D0948765

Fourier Series and Integrals

Probability and Mathematical Statistics

A Series of Monographs and Textbooks

Editors

Z. W. Birnbaum
University of Washington
Seattle, Washington

E. Lukacs
Catholic University
Washington, D.C.

FOURIER SERIES AND INTEGRALS

H. Dym

DEPARTMENT OF THEORETICAL MATHEMATICS
THE WEIZMANN INSTITUTE OF SCIENCE
REHOVOT, ISRAEL

H. P. McKean

COURANT INSTITUTE OF MATHEMATICAL SCIENCES
NEW YORK UNIVERSITY
NEW YORK, NEW YORK

1972

ACADEMIC PRESS   New York and London

ACADEMIC PRESS, INC.
111 Fifth Avenue, New York, New York 10003

United Kingdom Edition published by
ACADEMIC PRESS, INC. (LONDON) LTD.
24/28 Oval Road, London NW1

LIBRARY OF CONGRESS CATALOG CARD NUMBER: 73-187225
AMS (MOS) 1970 Subject Classification: 42-01, 42A16, 42A20
42A52, 42A64, 42A68, 42A76, 42A92, 42A96

PRINTED IN THE UNITED STATES OF AMERICA

To the memory of

TERRY MIRKIL

who started to write a book like this,
only better.

ולכבוד
יצחק בן אלקסנדר זישא הלוי,
חנה בת אליעזר ליפא ,
וחיה בת אליהו .

Contents

Chapter 3.   Fourier Integrals and Complex Function Theory

Chapter 4.   Fourier Series and Integrals on Groups

Additional Reading

Bibliography

Preface

SCOPE. The purpose of this book is to give a mathematical account of Fourier ideas on the circle and the line, on finite commutative groups, and on a few important noncommutative groups. Purely technical aspects of the subject, such as the pointwise convergence of Fourier series of wild functions, will be avoided. The emphasis is placed instead on the extraordinary power and flexibility of Fourier's basic series and integrals and on the astonishing variety of applications in which it is the chief tool. A brief sample of such applications will suffice here:

(1)   summation of series (1.7)
(2)   the isoperimetric problem (1.7)
(3)   Jacobi's identity for the theta function (1.7, 2.7, 2.11)
(4)   heat flow (1.8, 2.7)
(5)   wave motion, especially Huygens' principle (1.8, 2.7, 2.11)
(6)   random walks (1.10, 2.11)
(7)   the Poisson summation formula (2.7, 2.11, 4.5)
(8)   electrical circuits and filters (2.7, 3.5)
(9)   the central limit theorem (2.7, 2.11)
(10)   Heisenberg's inequality (2.8)
(11)   band- and time-limited signals (2.9)
(12)   Minkowski's theorem in the geometry of numbers (2.11)
(13)   steady-state radiation in stars (3.6)
(14)   Spitzer's identity (3.7)
(15)   polynomial approximation (1.7, 3.8, 3.9)
(16)   the distribution of primes (3.10)
(17)   Gauss' law of quadratic reciprocity (4.6)
(18)   the Zeeman effect in hydrogen (4.13)
(19)   representations of the rotation group $SO(3)$ (4.14)

PREREQUISITES. The level of preparation expected is a thorough knowledge of advanced calculus. Courant–John [1965] would be ideal. To this must be added a willingness to believe in (or to study up on) the Lebesgue integral, as outlined in 1.1. A smattering of complex function theory and groups would be helpful, though what is needed is explained (with proofs) in 3.1 and 4.1, respectively. A little physics would be nice too. Feynman [1963–1965] would be the ideal, but an attempt has been made to sketch everything needed in that line. The *most* important thing you can bring to this study is a real interest in applications and in the interconnections of different parts of mathematics.

The basic material was originally presented to an audience of junior and senior mathematics students in the course 18.49/MIT/Spring 1966; it has also been used for a similar audience at the City College of CUNY (Spring 1969) and as the format of a proseminar in collaboration with M. Schreiber at Rockefeller University (1969–1970). The book is now about double the original, what with new mathematical material, new applications, and a wide variety of exercises.

EXERCISES, usually with some hint as to their solution, are placed in nearly every section. Do not skip them! They are to be regarded as an integral part of the text.

USAGE. A function f on an open interval/region is "compact" if $f \equiv 0$ off a compact part of that interval/region. 1 is the unit function or the number "unity." 1_Q stands for the indicator function of Q, i.e., $1_Q(x) = 1$ if $x \in Q$ and 0 otherwise. R^n ($n \leqslant \infty$) is the n-dimensional real number space. C^n is the n-dimensional complex number space. Z^n is the lattice of integral points in R^n, in particular, Z^1 is just the integers 0, ± 1, ± 2, etc. The imaginary unit is i. Re and Im signify real and imaginary parts, * denotes complex conjugation, arg is the complex argument, and $|\ |$ is the complex norm: for $z = x + iy$,

$$\text{Re } z = x, \qquad \text{Im } z = y, \qquad z^* = x - iy,$$
$$\arg z = \tan^{-1}(y/x), \qquad |z| = \sqrt{x^2 + y^2}.$$

The phrase "if and only if" is abbreviated as iff.

NUMBERING. The reader will find that some sections are starred. This indicates more advanced (or merely more complicated) topics or side issues not needed in subsequent sections. The beginner should probably skip them all at a first reading.

1.2 means Section 2 of Chapter 1; Exercise 1.2.3. means Exercise 3 of that section, and it is the same with figures, subsections, etc. Brown [1969, p. 1] refers to page 1 of the item Brown [1969] listed in the bibliography.

ACKNOWLEDGMENTS. It is a special pleasure to thank Yehudah Barbut and Ruth Mandelbaum for drawing the figures and Mary Ellen O'Brien for cheerfully distilling an ocean of scrawl into an elegant typescript. We also thank C. Amitsur, S. Goldfinger, B. Loering and especially J. Schonfeld for reading portions of the manuscript and checking most of the exercises.

Historical Introduction

The basic idea of Fourier series is that "any" periodic function f of, say, time can be expressed as a "trigonometric sum" of sines and cosines of the same period T:

$$f(t) = \sum_{n=0}^{\infty} [\hat{f}_+(n) \cos(2\pi nt/T) + \hat{f}_-(n) \sin(2\pi nt/T)].$$

The idea comes up naturally in astronomical problems; in fact, Neugebauer [1952] discovered that the Babylonians used a primitive kind of Fourier series for the prediction of celestial events.

The history of the subject in more recent times begins with d'Alembert (1747) and his discussion of the oscillations of a violin string. The displacement $u = u(t, x)$ of the string, as a function of the time $t \geqslant 0$ and the place x, is a solution of

$$\partial^2 u/\partial t^2 = \partial^2 u/\partial x^2, \qquad t > 0, \quad 0 < x < 1, \quad \text{say,}$$

subject to the conditions

$$u(t, 0) = u(t, 1) = 0, \qquad t \geqslant 0;$$

$$\frac{\partial u}{\partial t}(0, x) = 0, \qquad 0 < x < 1.$$

1

The former expresses the fact that the ends of the string $[x = 0, 1]$ are tied down; the latter that it is at rest at time $t = 0$. The solution of this problem is the superposition of two waves traveling in opposite directions at speed 1, as expressed by d'Alembert's formula:

$$u(t, x) = \tfrac{1}{2}f(x+t) + \tfrac{1}{2}f(x-t),$$

in which f has to be an odd function of period 2, vanishing at $x = 0, \pm 1, \pm 2, \ldots$, so as to make $u = 0$ at the ends of the string. Euler (1748) proposed that such a function could be expanded into a sine series of period 2:

$$f(x) = \sum_{n=1}^{\infty} \hat{f}(n) \sin n\pi x$$

with the result that

$$u(t, x) = \sum_{n=1}^{\infty} \hat{f}(n) \cos n\pi t \sin n\pi x.$$

The "simple harmonic" $\cos n\pi t \sin n\pi x$ is interpreted as a "fundamental mode" of the string and the frequency $n/2$ as one of its "fundamental tones." The same idea was advanced by D. Bernouilli (1753) and Lagrange (1759) with much mutual criticism. The recipe

$$\hat{f}(n) = 2 \int_0^1 f(x) \sin n\pi x \, dx$$

for computing the coefficients, later associated with Fourier's name, first appears in a paper by Euler (1777).

Fourier's contributions begin in 1807 with his studies of the problem of heat flow

$$\partial u/\partial t = \tfrac{1}{2}(\partial^2 u/\partial x^2)$$

presented to the Académie des Sciences in 1811 and published (in part) as the celebrated *Théorie analytique de la chaleur* (1822). He made a serious attempt to prove that any piecewise smooth function f can be expanded into a trigonometric sum. A satisfactory proof of this fact was found soon afterwards by Dirichlet (1829). Riemann (1867) also made important contributions to this problem, but the rest of the 1800's were spent in what may now be judged to be spirited but unprofitable investigations of this kind. For additional information about the 18th and 19th centuries, see Coppel [1969] and Hobson [1926].

The key to further progress was provided by the new integral of Lebesgue (1904). The proper setting for Fourier series turned out to be the class of "Lebesgue measurable" functions f of period 1, say, with

$$\|f\|^2 = \int_0^1 |f(x)|^2 \, dx < \infty.$$

The crowning result of this period is the theorem of Riesz–Fischer (1907): *for such functions, the Fourier coefficients*

$$\hat{f}(n) = \int_0^1 f(x) e^{-2\pi i n x} dx, \qquad n \in Z^1$$

provide a $1:1$ *map of the function space onto the space of sequences* $\hat{f}(n)$ $(n = \ldots, -1, 0, 1, 2, \ldots)$ *with*

$$\|\hat{f}\|^2 = \sum_{n=-\infty}^{\infty} |\hat{f}(n)|^2 < \infty,$$

and not only does this map preserve the geometry:

$$\|f\| = \|\hat{f}\|;$$

but, the associated (formal) Fourier series

$$f(x) = \sum_{n=-\infty}^{\infty} \hat{f}(n) e^{2\pi i n x}$$

actually converges to f in the sense of that geometry:

$$\lim_{n \uparrow \infty} \int_0^1 |f(x) - \sum_{|k| \leqslant n} \hat{f}(k) e^{2\pi i k x}|^2 dx = 0.$$

A parallel development was carried out for the Fourier integral

$$\hat{f}(\gamma) = \int_{-\infty}^{\infty} f(x) e^{-2\pi i \gamma x} dx$$

for nonperiodic functions f, decaying sufficiently rapidly at $\pm \infty$, as summed up in the theorem of Plancherel (1910): *if f is Lebesgue measurable and if*

$$\|f\|^2 = \int_{-\infty}^{\infty} |f(x)|^2 dx < \infty,$$

then the same is true of $\hat{f}$ *(interpreting the integral with due caution),*

$$\|\hat{f}\| = \|f\|,$$

and f itself may be recovered from $\hat{f}$ *via the inverse Fourier integral*:

$$f(x) = \int_{-\infty}^{\infty} \hat{f}(\gamma) e^{2\pi i \gamma x} d\gamma.$$

Coming down to modern times, both Fourier series and integrals have been much advanced by the use of the powerful tools of complex function theory. Hardy, Littlewood, and Wiener took the lead in this new field. There has also been a very beautiful and fruitful interplay between Fourier series and integrals and commutative and noncommutative groups, beginning with Frobenius (1896–1903) and continuing down to the present day with the discoveries of Selberg, Weil, and Weyl. By now, the ideas of Fourier have made their way into every branch of mathematics and mathematical physics, from the theory of numbers to quantum mechanics—but that is the subject of this book.

Chapter 1 Fourier Series

1.1 THE LEBESGUE INTEGRAL

The conventional Riemann integral is a much less flexible tool than the integral of Lebesgue. The practical advantages have mostly to do with the interchange of the integral with other limiting operations, such as sums, other integrals, differentiation, and the like. The following outline is intended to convey the principal facts about Lebesgue integrals, as needed later. The reader who is content to take these statements as articles of faith may do so; nice proofs and additional information can be found in Munroe [1953] and Royden [1963]; for the more advanced student, Rudin [1966] and Hewitt and Stromberg [1965] are suggested. These are modern presentations; one of the oldest but most elementary and satisfactory explanations is given by de la Vallée-Poussin [1950] (the first edition appeared in 1916).

The principal difference between the integrals of Riemann and Lebesgue may be illustrated by pictures. Fig. 1 shows a positive-continuous function f defined on an interval $-\infty < a \leqslant x \leqslant b < \infty$, subdivided as for Riemann

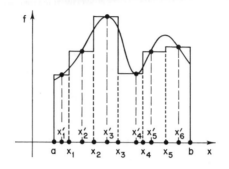

FIGURE 1

integration: You pick a series of points of subdivision

$$a = x_0 < x_1 < x_2 < \cdots < x_n = b,$$

form the Riemann sum

$$\sum_{k=1}^{n} f(x_k')\,(x_k - x_{k-1}),$$

in which x_k' is any point between x_{k-1} and x_k, and verify that this sum approaches a limit, namely the Riemann integral

$$\int_a^b f(x)\,dx,$$

as $n \uparrow \infty$ and the biggest of the lengths $x_k - x_{k-1}$ $(k \leqslant n)$ tends to 0. Lebesgue simply turned the recipe on its side and subdivided the range of the function instead of the domain. The idea, as indicated by the different shadings in Fig. 2, being to lump together the points at which the function takes on (approximately) the same values. This would appear to be a perfectly trivial

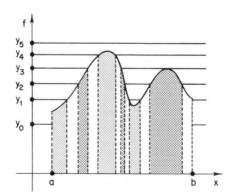

FIGURE 2

modification, but it has far-reaching consequences, as will be explained later. Lebesgue's recipe tells you first to subdivide the vertical axis by a series of points

$$\min f \geqslant y_0 < y_1 < \cdots < y_n \geqslant \max f,$$

next, to form the sum

$$\sum_{k=1}^{n} y_{k-1} \times \text{measure}(x: y_{k-1} \leqslant f(x) < y_k),$$

in which measure$(x: \cdots)$ is the sum of the lengths of the subintervals of $a \leqslant x \leqslant b$ on which the stated inequality takes place, and finally to verify that this sum approaches the *same* number

$$\int_a^b f(x)\, dx$$

as $n \uparrow \infty$ and the biggest of the lengths $y_k - y_{k-1}$ $(k \leqslant n)$ tends to 0. The point is that by now extending the idea of *measure* from unions of disjoint sub-intervals to the wider class of "measurable" subsets of the interval $a \leqslant x \leqslant b$, you can integrate a much wider class of functions by means of Lebesgue's recipe than you can by Riemann's.

The next item is a more formal explanation of the Lebesgue integral. Fix an interval $Q \subset R^1$, which may be bounded $(-\infty < a \leqslant x \leqslant b < \infty)$, or a half-line $(-\infty < a \leqslant x < \infty)$, or the whole line $(-\infty < x < \infty)$, and let us agree that *the measure of a (countable) union of nonoverlapping intervals is the sum of their lengths, finite or not*; especially, you ascribe to a single point or any countable family of points measure 0. This definition of measure may now be extended to the class of "Borel measurable" sets. This is the smallest collection of subsets of Q which contains all subintervals of Q and is closed under countable unions, countable intersections, and complementation. It turns out that if you require the extended measure of a countable union of disjoint "Borel measurable" sets to be the sum of their individual measures, then you can do this in only one way. A small additional extension (to the class of "Lebesgue measurable" sets) is made for technical convenience by throwing in any subset of a "Borel-measurable" set of measure 0 and ascribing to it measure 0 also. This second extension is the Lebesgue measure; from now on "measurable" *always* means "Lebesgue measurable." The most concrete way of expressing the Lebesgue measure is by the following recipe: *for any measurable set E,*

$$\text{measure}(E) = \inf \sum_{n=1}^{\infty} \text{length}(I_n),$$

in which the infemum is taken over the class of countable coverings of E by means of intervals I_n $[\bigcup_{n=1}^{\infty} I_n \supset E]$.

EXAMPLE 1.   To illustrate the idea, take $a = 0$ and $b = 1$. Any open sub-set G of $Q = [0, 1]$ is measurable, since it is the union of a countable number of nonoverlapping open intervals. The natural extension to compact sets K is to put

$$\text{measure}(K) = 1 - \text{measure}(G)$$

in which G is the (open) complement of K relative to Q. For example, the measure of

$$G = (\tfrac{1}{3}, \tfrac{2}{3})$$

$$\cup (\tfrac{1}{9}, \tfrac{2}{9}) \cup (\tfrac{7}{9}, \tfrac{8}{9})$$

$$\cup (\tfrac{1}{27}, \tfrac{2}{27}) \cup (\tfrac{7}{27}, \tfrac{8}{27}) \cup (\tfrac{19}{27}, \tfrac{20}{27}) \cup (\tfrac{25}{27}, \tfrac{26}{27})$$

$$\cup \quad \cdots$$

is equal to

$$\sum_{n=1}^{\infty} 2^{n-1} 3^{-n} = \tfrac{1}{3}(1 - \tfrac{2}{3})^{-1} = 1,$$

so that its complement K is of measure 0, though uncountable. K is the so-called Cantor set.

The most interesting properties of Lebesgue measure are listed below. A, B_1, B_2, etc. are measurable subsets of Q; for economy, measure (A) is now written $m(A)$.

(a) $0 \leqslant m(A) \leqslant m(Q) = b - a.$

(b) $m(A) \leqslant m(B)$ if $A \subset B.$

(c) $m(Q - A) + m(A) = m(Q).$

(d) $m\left(\bigcup_{n=1}^{\infty} B_n \right) \leqslant \sum_{n=1}^{\infty} m(B_n).$

(d′) $m\left(\bigcup_{n=1}^{\infty} B_n \right) = \sum_{n=1}^{\infty} m(B_n)$ if $B_i \cap B_j$ is empty for $i \neq j.$

(e) $m(B_n) \uparrow m(A)$ if $B_1 \subset B_2 \subset \cdots$ and $\bigcup_{n=1}^{\infty} B_n = A.$

(e′) $m(B_n) \downarrow m(A)$ if $B_1 \supset B_2 \supset \cdots$ and $\bigcap_{n=1}^{\infty} B_n = A,$
         provided $m(B_1) < \infty.$

The reader may find the following exercises instructive.

EXERCISE 1.   Check that each of (d′), (e), and (e′) implies the other two.

EXERCISE 2.   Check that if $f_1, f_2, \ldots$ are real continuous functions on the line and if $f = \lim_{n \uparrow \infty} f_n$ exists pointwise, then $(x: 0 \leqslant f(x) < 1)$ is measurable. *Hint*: $(x: f(x) < 1) = \bigcup_{k \geqslant 1} \bigcup_{m \geqslant 1} \bigcap_{n \geqslant m} (x: f_n(x) \leqslant 1 - 1/k).$

A (real) function f on Q is measurable if $(x: \alpha \leqslant f(x) < \beta)$ is a measurable set for every choice of α and β. The integral of a nonnegative measurable function $f \leqslant \infty$ is now defined by forming the Lebesgue-type sums

$$\sum_{k=0}^{\infty} k2^{-n} \times \text{measure}(x: k2^{-n} \leqslant f(x) < (k+1)2^{-n})$$

$$+ \infty \times \text{measure}(x: f(x) = \infty)$$

and making $n \uparrow \infty$, with the understanding that $0 \times \infty = 0$. As n increases, the subdivision $k2^{-n}: k \geqslant 0$ becomes finer and finer, and the sums increase to a finite or infinite limit, which is declared to be the Lebesgue integral of f:

$$\int_Q f = \int_a^b f = \int_a^b f(x)\, dx.$$

To sum up, *the Lebesgue integral of a nonnegative function always exists, although it may be $+\infty$*.

The Lebesgue integral of a signed (measurable) function f is now obtained by splitting f into a positive part $f^+ = \max(f, 0)$ and a negative part $f^- = \min(f, 0)$ and declaring the integral to be

$$\int_Q f = \int_Q f^+ - \int_Q (-f^-),$$

provided that at least one of $\int f^+$ and $\int(-f^-)$ is less than ∞; if both are less than ∞, f is said to be "summable." Notice that f is summable iff $|f| = f^+ - f^-$ is so. The integral of f times the indicator function 1_B of a measurable set $B \subset Q$ is declared to be the "integral of f over B":

$$\int_B f = \int_Q f 1_B.$$

EXERCISE 3.   Check that the Lebesgue integral $\int_0^1 f$ of the indicator function f of the rational numbers exists ($=0$) but that f is *not* integrable in the Riemann sense.

The class of measurable functions is closed under

(a) multiplication by real constants,
(b) addition (and subtraction),
(c) multiplication (and division) of functions (with self-evident precautions about $-\infty \times \infty$, and the like),
(d) (countable) infema, suprema, limits inferior and limits superior.

EXERCISE 4.   Check that any (piecewise) continuous function is measurable.

EXERCISE 5.   Check that if $f_1, f_2, \ldots$ are measurable, then $f = \liminf_{n\uparrow\infty} f_n$ is, too. Hint: $(x: f(x) > \alpha) = \bigcup_{\beta > \alpha} \bigcup_{m \geq 1} \bigcap_{n \geq m} (x: f_n(x) \geq \beta)$ in which the first union is over rational $\beta > \alpha$. Why?

The most important properties of the integral are listed below. $f, f_1, f_2, \ldots$ are real measurable functions; they are understood to be summable or nonnegative.

(a) $m(B) = \int_Q f$ if f is the indicator function of a measurable subset B of Q.

(b) $\alpha \times m(Q) \leq \int_Q f \leq \beta \times m(Q)$ if $\alpha \leq f \leq \beta$.

(c) $\int_{A \cup B} f = \int_A f + \int_B f$ if A and B are disjoint.

(d) $\int_Q (\text{constant} \times f) = \text{constant} \times \int_Q f$.

(e) $\int_Q (f_1 + f_2) = \int_Q f_1 + \int_Q f_2$.

(f)   Monotone convergence: $\lim_{n\uparrow\infty} \int_Q f_n = \int_Q f$ if $0 \leq f_1 \leq f_2 \leq \cdots$ and $\lim_{n\uparrow\infty} f_n = f$. The integrals exist automatically; both sides may be $+\infty$. The condition $0 \leq f_1$ may be replaced by the condition that f_1^- be summable, that is to say $\int f_1^- > -\infty$.

EXERCISE 6. $\lim_{n\uparrow\infty} \int f_n = \int f$ if $f_1 \geq f_2 \geq \cdots$, $\lim_{n\uparrow\infty} f_n = f$, and $\int f_1^+ < \infty$.

EXERCISE 7. $\int \sum_{n=1}^{\infty} f_n = \sum_{n=1}^{\infty} \int f_n$ if $f_n \geq 0$ for $n \geq 1$.

(g)   Dominated convergence: $\lim_{n\uparrow\infty} \int_Q f_n = \int_Q f$ if, for every $n \geq 1$, $|f_n|$ is bounded by a common summable function and $\lim_{n\uparrow\infty} f_n = f$. The special case of $|f_n| \leq \text{constant}$ and $m(Q) = b - a < \infty$ is called "bounded convergence."

EXERCISE 8.   Check that (f) implies (g) and vice versa. Hint: To pass from (f) to (g) look at $\inf_{k \geq n}(h \pm f_k)$ in case $|f_k| \leq h$.

EXERCISE 9. $\lim_{n\uparrow\infty} \int f_n \neq \int \lim_{n\uparrow\infty} f_n$ if $a = 0$, $b = \infty$, and $f_n = ne^{-nx}$. What goes wrong?

(h)   Fatou's lemma: $\int_Q f \leq \liminf_{n\uparrow\infty} \int_Q f_n$ if $f_n \geq 0$ for $n \geq 1$ and $\liminf_{n\uparrow\infty} f_n = f$.

The statements of (f), (g), and (h) can be amplified by permitting the conditions to fail on a set of measure 0. The phrase "almost everywhere" (abbreviated a.e.) is useful in this context; for example, $\lim_{n\uparrow\infty} f_n = f$ a.e. means that the set of points $x \in Q$ at which $\lim_{n \uparrow \infty} f_n$ either fails to exist, or else exists but fails to agree with $f(x)$, is of measure 0.

The integral of a complex function f is defined in the natural way. The function f is "summable" if $\int_Q |f| < \infty$. Then the real functions $\mathrm{Re} f$ and $\mathrm{Im} f$ are summable, and you put

$$\int_Q f = \int_Q \mathrm{Re} f + i \int_Q \mathrm{Im} f.$$

The preceding rules (c), (d), (e), and (g) still hold; (b) is replaced by

EXERCISE 10.   Check that $|\int_Q f| \leqslant \int_Q |f|$. *Hint*: $|\int_Q f| = \int \mathrm{Re}(f \times e^{-i\varphi})$ in which $\varphi = \arg(\int_Q f)$.

As an exercise in the use of Lebesgue measure, let us prove the so-called Borel–Cantelli lemma for future use. The statement is that *if*

$$\sum_{n=1}^{\infty} m(B_n) < \infty,$$

then almost no point $x \in Q$ belongs to an infinite number of the sets B_n ($n \geqslant 1$), that is

$$m\left(\bigcap_{n \geqslant 1} \bigcup_{k \geqslant n} B_k \right) = 0.$$

EXERCISE 11.   Check that $\bigcap_{n \geqslant 1} \bigcup_{k \geqslant n} B_k$ is really the set of points x that belong to B_n for infinitely many values of n. This set is customarily denoted by "B_n i.o."; for i.o., read *infinitely often*.

PROOF OF THE BOREL–CANTELLI LEMMA.

$$m\left(\bigcap_{n \geqslant 1} \bigcup_{k \geqslant n} B_k \right) \leqslant m\left(\bigcup_{k \geqslant n} B_k \right) \leqslant \sum_{k \geqslant n} m(B_k) \qquad \text{for any } n \geqslant 1.$$

This is the tail of a convergent sum, and can be made as small as you please by making $n \uparrow \infty$.

EXERCISE 12.   Prove Chebyshev's inequality:

$$m(x : |f(x)| \geqslant l) \leqslant l^{-2} \int_Q |f|^2.$$

Hint: $l^{-2} |f|^2 \geqslant 1$ on the set where $|f| \geqslant l$.

EXERCISE 13.   Check that if $f \geqslant 0$ and if $\int_Q f = 0$, then $f = 0$ a.e. *Hint*: $(x : f(x) > 0) = \bigcup_{n \geqslant 1} (x : f(x) > 1/n)$.

The Lebesgue integral is easily defined for functions on a region Q of the plane or of any finite-dimensional space R^n. The measure of sets is now built up from the measure (alias area or volume) of n-dimensional boxes much as before. The volume of the box

$$a_1 \leqslant x_1 \leqslant b_1, ..., a_n \leqslant x_n \leqslant b_n$$

is simply the customary product of its edges:

$$m(Q) = (b_1 - a_1) \times \cdots \times (b_n - a_n).$$

The integral is now defined first for $f \geqslant 0$, as a limit of Lebesgue sums, and then extended to complex f, just as before. The notations

$$\int_Q f = \int_Q f(x_1, ..., x_n) \, dx_1 \cdots dx_n = \int_Q f(x) \, d^n x$$

will be used interchangeably. The only really new point that comes up is

FUBINI'S THEOREM.   *The statement for the plane is that if f is a measurable function on the box $Q = I \times J$, then the two-dimensional integral*

$$\int_Q f(x) \, d^2 x = \int_Q f(x_1, x_2) \, dx_1 \, dx_2$$

*based upon area measure can also be evaluated as an iterated integral **in either order**, namely as*

$$\int_Q f \, d^2 x = \int_I dx_1 \left(\int_J f(x_1, x_2) \, dx_2 \right) = \int_J dx_2 \left(\int_I f(x_1, x_2) \, dx_1 \right),$$

***provided** f is summable in any of the three senses, i.e., if any of the integrals*

$$\int_Q |f| \, d^2 x, \qquad \int_I dx_1 \left(\int_J |f| \, dx_2 \right), \qquad \int_J dx_2 \left(\int_I |f| \, dx_1 \right)$$

*is less than ∞. For $f \geqslant 0$, even this condition is superfluous: **Either** all the integrals diverge to $+\infty$ or they all converge to the same value. A completely analogous statement holds for dimensions $n \geqslant 3$.*

Warning: Henceforth, all sets and functions are measurable.

1.2   THE GEOMETRY OF $\mathbf{L^2}(Q)$

The space $L^2(Q, dx)$, or $L^2(Q)$ for short, is the class of all complex (measurable) functions f on a subinterval Q of R^1, subject to

$$\|f\| = \|f\|_2 = \left(\int_Q |f|^2 \, dx \right)^{1/2} < \infty.$$

This is the class of functions with which Fourier series and/or integrals are most naturally associated and therefore plays a central role in everything to follow. The reader will notice the analogy with finite-dimensional space: If you picture f as a point in an infinite-dimensional space with coordinates $f(x): x \in Q$ and think of $\|f\|$ as the distance of this point from the origin, then the formula $\|f\|^2 = \int |f|^2$ appears as a kind of Pythagorean rule. The purpose of this section is to develop the elementary geometry of $L^2(Q)$ with this picture in mind.

Technical aside: Two functions from $L^2(Q)$ will be identified if they differ only on a set of measure 0; especially, f is identified with the 0-function if $f = 0$ a.e. This permits you to say that $f = 0$ iff $\|f\| = 0$. To be perfectly honest, you cannot uphold this convention unless you think of the points of $L^2(Q)$, not as individual functions, but as equivalence classes of functions; but it is simpler and not at all confusing to ignore the distinction.

The principal result of this section is that $L^2(Q)$ is a "Hilbert space." But first, the technical definition. To begin with, a Hilbert space H is *linear space*, i.e., a set of "points" $\alpha, \beta, \ldots$ that is closed under addition $[\alpha + \beta]$ and multiplication by complex constants $[c\alpha]$, subject to the usual rules $[\alpha + \beta = \beta + \alpha, c(\alpha + \beta) = c\alpha + c\beta, \ldots]$. Besides this, H is endowed with an inner product, which associates a complex number (α, β) to each pair of points α and β in H, subject to the rules

$$(\alpha, \beta) = (\beta, \alpha)^*,$$

$$(\alpha + \beta, \gamma) = (\alpha, \gamma) + (\beta, \gamma),$$

$$(c\alpha, \beta) = c(\alpha, \beta),$$

$$(\alpha, \alpha) \geqslant 0 \qquad \text{with equality iff} \quad \alpha = 0.$$

The distance between two points α and β in H is declared to be

$$\|\alpha - \beta\| = (\alpha - \beta, \alpha - \beta)^{\frac{1}{2}}.$$

Notice that $\|\alpha - \beta\| = 0$ iff $\alpha = \beta$. The Hilbert space H is also required to be "complete." This means that if $\|\alpha_m - \alpha_n\| \to 0$ as m and n tend to ∞ then the points α_n must actually converge to some point $\beta \in H$ in the sense that

$$\|\alpha_n - \beta\| \to 0$$

as $n \uparrow \infty$. *In this book*, H is also required to be "separable." This means that there is a countable family $\alpha_n : n \geqslant 1$ of points of H which is "dense," that is, any point β of H can be approximated arbitrarily well by a member of this family:

$$\inf_{n \geqslant 1} \|\alpha_n - \beta\| = 0.$$

Warning: The assumption of separability is *not* standard. For an example of a nonseparable Hilbert space see Akhiezer and Glazman [1961, p. 29].

The finite-dimensional complex-number space C^n, with points $\alpha = (a_1, ..., a_n)$, $\beta = (b_1, ..., b_n)$, ..., is a Hilbert space under the inner product

$$(\alpha, \beta) = \sum_{i=1}^{n} a_i b_i^*,$$

but that is not the only possibility. As you will see shortly, $L^2(Q)$ is also a Hilbert space, albeit an infinite-dimensional one. The fact is that there is nothing in the definition to hold the dimension down. However, all infinite-dimensional (separable) Hilbert spaces can be identified with the infinite-dimensional complex number space C^∞, as will be explained in the next section.

A few simple geometrical remarks about H may be helpful. The chief point is that if you are dealing with $n < \infty$ points in H, you might as well be in C^n; especially, all the familiar rules of trignometry apply!

The first item of business is *Schwarz's inequality*, which states that

$$|(\alpha, \beta)| \leqslant \|\alpha\| \|\beta\|.$$

PROOF.   Take a real number x and an angle $0 \leqslant \theta < 2\pi$. Then $\alpha - x e^{i\theta} \beta$ belongs to H, and

$$\|\alpha - x e^{i\theta} \beta\|^2 = (\alpha - x e^{i\theta} \beta, \alpha - x e^{i\theta} \beta)$$

$$= (\alpha, \alpha) - (x e^{i\theta} \beta, \alpha) - (\alpha, x e^{i\theta} \beta) + (x e^{i\theta} \beta, x e^{i\theta} \beta)$$

$$= \|\alpha\|^2 - 2 \operatorname{Re}[x e^{-i\theta} (\alpha, \beta)] + x^2 \|\beta\|^2.$$

Pick θ equal to the argument of (α, β). Then

$$\|\alpha\|^2 - 2x|(\alpha, \beta)| + x^2 \|\beta\|^2 = \|\alpha - x e^{i\theta} \beta\|^2 \geqslant 0$$

is a polynomial of degree 2 in the variable x, and since it is nonnegative, its discriminant

$$4[|(\alpha, \beta)|^2 - \|\alpha\|^2 \|\beta\|^2]$$

must be zero or less. But this is precisely the Schwarz inequality.

EXERCISE 1.   Check that equality prevails in Schwarz's inequality, i.e., $|(\alpha, \beta)| = \|\alpha\| \|\beta\|$, iff α and β are proportional.

EXERCISE 2.   Check that for fixed β, the inner product (α, β) is a continuous function of α. *Hint:* $(\alpha_1, \beta) - (\alpha_2, \beta) = (\alpha_1 - \alpha_2, \beta)$; now use Schwarz's inequality.

EXERCISE 3.   Prove the *triangle inequality*

$$\|\alpha \pm \beta\| \leqslant \|\alpha\| + \|\beta\|.$$

A glance at Fig. 1 suggests the geometrical content of this: It says that the far side of a triangle is shorter than the sum of the lengths of the 2 adjacent sides.   *Hint*: $\|\alpha \pm \beta\|^2 = \|\alpha\|^2 \pm 2\,\mathrm{Re}(\alpha, \beta) + \|\beta\|^2$;   now   use   Schwarz's   inequality.

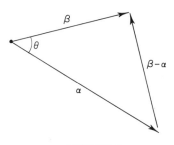

FIGURE 1

EXERCISE 4.   Define the angle between $\alpha \neq 0$ and $\beta \neq 0$ by the rule

$$\cos\theta = \mathrm{Re}(\alpha, \beta)/(\|\alpha\|\,\|\beta\|).$$

Check the "law of cosines":

$$\|\alpha - \beta\|^2 = \|\alpha\|^2 - 2\,\|\alpha\|\,\|\beta\|\,\cos\theta + \|\beta\|^2.$$

α and β are declared to be *perpendicular* if $(\alpha, \beta) = 0$. In this circumstance, the law of cosines becomes the "Pythagorean rule":

$$\|\alpha + \beta\|^2 = \|\alpha\|^2 + \|\beta\|^2.$$

Check the extended Pythagorean rule:

$$\left\|\sum_{k=1}^{n}\alpha_k\right\|^2 = \sum_{k=1}^{n}\|\alpha_k\|^2 \qquad \text{if}\quad (\alpha_i, \alpha_j) = 0 \quad \text{for}\quad i \neq j.$$

So much for elementary geometry. The proof that $L^2(Q)$ is a Hilbert space will now be presented in a series of simple steps with explanatory asides.

Step 1:   Check that $L^2(Q)$ is closed under multiplication by complex numbers and under addition. The first is self-evident, while the second follows from the elementary bound

$$|f + g|^2 \leqslant 2|f|^2 + 2|g|^2$$

upon integrating over Q.

Step 2:   For any f and g from $L^2(Q)$, the elementary bound

$$2|fg^*| = 2|f||g| \leqslant |f|^2 + |g|^2$$

shows that fg^* is a summable function, so that the *inner product*

$$(f,g) = \int_Q fg^*$$

makes sense. The reader will have no difficulty in checking that all the customary rules for inner products hold; after all, from an algebraic standpoint,
$\int fg^*$ is just the same as the inner product for C^n ($n < \infty$), only now you have
an integral in place of a finite sum. An automatic consequence is the Schwarz
inequality:

$$|(f,g)|^2 = \left| \int_Q fg^* \right|^2 \leqslant \|f\|^2 \|g\|^2 = \int_Q |f|^2 \int_Q |g|^2 .$$

EXERCISE 5.   State precisely when equality holds in the present context, and compare with Exercise 1.

EXERCISE 6.   Check *Minkowski's inequality*:

$$\left(\int_Q |f+g|^2 \right)^{1/2} \leqslant \left(\int_Q |f|^2 \right)^{1/2} + \left(\int_Q |g|^2 \right)^{1/2} .$$

Step 3:   Prove that $L^2(Q)$ is complete. This would *not* have been the
case if $L^2(Q)$ had been defined by means of the Riemann integral; therein
lies one of the chief advantages of the Lebesgue viewpoint. Before entering
into the details of the proof, a short aside about different modes of convergence of functions may be helpful.

Aside: f_n converges to f in the sense of distance in $L^2(Q)$ iff

$$\lim_{n\uparrow\infty} \|f - f_n\|^2 = \lim_{n\uparrow\infty} \int_Q |f - f_n|^2 = 0 .$$

This mode of convergence is to be compared with *pointwise convergence
almost everywhere* (a.e.). The latter phrase means that

$$m(x: f_n(x) \text{ fails to approach } f(x) \text{ as } n\uparrow\infty) = 0 .$$

Neither type of convergence implies the other as simple examples show.

EXAMPLE 1. $f_n = \sqrt{n} \times$ the indicator function of $0 \leqslant x \leqslant 1/n$ converges pointwise to 0 a.e., but

$$\int_0^1 |f_n|^2 = n \times 1/n = 1 \qquad \text{for every } \; n \geqslant 1,$$

so f_n does not converge to $f = 0$ in $L^2[0, 1]$.

   EXAMPLE 2. Take $L^2[0, 1]$ again, and identify 0 and 1 so that the interval $0 \leqslant x \leqslant 1$ is pictured as a circle. Now think of a rectangular "whisker" of height 1 that rotates at speed 1 counterclockwise around the circle, becoming thinner and thinner as time goes on, as in Fig. 2. The whisker at time $0 \leqslant t < \infty$ is described by the indicator function

$$f_t(x) = \begin{cases} 1 & \text{for } \; t \leqslant x \leqslant t + \Delta, \\ 0 & \text{otherwise,} \end{cases}$$

in which $\Delta = e^{-t}$, say, is the width and x is counted modulo 1. Clearly, $\|f_t\|^2 = \Delta$ tends to 0 so that f_t approaches 0 in $L^2[0, 1]$, but no matter where you sit on the circle, the whisker keeps on coming by, and you see it for arbitrarily large times; in fact,

$$m(x : f_t(x) \text{ converges to 0 as } t \uparrow \infty) = 0.$$

The reader may rightly object that the functions f_t have a continuous parameter $t \geqslant 0$ rather than an integral parameter $n \geqslant 1$.

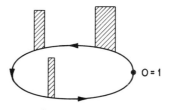

FIGURE 2

   EXERCISE 7. Overcome this objection by checking that if you take

$$
\begin{aligned}
t_1 &= 0, & \Delta_1 &= 1, \\
t_2 &= 1, & \Delta_2 &= \tfrac{1}{2}, \\
t_3 &= \tfrac{3}{2}, & \Delta_3 &= \tfrac{1}{2}, \\
t_4 &= 2, & \Delta_4 &= \tfrac{1}{4}, \\
t_5 &= \tfrac{9}{4}, & \Delta_5 &= \tfrac{1}{4}, \\
t_6 &= \tfrac{10}{4}, & \Delta_6 &= \tfrac{1}{4}, \\
t_7 &= \tfrac{11}{4}, & \Delta_7 &= \tfrac{1}{4}, \\
t_8 &= 3, & \Delta_8 &= \tfrac{1}{8}, \\
&\;\vdots & &\;\vdots
\end{aligned}
$$

then the corresponding functions $f_1 = f_{t_1}, f_2 = f_{t_2}, \ldots$ converge to 0 in $L^2[0,1]$, but

$$m(x: f_n(x) \text{ converges to 0 as } n \uparrow \infty) = 0.$$

Despite these examples, there *is* a relation between the two modes of convergence: *To wit, if f_n converges to f in $L^2(Q)$, then you can find a subsequence $n_1 < n_2 < \cdots \uparrow \infty$ so that f_{n_k} converges to f pointwise a.e. as $k \uparrow \infty$.*

PROOF.   Because $\lim_{n \uparrow \infty} \|f_n - f\| = 0$, you can pick $n_1 < n_2 < \cdots$ so that

$$\|f_{n_k} - f\|^2 \leqslant 2^{-k} \qquad \text{for} \quad k = 1, 2, \quad \text{etc.}$$

By the theorem of monotone convergence,

$$\int_Q \sum_{k=1}^{\infty} |f_{n_k} - f|^2 = \sum_{k=1}^{\infty} \int_Q |f_{n_k} - f|^2 \leqslant \sum_{k=1}^{\infty} 2^{-k} = 1 < \infty,$$

so that

$$\sum_{k=1}^{\infty} |f_{n_k} - f|^2 < \infty \quad \text{a.e.,}$$

and this happens only if the general term of the sum approaches 0 a.e., that is, only if

$$\lim_{n \uparrow \infty} f_{n_k}(x) = f(x) \quad \text{a.e.}$$

The proof is finished.

EXERCISE 8.   Check that if f_n converges to f in $L^2(Q)$ and if it also converges to f^0 pointwise a.e., then $f = f^0$ a.e.

EXERCISE 9.   Check that if f_n converges to f pointwise a.e. and if $\sup_{n \geqslant 1} |f_n|^2$ is summable, then f_n also converges to f in $L^2(Q)$. *Hint*:   Use dominated convergence.

Step 3 (concluded)*:*   The proof of completeness can now be made. The problem is to check that if $\|f_n - f_m\|$ tends to 0 as n and $m \uparrow \infty$, then there is an actual function $f \in L^2(Q)$ such that

$$\lim_{n \uparrow \infty} \|f_n - f\| = 0.$$

PROOF.   The proof is a little more elaborate than the preceding one. The first task is to produce a plausible candidate for f. Pick $n_1 < n_2 < \cdots$ in order to make

$$\|f_{n_i} - f_{n_j}\|^2 \leqslant 2^{-i} \qquad \text{for all} \quad j \geqslant i;$$

the simplest thing to do is to pick

$n_1 \geq 1$ in order to make $\|f_n - f_{n_1}\|^2 \leq \frac{1}{2}$ for all $n \geq n_1$,

$n_2 \geq n_1$ in order to make $\|f_n - f_{n_2}\|^2 \leq \frac{1}{4}$ for all $n \geq n_2$,

and so on.
Define

$$A_j = (x: |f_{n_{j+1}}(x) - f_{n_j}(x)| \geq 2^{-j/3})$$
$$B = \bigcap_{i \geq 1} \bigcup_{j \geq i} A_j = A_j \quad \text{i.o.,}$$

in which i.o. stands for "infinitely often" [see Exercise 1.1.11].

EXERCISE 10.   Check that if $x \in Q - B$, then $f_{n_j}(x)$ converges to a finite limit as $j \uparrow \infty$. *Hint:* $x \in Q - B$ means that $|f_{n_{j+1}}(x) - f_{n_j}(x)| < 2^{-j/3}$ for all sufficiently large j.

By the Chebyshev inequality [Exercise 1.1.12],

$$m(A_j) \leq 2^{2j/3} \|f_{n_{j+1}} - f_{n_j}\|^2 \leq 2^{-j/3},$$

and since this is the general term of a convergent sum, the Borel–Cantelli lemma, just before Exercise 1.1.11 tells you that

$$m(B) = m(A_j \quad \text{i.o.}) = 0.$$

Therefore,

$$C = (x: f_{n_j}(x) \text{ fails to converge as } j \uparrow \infty) \subset B$$

is of measure 0, too [see Exercise 10], and you can define a measurable function f by the rule

$$f(x) = \begin{cases} 0 & \text{if } x \in C, \\ \lim_{j \uparrow \infty} f_{n_j}(x) & \text{otherwise}. \end{cases}$$

The claim is that f_n converges to this function f in $L^2(Q)$. Because $m(C) = 0$, f_{n_j} converges to f pointwise a.e., and by Fatou's lemma

$$\|f_n - f\|^2 = \int_Q \lim_{j \uparrow \infty} |f_n - f_{n_j}|^2 \leq \liminf_{j \uparrow \infty} \int_Q |f_n - f_{n_j}|^2$$

$$= \liminf_{j \uparrow \infty} \|f_n - f_{n_j}\|^2 \leq 2^{-i} \quad \text{if } n \geq n_i.$$

From this you learn two things: first, that $f \in L^2(Q)$, since

$$\|f\| \leq \|f_{n_1}\| + \|f_{n_1} - f\| \leq \|f_{n_1}\| + 2^{-\frac{1}{2}} < \infty,$$

and second, that f_n approaches f in $L^2(Q)$ since $2^{-i}\downarrow 0$ as $i\uparrow\infty$. The proof is finished.

Step 4: Verify that $L^2(Q)$ is separable. This will finish the proof that it is a Hilbert space. The problem is to show that you can come as close as you please to any function $f \in L^2(Q)$ by functions from a fixed countable family $K \subset L^2(Q)$. Clearly, it suffices to deal with *real* functions. For any such function f, any integral i, j, and $k \geqslant 1$, let

$$f_1 = \begin{cases} f & \text{if } |x| \leqslant i, \\ 0 & \text{if } |x| > i; \end{cases}$$

$$f_2 = \begin{cases} f_1 & \text{if } |f_1| \leqslant j, \\ 0 & \text{if } |f_1| > j; \end{cases}$$

and

$$f_3 = k^{-1}[kf_2],$$

in which $[f]$ stands for the largest integer $\leqslant f$. Each of these functions is measurable and

$$\|f - f_1\|^2 = \int_{Q \cap (x:|x|>i)} |f|^2,$$

$$\|f_1 - f_2\|^2 = \int_{Q \cap (x:|f_1|>j)} |f_1|^2 \leqslant \int_{Q \cap (x:|f|>j)} |f|^2,$$

$$\|f_2 - f_3\|^2 \leqslant \int_{Q \cap (x:|x| \leqslant i)} k^{-2} \leqslant 2ik^{-2},$$

so that you can make

$$\|f - f_3\| \leqslant \|f - f_1\| + \|f_1 - f_2\| + \|f_2 - f_3\|$$

as small as you like by taking i, j, and k large, in that order. Bring in the family K of piecewise constant functions f, which vanish far out, take rational values only, and jump only at a finite number of rational points. K is a countable subfamily of $L^2(Q)$ and is closed under finite sums with rational coefficients. Now $f_3 = l/k$ on the set

$$A = Q \cap (x: |x| \leqslant i) \cap (x: l/k \leqslant f_2(x) < (l+1)/k),$$

so f_3 is the sum (with coefficients l/k) of indicator functions 1_A of sets of measure $\leqslant 2i < \infty$. Therefore, to finish the proof, you only have to check that such an indicator function can be well approximated by a function from K. By the recipe for Lebesgue measure of Section 1.1, you can cover such a set A by a *countable* family of intervals I_k, so closely as to make

$$0 \leqslant \sum_{k=1}^{\infty} m(I_k) - m(A)$$

as small as you please. Especially, you can find a finite union $B = \bigcup_{k=1}^{n} I_k$ so as to make

$$m(A-B) + m(B-A) = \int_{A-B} 1 + \int_{B-A} 1 = \|1_A - 1_B\|^2$$

as small as you like, and you can even modify B to make the endpoints of the intervals rational, at the expense of a small change in $\|1_A - 1_B\|$. But then $1_B \in K$, and the proof is finished.

EXERCISE 11. Amplify the proof of Step 4 to verify that the class $C^{\infty}(Q)$ of infinitely differentiable functions f with $\int_Q |f|^2 < \infty$ is dense in $L^2(Q)$. Check that if Q is unbounded ($Q = R^1$, say), you can even make do with compact *infinitely* differentiable functions. The adjective "compact" means that f has compact support, i.e., $f = 0$ far out. A special case is the fact that compact and/or bounded functions are dense in $L^2(Q)$. *Hint:* By the proof, any real function $f \in L^2(Q)$ can be well approximated by a step

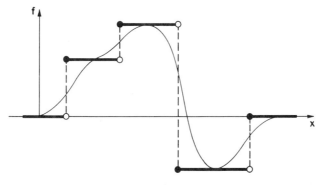

FIGURE 3

function, as pictured in Fig. 3 (heavy line), and you have only to smooth out the jumps in an infinitely differentiable way as indicated by the lighter curve. The prototype of this smoothing is the function

$$f(x) = \begin{cases} 0 & \text{for } x \leqslant 0, \\ \exp[-x^{-1} \exp(x-1)^{-1}] & \text{for } 0 < x < 1, \\ 1 & \text{for } x \geqslant 1, \end{cases}$$

which interpolates between the constant values 0 for $x \leqslant 0$ and 1 for $x \geqslant 1$ in an infinitely differentiable way.

EXERCISE 12. Check that for any nonnegative measurable function Δ on Q, the class $\mathsf{L}^2(Q, \Delta(x)\,dx)$ of measurable functions with

$$\|f\|^2 = \int_Q |f(x)|^2 \Delta(x)\,dx < \infty$$

is likewise a Hilbert space.

EXERCISE 13. The infinite-dimensional complex number space C^∞ is the class of all sequences $c = (c_1, c_2, \ldots)$ of complex numbers with

$$\|c\| = \left(\sum |c_n|^2\right)^{\frac{1}{2}} < \infty.$$

Check that C^∞ is a Hilbert space under the inner product

$$(a, b) = \sum a_n b_n^*.$$

The proof is the same as for $\mathsf{L}^2(Q)$ only easier. Another name for C^∞ is $\mathsf{L}^2(Z^+)$, Z^+ being the integers $n \geqslant 1$; this brings out the analogy with $\mathsf{L}^2(Q)$. The space $\mathsf{L}^2(Z^1)$ of complex functions on the integers $n = 0, \pm 1, \pm 2, \ldots$ $(= Z^1)$ is defined similarly.

1.3   THE GEOMETRY OF $\mathsf{L}^2(Q)$ CONTINUED

The principal task of this section is to introduce coordinates into $\mathsf{L}^2(Q)$ [similar to the familiar perpendicular coordinates of the finite-dimensional complex number space C^n $(n < \infty)$], with a view to showing that $\mathsf{L}^2(Q)$ is isomorphic to the infinite-dimensional complex number space $C^\infty = \mathsf{L}^2(Z^+)$ described in Exercise 1.2.13 (see Theorem 3 for the precise statement). The fact is, that, up to isomorphism, C^∞ *is the only infinite-dimensional Hilbert space there is!* But do not get a false impression from that statement: The principal interest of $\mathsf{L}^2(R^1)$, for example, is *not* that it is a Hilbert space, but that it is populated by measurable functions on the line, so that differential operators and the like can come into play; that the line sits in the complex plane, so that complex function theory can be brought to bear; and so on. All that is *accidental* from the Hilbert-space viewpoint, but *of the first importance* for applications. This point will become clearer as you proceed. Now, back to the geometry of $\mathsf{L}^2(Q)$.

A pair of functions f_1 and f_2 from $\mathsf{L}^2(Q)$ is *perpendicular*[1] if their inner product is 0, i.e., if $(f_1, f_2) = 0$. A family of functions $f_n: n \geqslant 1$ from $\mathsf{L}^2(Q)$ is a *perpendicular family* if $(f_i, f_j) = 0$ for $i \neq j$; if, in addition, the functions are of unit length $[\|f_n\| = 1]$, it is a *unit-perpendicular family*.

---

[1] The customary adjective is "orthogonal"; "perpendicular" is more geometrical and therefore preferable (at least it seemed to be at one time).

EXAMPLE 1.   The functions

$$e_n(x) = e^{2\pi inx}, \qquad n \in Z^1, \quad 0 \leqslant x \leqslant 1$$

form a unit-perpendicular family in $L^2[0,1]$.

EXERCISE 1.   Check that for any unit-perpendicular family with $n \leqslant \infty$ members

$$\|e_i - e_j\| = \sqrt{2} \qquad \text{for} \quad i \neq j.$$

The geometrical content is that on the surface of the unit sphere $(f: \|f\| = 1)$ there lie $n \leqslant \infty$ points at mutual distances $\sqrt{2}$! Notice especially, that *no* subsequence of the e_n's converges in $L^2(Q)$.

EXERCISE 2.   Check that for any unit-perpendicular family $e_n: n \geqslant 1$ and any $c \in L^2(Z^+)$, the sum $\sum_{n=1}^{\infty} c_n e_n$ converges in $L^2(Q)$ to a function f, i.e., $\|f - \sum_{k=1}^{n} c_k e_k\|$ tends to 0 as $n \uparrow \infty$. Prove the Pythagorean rule

$$\|f\|^2 = \sum_{n=1}^{\infty} |c_n|^2 = \|c\|^2.$$

A family of functions $f_n: n \geqslant 1$ *spans* $L^2(Q)$ if *finite* sums $\sum_{k=1}^{n} c_k f_k$ with complex coefficients are dense in the latter, that is, if for any $f \in L^2(Q)$, you can make

$$\left\| f - \sum_{k=1}^{n} c_k f_k \right\|$$

as small as you please by proper choice of $1 \leqslant n < \infty$ and the coefficients $c_1, \ldots, c_n$. Naturally, the choice of coefficients depends upon f and may change as n is increased: As closer approximations to f are sought by increasing the number of summands, the earlier coefficients may need to be adjusted. An important exception to this self-evident observation occurs in the case of a unit-perpendicular family. This fact is at the basis of standard curve-fitting recipes and is the content of the following theorem.

THEOREM 1.   *Let $e_n: n \geqslant 1$ be any unit-perpendicular family from $L^2(Q)$. Then for any $f \in L^2(Q)$, any $n \geqslant 1$, and any complex numbers $c_1, \ldots, c_n$,*

$$\left\| f - \sum_{k=1}^{n} \hat{f}(k) e_k \right\| \leqslant \left\| f - \sum_{k=1}^{n} c_k e_k \right\|,$$

in which $\hat{f}(k)$ is the so-called Fourier coefficient:

$$\hat{f}(k) = (f, e_k) = \int_Q f e_k^*.$$

The lower bound on the left-hand side is attained iff

$$c_k = \hat{f}(k) \qquad \text{for every } k \leqslant n.$$

PROOF.   To see this, expand the right-hand side as follows:

$$\left\| f - \sum_{k=1}^{n} c_k e_k \right\|^2 = \left\| f - \sum_{k=1}^{n} \hat{f}(k) e_k + \sum_{k=1}^{n} [\hat{f}(k) - c_k] e_k \right\|^2$$

$$= \left\| f - \sum_{k=1}^{n} \hat{f}(k) e_k \right\|^2$$

$$+ 2 \operatorname{Re}\left(f - \sum_{k=1}^{n} \hat{f}(k) e_k, \sum_{j=1}^{n} [\hat{f}(j) - c_j] e_j \right)$$

$$+ \left\| \sum_{k=1}^{n} [\hat{f}(k) - c_k] e_k \right\|^2.$$

The large inner product in the middle expression is a sum of constant multiples of terms like

$$\left(f - \sum_{k=1}^{n} \hat{f}(k) e_k, e_j \right) = (f, e_j) - \sum_{k=1}^{n} \hat{f}(k) (e_k, e_j)$$

$$= (f, e_j) - \hat{f}(j) = 0,$$

so you are left with

$$\left\| f - \sum_{k=1}^{n} c_k e_k \right\|^2 = \left\| f - \sum_{k=1}^{n} \hat{f}(k) e_k \right\|^2 + \sum_{k=1}^{n} |\hat{f}(k) - c_k|^2.$$

The proof is finished.

A (unit-perpendicular) *basis* of $L^2(Q)$ is a unit-perpendicular family which spans the whole space.

EXERCISE 3.   Check that a unit-perpendicular family is a basis if

$$f = \sum_{n=1}^{\infty} \hat{f}(n) e_n$$

for any $f \in L^2(Q)$. The sum is to converge in $L^2(Q)$; it is the so-called Fourier series of f.

EXERCISE 4.   Check that for any unit-perpendicular family and any $f \in L^2(Q)$,

$$0 \leqslant \left\| f - \sum_{k=1}^{n} \hat{f}(k) e_k \right\|^2 = \|f\|^2 - \sum_{k=1}^{n} |\hat{f}(k)|^2.$$

Deduce *Bessel's inequality*,

$$\sum_{n=1}^{\infty} |\hat{f}(n)|^2 \leqslant \|f\|^2$$

and conclude that $e_n : n \geqslant 1$ is a basis iff the *Plancherel identity* holds:

$$\sum_{n=1}^{\infty} |\hat{f}(n)|^2 = \|f\|^2$$

for every $f \in L^2(Q)$.

EXERCISE 5.   A unit-perpendicular family spans $L^2(Q)$ iff the only function $f \in L^2(Q)$, which is perpendicular to the whole family is $f \equiv 0$. *Hint*:   To what is $f - \sum \hat{f}(n) e_n$ perpendicular?

THEOREM 2. $L^2(Q)$ *has a unit-perpendicular basis.*

PROOF. $L^2(Q)$ is separable, so you can find a countable dense family of functions, which may be listed as $f_1, f_2, \ldots$. The function f_n is rejected if it can be expressed as a finite sum of complex multiples of the preceding functions $f_k : k < n$. The functions that survive this weeding out are relisted as $g_1, g_2, \ldots$. A unit-perpendicular family is now constructed by the so-called Gram–Schmidt recipe:

$$e_1 = \frac{g_1}{\|g_1\|},$$
$$\vdots$$
$$e_n = \frac{g_n - \sum_{k<n} (g_n, e_k) e_k}{\|g_n - \sum_{k<n} (g_n, e_k) e_k\|},$$
$$\vdots$$

The weeding out ensures that none of the lengths by which you have to divide is 0, and it is self-evident that the $e_n : n \geqslant 1$ form a unit-perpendicular family. The geometrical content of the recipe for e_n is that you project out all the components of g_n in the directions $e_k : k < n$, so that what is left over is perpendicular to $e_k : k < n$. The process is illustrated for $n = 2$ in Fig. 1.

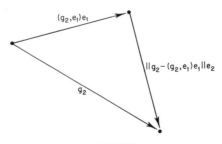

FIGURE 1

Now it is easy to see that the family $e_n: n \geqslant 1$ spans $L^2(Q)$, since each f from the original list can be expressed as a finite sum of complex multiples of the e's. This finishes the proof.

EXERCISE 6.   Show that $L^2(Q)$ cannot have a finite basis; this is what is meant by saying that $L^2(Q)$ is "infinite-dimensional." *Hint:* If A and B are disjoint measurable subsets of Q of finite measure, then 1_A and 1_B are perpendicular in $L^2(Q)$.

EXERCISE 7.   The functions $x^n: n \geqslant 0$ span $L^2[-1, 1]$; see Subsection 1.7.3 for the proof. Compute the first few functions of the corresponding unit-perpendicular basis by the Gram–Schmidt recipe. Apart from indexing and multiplicative constants, these are the so-called Legendre polynomials which play an important role in Sections 4.8–4.14 in connection with the rotation group.

Answer:

$$e_1 = \frac{1}{\sqrt{2}}, \qquad\qquad e_2 = \frac{\sqrt{3}}{\sqrt{2}} x,$$

$$e_3 = \frac{\sqrt{5}}{\sqrt{2}}\left(\frac{3}{2} x^2 - \frac{1}{2}\right), \qquad e_4 = \frac{\sqrt{7}}{\sqrt{2}}\left(\frac{5}{2} x^3 - \frac{3}{2} x\right).$$

The next item of business is the Riesz–Fischer theorem:

THEOREM 3. $L^2(Q)$ *and* $L^2(Z^+)$ *are isomorphic, i.e., there is a* 1:1, *linear, distance-preserving map of* $L^2(Q)$ *onto* $L^2(Z^+)$. *The adjective "linear" means that the map respects multiplication by complex numbers and addition.*

To be more specific, if $e_n: n \geqslant 1$ is any unit-perpendicular basis of $L^2(Q)$ and if $\hat{f}(n) = (f, e_n)$, then the map $f \to \hat{f}$ is such an *isomorphism*.

Amplification: Hilbert spaces can be classified by their dimension. Any unit-perpendicular basis of a fixed Hilbert space H contains the same number n ($\leqslant \infty$) of members; this number is the dimension of H, and it is a general fact that H is isomorphic to C^n in the sense used just above. The proof is exactly the same as for $L^2(Q)$.

PROOF.   By Exercise 4, the Fourier coefficients $\hat{f}(n) = (f, e_n)$ satisfy the Plancherel identity

$$\|f\|^2 = \sum_{n=1}^{\infty} |\hat{f}(n)|^2.$$

Therefore, $f \to \hat{f}$ is a length-preserving map of $L^2(Q)$ into $L^2(Z^+)$. The fact that the map is $1:1$ and linear is self-evident. Finally, the map is onto $L^2(Z^+)$: Namely, by Exercise 2, if $c \in L^2(Z^+)$, then $\sum_{n=1}^{\infty} c_n e_n$ converges to a function $f \in L^2(Q)$, and

$$\hat{f}(k) = (f, e_k) = \left(\lim_{n \uparrow \infty} \sum_{j=1}^{n} c_j e_j, e_k \right)$$

$$= \lim_{n \uparrow \infty} \sum_{j=1}^{n} c_j (e_j, e_k)$$

$$= c_k,$$

for every $k \geqslant 1$, that is, $\hat{f} = c$. The proof is finished.

EXERCISE 8.   Check that the map $\wedge$ automatically preserves inner products:

$$(f_1, f_2) = \int_Q f_1 f_2^* = (\hat{f}_1, \hat{f}_2) = \sum_{n=1}^{\infty} \hat{f}_1(n) \hat{f}_2^*(n).$$

This fact is the so-called *Parseval identity*. *Hint*: $\|f_1 - f_2\| = \|(f_1 - f_2)^{\wedge}\| = \|\hat{f}_1 - \hat{f}_2\|$.

A little glossary of the chief results of this section may be helpful; $e_n: n \geqslant 1$ is any unit-perpendicular basis of $L^2(Q)$ and $\hat{f}(n) = (f, e_n)$.

Bessel: $\|f\|^2 = \int_Q |f|^2 \geqslant \|\hat{f}\|^2 = \sum |\hat{f}|^2.$

Plancherel: $\|f\|^2 = \|\hat{f}\|^2.$

Parseval[2]: $(f_1, f_2) = \int_Q f_1 f_2^* = (\hat{f}_1, \hat{f}_2) = \sum \hat{f}_1 \hat{f}_2^*.$

Riesz–Fischer: $\wedge$ is an isomorphism from $L^2(Q)$ onto $L^2(Z^+)$.

EXERCISE 9.   Check that the so-called Haar functions

$$e_0^0(x) = 1 \qquad \text{for } 0 \leqslant x \leqslant 1,$$

$$e_n^k(x) = \begin{cases} 2^{n/2} & \text{for } \dfrac{k-1}{2^n} \leqslant x < \dfrac{k-\frac{1}{2}}{2^n}, \\[2mm] -2^{n/2} & \text{for } \dfrac{k-\frac{1}{2}}{2^n} \leqslant x < \dfrac{k}{2^n}, \\[2mm] 0 & \text{otherwise}, \end{cases}$$

---

[2] The distinction drawn here between Parseval and Plancherel is for convenience and is not historically accurate. Parseval is generally credited with (a primitive version of) both formulas for Fourier series, while Plancherel is more properly associated with the Fourier integral of Section 2.4.

defined for $1 \leqslant k \leqslant 2^n$ and $n \geqslant 1$, form a unit-perpendicular basis of $L^2[0,1]$. Draw pictures of the first few functions. *Hint:* To prove that the functions span, verify that if f is perpendicular to them all, then its indefinite integral $\int_0^x f$ vanishes for every x of the form $k/2^n$ with $1 \leqslant k \leqslant 2^n$ and $n \geqslant 0$, and so $\int_B f = 0$ for every measurable $B \subset [0,1]$.

The following exercises have to do with a *closed subspace* A of $L^2(Q)$. The word *subspace* means that A is closed under addition and under multiplication by complex constants; the modifier *closed* signifies that A is also closed under limits in the sense of $L^2(Q)$. The subspace A is *not* assumed to be separable; this property is inherited from $L^2(Q)$, as will appear in Exercise 13.

EXERCISE 10. The annihilator $B(= A^0)$ of A is the class of functions from $L^2(Q)$ that are perpendicular to every function from A. Check that B is also a closed subspace. Is this true if the subspace A is not closed?

EXERCISE 11. For any $f \in L^2(Q)$, there is a point Pf in A which is closest to f, i.e.,

$$\|f - Pf\| \leqslant \|f - g\| \qquad \text{for any} \quad g \in A.$$

Hint: Pick $g_n \in A$ so as to make $\|f - g_n\| \downarrow \inf_A \|f - g\|$. Then apply the Gram–Schmidt recipe to convert $g_n : n \geqslant 1$ into a unit-perpendicular family $e_n : n \geqslant 1$ and put $Pf = \sum (f, e_n) e_n$; see Theorem 1 for help with the rest.

EXERCISE 12. $f - P$ belongs to $B = A^0$. Check this and verify that $f = Pf + (f - Pf)$ is the *only* way of splitting f into a piece from A and a piece from B. This state of affairs is summed up by saying that $L^2(Q)$ is the (perpendicular) sum of A and B:

$$L^2(Q) = A \oplus B.$$

Hint: Pick $g \in A$. Then $k(\varepsilon) = \|f - Pf + \varepsilon g\|^2$ is a polynomial of degree 2 and is least for $\varepsilon = 0$. Compute $k'(0)$.

EXERCISE 13. The so-called projection $f \to Pf$ of Exercise 12 is a linear map of $L^2(Q)$ into itself, i.e., it respects addition and multiplication by complex constants. Why? Besides, as you will verify,

(a) $P^2 = P$,
(b) $(Pf_1, f_2) = (f_1, Pf_2)$,
(c) $\|Pf\| \leqslant \|f\|$,
(d) $P = 1$ on A and 0 on B.

Item (d) justifies calling P *the* projection upon A. Now check that A inherits the separability of $\mathsf{L}^2(Q)$.

EXERCISE 14. Show that the family f_n: $n \geqslant 1$ spans $\mathsf{L}^2(Q)$ iff $(f, f_n) = 0$ for every $n \geqslant 1$ implies $f \equiv 0$. *Hint:* What is the annihilator of the family f_n: $n \geqslant 1$?

EXERCISE 15. Show that any linear map l of $\mathsf{L}^2(Q)$ into the complex numbers which is bounded in the sense that

$$|l(f)| \leqslant \text{constant} \times \|f\|$$

where the constant is independent of f, can be expressed as an inner product:

$$l(f) = (f, g)$$

for some $g \in \mathsf{L}^2(Q)$. This is the so-called *Riesz representation theorem*. *Hint.* Suppose $l \not\equiv 0$ and let $\mathsf{A} = (f: l(f) = 0)$. Check that $\mathsf{B} = \mathsf{A}^0$ is of dimension 1 and find a function $g \in \mathsf{L}^2(Q)$ so that $l(f) = 0$ iff $(f, g) = 0$. Then

$$l(f) = \|g\|^{-2} l(g)(f, g) = \text{constant} \times (f, g).$$

Why?

There are many excellent books on Hilbert space; among the best at an advanced level are Akhiezer and Glazman [1961–1963] and Riesz and Sz.-Nagy [1955]; at an elementary level, Berberian [1961] is recommended.

1.4 SQUARE SUMMABLE FUNCTIONS ON THE CIRCLE AND THEIR FOURIER SERIES

Attention is now focused on the space $\mathsf{L}^2(S^1)$, where S^1 is a unit circular circumference. This may be thought of as the interval $0 \leqslant x \leqslant 1$ with the endpoints 0 and 1 identified, and you may think of functions on S^1 as periodic functions on R^1 of period 1, so that $f(x+1) = f(x)$. A function f is continuous on the circle only if $f(0) = f(0+) = f(1) = f(1-)$. You can also picture the circle as the interval $-\frac{1}{2} \leqslant x \leqslant \frac{1}{2}$ with identification of endpoints; this attitude will be helpful occasionally. The space $\mathsf{L}^2(S^1)$ is the Hilbert space of (complex) measurable functions f on S^1 with

$$\|f\| = \|f\|_2 = \left(\int_0^1 |f|^2 \right)^{1/2} < \infty.$$

The inner product is

$$(f,g) = \int_0^1 fg^*,$$

and there is a self-evident isomorphism with $L^2[0,1]$. The space $C^n(S^1)$ of $n\,(<\infty)$ times continuously differentiable functions on the circle will also come into play, as well as the space $C^\infty(S^1)$ of infinitely differentiable functions. The space $C^\infty(S^1)$ is dense in $L^2(S^1)$ by Exercise 1.2.11. The space $C^0(S^1) = C(S^1)$ is just the space of continuous functions; for such functions, the new length

$$\|f\|_\infty = \max_{0 \leqslant x < 1} |f(x)|$$

is often convenient. The subscript ∞ will distinguish this notion of length from

$$\|f\|_2 = \left(\int_0^1 |f|^2\right)^{1/2} \leqslant \|f\|_\infty.$$

THEOREM 1.   *The unit-perpendicular family*

$$e_n(x) = e^{2\pi i n x}, \qquad n \in Z^1$$

of Example 1.3.1 is a basis for $L^2(S^1)$, *that is, any function* $f \in L^2(S^1)$ *can be expanded into a Fourier series*

$$f = \sum_{n=-\infty}^\infty \hat{f}(n)\, e_n$$

with coefficients

$$\hat{f}(n) = (f, e_n) = \int_0^1 f e_n^* = \int_0^1 f(x)\, e^{-2\pi i n x}\, dx,$$

the sum being understood in the sense of distance in $L^2(S^1)$. *By Theorem 1.3.3, the map* $f \to \hat{f}$ *is therefore an isomorphism of* $L^2(S^1)$ *onto* $L^2(Z^1)$ *[see Exercise 1.2.13], and there is a Plancherel identity:*

$$\|f\|_2^2 = \int_0^1 |f|^2 = \|f\|^2 = \sum_{n=-\infty}^\infty |\hat{f}(n)|^2.$$

Warning: Until further notice e_n *will always mean* $e^{2\pi i n x}$; *and "Fourier series" will refer to this particular family!*

EXERCISE 1. $1, \sqrt{2}\cos 2\pi n x : n \geqslant 1$, and $\sqrt{2}\sin 2\pi n x : n \geqslant 1$ form a unit-perpendicular family in $L^2(S^1)$. Check this and deduce that the (complex) Fourier series for f can also be expressed in the real form

$$f = \hat{f}_{\text{even}}(0) + \sum_{n=1}^\infty [\hat{f}_{\text{even}}(n)\sqrt{2}\cos 2\pi n x + \hat{f}_{\text{odd}}(n)\sqrt{2}\sin 2\pi n x],$$

with coefficients

$$\hat{f}_{\text{even}}(0) = \int_0^1 f(x)\, dx,$$

$$\hat{f}_{\text{even}}(n) = \sqrt{2}\int_0^1 f(x)\cos 2\pi nx\, dx, \qquad n \geqslant 1,$$

$$\hat{f}_{\text{odd}}(n) = \sqrt{2}\int_0^1 f(x)\sin 2\pi nx\, dx, \qquad n \geqslant 1.$$

The key step in proving that the exponentials $e_n \colon n \in Z^1$ span $L^2(S^1)$ is to check that the Fourier series of a smooth function f actually converges (to f!). This is the content of the following theorem.

THEOREM 2.   *For any $1 \leqslant p < \infty$ and any $f \in C^p(S^1)$, the partial sums*

$$S_n = S_n(f) = \sum_{|k| \leqslant n} \hat{f}(k)\, e_k$$

converge to f, uniformly as $n \uparrow \infty$; in fact, $\|S_n - f\|_\infty$ is bounded by a constant multiple of $n^{-p+\frac{1}{2}}$.

Amplification: The bound on $\|S_n - f\|_\infty$ indicates that the speed of convergence of a Fourier series improves with the smoothness of f. This reflects the fact that *local* features of f (such as smoothness) are reflected in *global* features of $\hat{f}$ (such as rapid decay at $n = \pm\infty$). This local–global duality is one of the major themes of Fourier series and integrals, as you will see later.

PROOF   (for $p = 1$).   Bring in the so-called *Dirichlet kernel*

$$D_n(x) = \sum_{|k| \leqslant n} e_k(x) = \sum_{|k| \leqslant n} e^{2\pi ikx}$$

$$= e^{-2\pi inx} \sum_{k=0}^{2n} e^{2\pi ikx}$$

$$= e^{-2\pi inx}\, \frac{e^{2\pi i(2n+1)x} - 1}{e^{2\pi ix} - 1}$$

$$= \frac{\sin \pi(2n+1)x}{\sin \pi x},$$

with the understanding that $D_n(0) = (2n+1)$, and note that you obtain the value of the forbidding-looking integral

$$\int_0^1 D_n = \int_0^1 \frac{\sin \pi(2n+1)x}{\sin \pi x}\, dx$$

as a fringe benefit:

$$\int_0^1 D_n = \sum_{|k| \leqslant n} \int_0^1 e_k = 1.$$

The introduction of D_n is motivated by the desire to express S_n in a more transparent way:

$$S_n(x) = \sum_{|k| \leqslant n} \hat{f}(k) e_k(x) = \sum_{|k| \leqslant n} e_k(x) \int_0^1 f(y) e_k^*(y) \, dy$$

$$= \int_0^1 \sum_{|k| \leqslant n} e_k(x-y) f(y) \, dy$$

$$= \int_0^1 D_n(x-y) f(y) \, dy$$

$$= \int_{-\frac{1}{2}}^{\frac{1}{2}} f(x+y) D_n(y) \, dy.$$

To achieve the final expression, make the substitution $y - x \to y$, and use the fact that D_n is even:

$$\int_0^1 D_n(x-y) f(y) \, dy = \int_{-x}^{1-x} f(x+y) D_n(-y) \, dy = \int_{-x}^{1-x} f(x+y) D_n(y) \, dy;$$

then notice that $f(x + \cdot) D_n$ is a periodic function of period 1, so that you can replace the integral over $-x \leqslant y \leqslant 1 - x$ by the integral over *any* period you like, for example, $-\frac{1}{2} \leqslant x \leqslant \frac{1}{2}$.

The problem of verifying that S_n is a good approximation to f can now be better understood with the aid of a picture of D_n; see Fig. 1 for a sketch of D_8. The peak tends to ∞ with n. At the same time, the oscillations to either side become increasingly rapid, and while they do not die away, you can

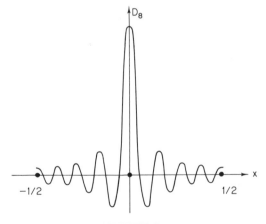

FIGURE 1

hope that they will, on the average, knock each other out, with the result that the major contribution to the integral comes from a small neighborhood of $y = 0$. The plan is to take (indirect) advantage of this phenomenon.

To begin with, since $f \in C^1(S^1)$,

$$(f')^\wedge(n) = \int_0^1 f' e_n^* = -\int_0^1 f e_n'^* = 2\pi i n \hat{f}(n)$$

by partial integration. Therefore, by the inequalities of Schwarz and Bessel, if $n \leqslant n' < \infty$, then

$$|S_n - S_{n'}| \leqslant \sum_{|k|>n} |\hat{f}(k)| = \sum_{|k|>n} |(f')^\wedge(k)| \, |2\pi k|^{-1}$$

$$\leqslant \left(\sum_{|k|>n} |(f')^\wedge(k)|^2 \right)^{1/2} \left(\sum_{|k|>n} (2\pi k)^{-2} \right)^{1/2}$$

$$\leqslant \|f'\|_2 \times \text{a constant multiple of } n^{-1/2}.$$

This shows that S_n converges uniformly, at the advertised speed, to *something*. The only question is: *to what*?

Because

$$\int_{-\frac{1}{2}}^{\frac{1}{2}} D_n = 1,$$

the discrepancy between f and the partial sum can be expressed as

$$S_n(x) - f(x) = \int_{-\frac{1}{2}}^{\frac{1}{2}} [f(x+y) - f(x)] D_n(y) \, dy$$

$$= \int_{-\frac{1}{2}}^{\frac{1}{2}} Q(x,y) \sin \pi(2n+1) y \, dy,$$

with the understanding that

$$Q(x,y) = [f(x+y) - f(x)]/\sin \pi y$$

stands for $\pi^{-1} f'(x)$ at $y = 0$. Fix $-\frac{1}{2} \leqslant x < \frac{1}{2}$. As a function of $-\frac{1}{2} \leqslant y < \frac{1}{2}$, Q belongs to $L^2[-\frac{1}{2}, \frac{1}{2})$, and

$$S_n(x) - f(x) = \int_{-\frac{1}{2}}^{\frac{1}{2}} Q(x,y) \frac{e^{2\pi i n y} e^{\pi i y} - e^{-2\pi i n y} e^{-\pi i y}}{2i} \, dy$$

$$= (2i)^{-1} (Q^+)^\wedge(-n) - (2i)^{-1} (Q^-)^\wedge(n)$$

in which $Q^\pm = Q \exp(\pm \pi i y)$. By a second application of Bessel's inequality

$$\sum |(Q^\pm)^\wedge(n)|^2 \leqslant \|Q\|_2^2 < \infty,$$

so $(Q^\pm)^\wedge(n)$ approaches 0 as $|n|\uparrow\infty$, and $\lim_{n\uparrow\infty} S_n = f$ for each fixed $-\frac{1}{2} \leqslant x < \frac{1}{2}$, separately. Now make $n'\uparrow\infty$ in the preceding estimate for $|S_n - S_{n'}|$ to verify that $\|S_n - f\|_\infty$ is bounded by a constant multiple of $n^{-1/2}$. This completes the proof for the case $p = 1$.

EXERCISE 2. Finish the proof of Theorem 1. *Hint:* The space $\mathbf{C}^1(S^1)$ is dense in $\mathbf{L}^2(S^1)$ by Exercise 1.2.11.

EXERCISE 3. Finish the proof of Theorem 2 for $2 \leqslant p < \infty$. *Hint:* $(f')^\wedge(n) = 2\pi in\hat{f}(n)$.

EXERCISE 4. Check that $f \in \mathbf{C}^\infty(S^1)$ iff $\hat{f}$ is *rapidly decreasing* in the sense that $n^p\hat{f}(n)$ approaches 0 as $|n|\uparrow\infty$ for every $p < \infty$, separately. *Hint:* For rapidly decreasing $\hat{f}$, $\sum \hat{f}(n)e_n'$ converges uniformly to a periodic function f_1, and

$$\int_0^x f_1 = \sum \hat{f}(n) \int_0^x e_n' = \sum \hat{f}(n) [e_n(x) - e_n(0)] = f(x) - f(0).$$

The pointwise convergence of the Fourier series of a function from $\mathbf{L}^2(S^1)$ is a very complicated business in general. Carleson [1966] proved that it must converge a.e., but there are examples with $f \in \mathbf{C}(S^1)$ in which the sum diverges at uncountably many points. The situation for summable functions $[\int_0^1 |f| < \infty]$ is even more horrible; the most famous example is that of Kolmogorov [1926] in which the sum diverges everywhere! The proofs by Carleson and Kolmogorov are complicated and cannot be presented here; in fact, these remarks are merely meant to point out the very attractive simplicity of the following result of Fejér [1904].

THEOREM 3. *For functions f of class $\mathbf{C}(S^1)$, the arithmetic means $n^{-1}(S_0 + \cdots + S_{n-1})$ of the partial sums $S_n = \sum_{|k| \leqslant n} \hat{f}(k)e_k$ converge uniformly to f.*

EXERCISE 5. Check that the arithmetic means $n^{-1}(x_0 + \cdots + x_{n-1})$ of the numerical series $x_0, x_1, \ldots$ converge to y if $\lim_{n\uparrow\infty} x_n = y$. Give an example to show that $\lim_{n\uparrow\infty} n^{-1}(x_0 + \cdots + x_{n-1})$ can exist even if $\lim_{n\uparrow\infty} x_n$ does not.

PROOF OF THEOREM 3. The proof makes use of the arithmetical means of the Dirichlet kernel:

$$F_n(x) = \frac{1}{n}(D_0 + \cdots + D_{n-1})$$

$$= \frac{1}{n}\sum_{k=0}^{n-1} \frac{\sin \pi(2k+1)x}{\sin \pi x}$$

$$= \frac{1}{n}\left[\frac{\sin n\pi x}{\sin \pi x}\right]^2.$$

EXERCISE 6.   Check the summation.   *Hint:*   Think of $\sin \pi(2k+1)x$ as the imaginary part of the complex exponential and sum the resulting geometrical series.

F_n is the so-called *Fejér kernel.* Note for future use that F_n is nonnegative and that

$$\int_{-\frac{1}{2}}^{\frac{1}{2}} F_n = n^{-1} \sum_{k=0}^{n-1} \int_{-\frac{1}{2}}^{\frac{1}{2}} D_k = 1.$$

The proposal is to check that the discrepancy

$$n^{-1} \sum_{k=0}^{n-1} S_k(x) - f(x) = n^{-1} \sum_{k=0}^{n-1} \int_{-\frac{1}{2}}^{\frac{1}{2}} f(x+y) D_k(y) \, dy - f(x)$$

$$= \int_{-\frac{1}{2}}^{\frac{1}{2}} f(x+y) F_n(y) \, dy - f(x)$$

$$= \int_{-\frac{1}{2}}^{\frac{1}{2}} [f(x+y) - f(x)] F_n(y) \, dy$$

is uniformly small. This is much easier for the present Fejér kernel F_n than it was for the Dirichlet kernel D_n since now the tails are small for large n and not just negligible due to rapid oscillation; see Fig. 2 for a sketch of F_8.

To make the actual estimates pick a small positive number $\delta < \frac{1}{2}$ and divide the interval of integration into two parts according as $|y| < \delta$ or $|y| \geq \delta$. The first piece is bounded as follows:

$$\left| \int_{|y| < \delta} \right| \leq \int_{-\delta}^{\delta} |f(x+y) - f(x)| F_n(y) \, dy$$

$$\leq \max_{|y| \leq \delta} \max_{|x| \leq \frac{1}{2}} |f(x+y) - f(x)| \int_{|y| < \delta} F_n(y) \, dy$$

$$\leq \max_{|y| \leq \delta} \| f_y - f \|_\infty,$$

in which f_y is the translated function $f_y(x) = f(x+y)$. This can be made as small as you please by making $\delta \downarrow 0$, since f is uniformly continuous. As to the other piece,

$$\left| \int_{|y| \geq \delta} \right| \leq \frac{4}{n} \| f \|_\infty \int_{\delta}^{\frac{1}{2}} \left[\frac{\sin n\pi x}{\sin \pi x} \right]^2 dx$$

$$\leq \frac{2}{n} (\sin \pi\delta)^{-2} \| f \|_\infty,$$

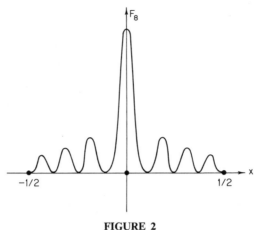

FIGURE 2

and this can be made small for fixed $0 < \delta < \frac{1}{2}$ by making n large enough. The proof is finished.

EXERCISE 7.   Check that for any $f \in L^2(S^1)$

$$\lim_{n\uparrow\infty} \frac{1}{n} \sum_{k=0}^{n-1} f\left(x + \frac{k}{n}\right) = \hat{f}(0) = \int_0^1 f$$

in the sense of distance in $L^2(S^1)$. *Hint:*   Compute the Fourier coefficients of the sum and use the Plancherel identity.

EXERCISE 8.   Check that for any $f \in C(S^1)$ and $0 \leqslant r < 1$,

$$\sum \hat{f}(n)\, r^{|n|}\, e_n(x) = \int_0^1 \frac{1 - r^2}{1 - 2r\cos 2\pi(x-y) + r^2} f(y)\, dy.$$

Hint: Express $\hat{f}(n)$ as an integral and bring the sum under the integral sign.

EXERCISE 9.   Use Exercise 8 to prove the following variant of Theorem 3: For $f \in C(S^1)$,

$$\lim_{r\uparrow 1} \left\| \sum \hat{f}(n)\, r^{|n|}\, e_n - f \right\|_\infty = 0.$$

The fact goes back to Poisson; the formula of Exercise 8 is known by his name. *Hint:*

$$\int_0^1 \frac{1 - r^2}{1 - 2r\cos 2\pi y + r^2}\, dy = \sum \hat{1}(n)\, r^{|n|}\, e_n(0) = 1.$$

1.5   SUMMABLE FUNCTIONS AND THEIR FOURIER SERIES

The next topic is the space $L^1(S^1)$ of (complex) summable functions on the circle. This space, or more generally, the space $L^1(Q)$ of summable functions on an interval Q, is provided with a length

$$\|f\|_1 = \int_Q |f| = \int_Q |f(x)|\, dx,$$

permitting you to define a distance much as for $L^2(Q)$. Clearly, the triangle inequality,

$$\|f+g\|_1 = \int_Q |f+g| \leqslant \int_Q |f| + \int_Q |g| = \|f\|_1 + \|g\|_1,$$

is satisfied, so that $L^1(Q)$ is closed under addition of functions, and it goes without saying that it is also closed under multiplication by complex numbers.

Technical point: The reader should bear in mind that, just as for $L^2(Q)$ the things that live in $L^1(Q)$ are actually *equivalence classes* of functions; especially, you identify f with the 0-function if $f = 0$ a.e. so as to make $f = 0$ iff $\|f\|_1 = 0$. As before, it is simpler, and not at all confusing, to ignore this point most of the time and to speak of $f \in L^1(Q)$ as a *function*.

EXERCISE 1.   Check that the space $L^1(Q)$ is *not* a Hilbert space. *Hint:* In any Hilbert space $\|\alpha + \beta\|^2 + \|\alpha - \beta\|^2 = 2\|\alpha\|^2 + 2\|\beta\|^2$. Try this out for $Q = [0, 1]$, $\alpha = 1$, and $\beta = x$. What happens? Try again.

EXERCISE 2.   Show that the space $L^1(Q)$ is *complete* and *separable*. The definitions are as for $L^2(Q)$, but relative to the distance of $L^1(Q)$. *Hint:*   The proof for $L^2(Q)$ is easily adapted from Section 1.2 with minor changes only.

EXERCISE 3.   Show that the class of compact functions belonging to $C^\infty(Q)$ is dense in $L^1(Q)$. *Hint:*   See Exercise 1.2.11. There is an inclusion between $L^1(Q)$ and $L^2(Q)$ if Q is bounded: Namely, by Schwarz's inequality,

$$\|f\|_1^2 = \left(\int_Q |f| \times 1\right)^2 \leqslant \int_Q |f|^2 \int_Q 1^2 = \|f\|_2^2 m(Q),$$

so that $L^2(Q)$ is included in $L^1(Q)$. In particular, this holds for functions on the circle, so that everything to be proved later about Fourier series of summable functions on the circle also holds for functions of class $L^2(S^1)$.

EXERCISE 4.   Check by example that the inclusion of $L^2[0, 1]$ in $L^1[0, 1]$ is *proper*, that is, find a summable function f with $\int_0^1 |f|^2 = \infty$.

EXERCISE 5. Check by examples that neither $L^1(R^1)$ nor $L^2(R^1)$ is included in the other.

EXERCISE 6. Show that any linear map l of $L^1(Q)$ into the complex numbers, subject to $|l(f)| \leqslant$ constant $\times \|f\|_1$, with a constant independent of f, can be expressed as $l(f) = \int_Q fg^*$ for some bounded measurable function g. *Hint:* $L^2(Q) \subset L^1(Q)$ if Q is bounded. Now use Exercise 1.3.15 to find such a function $g \in L^2(Q)$ and check that $\int_b^a |g| \leqslant$ constant $\times (b-a)$ for any interval $a \leqslant x \leqslant b$.

Given $f \in L^1(S^1)$, the function fe_n^* is summable, so you can set up a (formal) Fourier series for f by the customary recipe:

$$f = \sum \hat{f}(n) e_n$$

with coefficients

$$\hat{f}(n) = \int_0^1 fe_n^* = \int_0^1 f(x) e^{-2\pi i n x} dx.$$

The principal fact about such Fourier series is contained in

THEOREM 1. *The arithmetical means* $n^{-1}(S_0 + \cdots + S_{n-1})$ *of the partial sums* $S_n = \sum_{|k| \leqslant n} \hat{f}(k) e_k$ *converge to* f *in the sense of distance in* $L^1(S^1)$:

$$\lim_{n \uparrow \infty} \|n^{-1}(S_0 + \cdots + S_{n-1}) - f\|_1 = 0;$$

in particular, the map $f \to \hat{f}$ *is* 1:1.

PROOF. The discrepancy between $n^{-1}(S_0 + \cdots + S_{n-1})$ and f may be expressed by means of Fejér's kernel F_n, as in the proof of Theorem 1.4.3:

$$n^{-1}(S_0 + \cdots + S_{n-1}) - f = \int_{-\frac{1}{2}}^{\frac{1}{2}} [f(x+y) - f(x)] F_n(y) \, dy,$$

and the length of the discrepancy can be bounded as follows:

$$\|n^{-1}(S_0 + \cdots + S_{n-1}) - f\|_1 \leqslant \int_{-\frac{1}{2}}^{\frac{1}{2}} dx \int_{-\frac{1}{2}}^{\frac{1}{2}} |f(x+y) - f(x)| F_n(y) \, dy$$

$$= \int_{-\frac{1}{2}}^{\frac{1}{2}} \left(\int_{-\frac{1}{2}}^{\frac{1}{2}} |f(x+y) - f(x)| \, dx \right) F_n(y) \, dy$$

$$= \int_{-\frac{1}{2}}^{\frac{1}{2}} \|f_y - f\|_1 F_n(y) \, dy,$$

in which f_y is the translated function $f_y(x) = f(x+y)$. The rest of the proof runs parallel to that of Theorem 1.4.3. The only new ingredient is contained in

EXERCISE 7. Show that the map $f \to f_y$ is continuous in the distance of $L^1(S^1)$, i.e.,

$$\lim_{y \to 0} \|f_y - f\|_1 = 0.$$

Hint: $\|f_y - f\|_1 \leqslant \|f_y - f\|_\infty$, for f from $C(S^1)$ and the latter is dense in $L^1(S^1)$.

The Fourier coefficients of a summable function f do not satisfy

$$\sum |\hat{f}(n)|^2 < \infty;$$

that happens only if $f \in L^2(S^1)$, but they are *bounded*:

$$|\hat{f}(n)| = \left| \int_0^1 f e_n^* \right| \leqslant \int_0^1 |f| = \|f\|_1.$$

This crude estimate is much improved upon by the so-called *Riemann–Lebesgue lemma*:

THEOREM 2. *The Fourier coefficient $\hat{f}(n)$ of any function $f \in L^1(S^1)$ tends to 0 as $|n| \uparrow \infty$.*

PROOF. $\hat{f}(n) = \int f e_n^*$ can also be expressed as

$$\hat{f}(n) = -\int_0^1 f(x) \exp[-2\pi i n(x - (2n)^{-1})] = -\int_0^1 f(x + (2n)^{-1}) e_n^*(x)\, dx,$$

since $e^{\pi i} = -1$. Now average the two expressions for $\hat{f}(n)$:

$$\hat{f}(n) = \tfrac{1}{2} \int_0^1 f(x) e_n^*(x)\, dx - \tfrac{1}{2} \int_0^1 f(x + (2n)^{-1}) e_n^*(x)\, dx$$

$$= \tfrac{1}{2} \int_0^1 [f(x) - f(x + (2n)^{-1})] e_n^*(x)\, dx$$

and estimate as follows:

$$|\hat{f}(n)| \leqslant \tfrac{1}{2} \int_0^1 |f(x) - f(x + (2n)^{-1})|\, dx = \tfrac{1}{2} \|f - f_{1/2n}\|_1.$$

By Exercise 7, this approaches 0 as $|n| \uparrow \infty$. The proof is finished.

An important application of the Riemann–Lebesgue lemma is to verify that the *local* convergence of the Fourier series of a summable function depends only upon the *local* behavior of the function. This is the content of

THEOREM 3. *Take $f \in L^1(S^1)$ vanishing near $x = 0$. Then S_n approaches 0 as $n \uparrow \infty$, uniformly near $x = 0$.*

Amplification: Obviously, the point $x=0$ is in no way special; in fact, if f and g are summable functions and if $f=g$ near some fixed point x_0 of the circle, then their partial Fourier sums behave in the same way in the vicinity of x_0: Namely, $S_n(f)-S_n(g)=S_n(f-g)$ tends to 0 uniformly, near x_0.

PROOF OF THEOREM 3.   Suppose that $f=0$ for $|x|\leqslant\delta$ and express the partial sum S_n by means of the Dirichlet kernel of Section 1.4:

$$S_n(x)=\int_{-\frac{1}{2}}^{\frac{1}{2}}f(x+y)\frac{\sin(2n+1)\pi y}{\sin\pi y}\,dy.$$

Because $f(x+y)=0$ if both $|x|$ and $|y|$ are $\leqslant\delta/2$, the functions

$$f^{\pm}(y)=\frac{f(x+y)e^{\pm i\pi y}}{2i\sin\pi y}$$

are summable if $|x|\leqslant\delta/2$, and therefore

$$S_n(x)=\int_{-\frac{1}{2}}^{\frac{1}{2}}\frac{f(x+y)}{\sin\pi y}\frac{e^{\pi iy}e^*_{-n}(y)-e^{-\pi iy}e_n{}^*(y)}{2i}\,dy$$

$$=(f^{+})^{\wedge}(-n)-(f^{-})^{\wedge}(n)$$

tends to 0 as $n\uparrow\infty$ for any fixed $|x|\leqslant\delta/2$, by the Riemann–Lebesgue lemma. At the same time, for any $|x_1|\leqslant\delta/2$ and $|x_2|\leqslant\delta/2$,

$$|S_n(x_2)-S_n(x_1)|\leqslant\int_{\frac{1}{2}\geqslant|y|\geqslant\delta/2}|f(x_2+y)-f(x_1+y)|\left|\frac{\sin(2n+1)\pi y}{\sin\pi y}\right|dy$$

$$\leqslant\left(\sin\frac{\pi\delta}{2}\right)^{-1}\|f_{x_2}-f_{x_1}\|_1$$

$$=\left(\sin\frac{\pi\delta}{2}\right)^{-1}\|f_y-f\|_1$$

for $y=x_2-x_1$. By Exercise 7, this makes S_n continuous for $|x|\leqslant\delta/2$, uniformly in n, and a moment's reflection will convince you that this forces the convergence of S_n for $|x|\leqslant\delta/2$ to be uniform. The proof is finished.

EXERCISE 8.   Dini's test states that if $f\in L^1(S^1)$ and if for fixed $|x|\leqslant\frac{1}{2}$, the function $y^{-1}[f(x+y)-f(x)]$ is summable, then $\lim_{n\uparrow\infty}S_n(x)=f(x)$. Prove it. *Hint:* Use the Dirichlet kernel and the Riemann–Lebesgue lemma, as earlier.

A new feature of summable functions now comes into play: The space

$L^1(S^1)$ is an algebra under the multiplication defined by the "convolution" product

$$f \circ g = \int_0^1 f(x-y)g(y)\,dy.$$

For fixed $0 \leqslant x < 1$, the product $f(x-\cdot)g$ may be nonsummable, so it is necessary to check that $f \circ g$ makes sense. To do this with all the proper technical flourishes, look first at the plane integral

$$I = \int_0^1 \int_0^1 |f(x-y)g(y)|\,dx\,dy.$$

The integrand $|f(x-y)g(y)|$ is a nonnegative (plane) measurable function, so the integral makes sense $[I \leqslant \infty]$, and by Fubini's theorem, you can evaluate it as an iterated integral:

$$I = \int_0^1 |g(y)|\,dy \int_0^1 |f(x-y)|\,dx = \int_0^1 |g| \int_0^1 |f| = \|f\|_1 \|g\|_1 < \infty.$$

Fubini is used once more to conclude from

$$I = \int_0^1 dx \int_0^1 |f(x-y)g(y)|\,dy < \infty$$

that $f(x-\cdot)g$ is summable for almost every $0 \leqslant x < 1$. Thus, $f \circ g(x)$ is given by an honest Lebesgue integral for almost every $0 \leqslant x < 1$ and is itself a (periodic) summable function:

$$\|f \circ g\|_1 \leqslant I = \|f\|_1 \|g\|_1 < \infty.$$

This kind of finicky proof is not very interesting, but it is important to understand precisely what is involved.

EXERCISE 9. Check that the product $f \circ g$ is associative and commutative, putting in all the technical details.

EXERCISE 10. Check that $L^2(S^1)$ is an ideal in $L^1(S^1)$. This means that $f \circ g$ belongs to $L^2(S^1)$ as soon as one of the factors does.

EXERCISE 11. Check that the Dirichlet and Fejér formulas for partial sums can be expressed as

$$S_n = f \circ D_n$$

and

$$n^{-1}(S_0 + \cdots + S_{n-1}) = f \circ F_n.$$

A pretty interplay takes place between the product $f \circ g$ and Fourier series, stemming from

$$(f \circ g)^{\wedge} = \hat{f}\hat{g}\,;$$

this simple formula plays a very important role in applications, as you will see later.

PROOF.   By Fubini's theorem,

$$(f \circ g)^{\wedge}(n) = \int_0^1 \left[\int_0^1 f(x-y)g(y)\, dy \right] e_n^*(x)\, dx$$

$$= \int_0^1 \int_0^1 f(x-y)\, e_n^*(x-y) g(y) e_n^*(y)\, dx\, dy$$

$$= \int_0^1 g(y) e_n^*(y)\, dy \left[\int_0^1 f(x-y) e_n^*(x-y)\, dx \right]$$

$$= \int_0^1 g e_n^* \int_0^1 f e_n^*$$

$$= \hat{f}(n)\,\hat{g}(n).$$

EXERCISE 12.   Check that $L^1(S^1)$ does not have a multiplicative identity. Hint: A multiplicative identity e would satisfy $e \circ f = f$. Now look at $\hat{e}$ keeping the Riemann–Lebesgue lemma in mind.

EXERCISE 13.   Define f^n to be the n-fold product $f \circ \cdots \circ f$ of a summable function f. Prove that

$$\lim_{n \uparrow \infty} (\|f^n\|_1)^{1/n} = \|\hat{f}\|_\infty = \max_{|n| < \infty} |\hat{f}(n)|,$$

under the extra assumption that f belongs to $L^2(S^1)$.[1] Hint: The fact that $\|\hat{f}\|_\infty$ does not exceed the left-hand side is self-evident. To finish the proof use

$$\|f^n\|_1 = \int f^n \{\exp[i\arg(f^n)]\}^* = \sum \hat{f}^n \{\exp[i\arg(f^n)]\}^{\wedge *}.$$

The map $f \to \hat{f}$ maps $L^2(S^1)$ onto $L^2(Z^1)$, and

$$\|f\|_2 = \|\hat{f}\|_2 = \left(\sum |\hat{f}|^2 \right)^{1/2} < \infty.$$

The situation for $L^1(S^1)$ is very much more complicated. The information

---

[1] For a proof of this formula without the extra assumption see, for example, Edwards [1967, article 11.4.14].

at hand about the class A of Fourier coefficients $\hat{f}$ of summable functions may be summarized as follows:

(a)   A is populated by bounded functions $\hat{f}$:

$$\|\hat{f}\|_\infty = \lim_{n\uparrow\infty}(\|f^n\|_1)^{1/n} \leqslant \|f\|_1.$$

(b) $\hat{f} = 0$ iff $f = 0$, i.e., $\wedge$ is a $1:1$ map.

(c) $\hat{f} = 0$ at $\pm\infty$, i.e., $\lim_{|n|\uparrow\infty} \hat{f}(n) = 0$.

(d)   A is an algebra: namely,

$$(\hat{f}\hat{g})(n) = \hat{f}(n)\hat{g}(n) = (f \circ g)^\wedge(n).$$

Unfortunately, (a) and (c) do not suffice to single out precisely which functions $\hat{f}$ arise as Fourier coefficients of summable functions. The situation is thus entirely different from $L^2(S^1)$ where the condition

$$\|\hat{f}\|_2 = (\sum|\hat{f}|^2)^{1/2} < \infty$$

is decisive. The best information currently available indicates that A does not have *any* neat description.

EXERCISE 14.   Check that the class B of summable functions $\hat{f}$ is a subalgebra of A. The adjective "summable" means that $\|\hat{f}\|_1 = \sum|\hat{f}(n)| < \infty$.

EXERCISE 15.   Check that $B^\vee = (f = \sum \hat{f}(n)\,e_n : \hat{f} \in B)$ is a subalgebra of $C(S^1)$ under the *ordinary* multiplication of functions. *Hint:* $(fg)^\wedge(n) = \sum \hat{f}(n-k)\hat{g}(k)$.

A celebrated theorem of Wiener [1933, p. 91] states that if $f \in B^\vee$ is root-free $[f \neq 0]$, then also $1/f \in B^\vee$. A wide variety of fascinating and delicate results about A and B have been obtained since that date; the advanced student will find a nice account in Edwards [1967].

1.6★   GIBBS' PHENOMENON

Thus far, the object has been to show how *well* Fourier series converge. Gibbs' phenomenon has to do with how poorly they converge in the vicinity of a jump of f. The statement is that in the vicinity of a simple jump of the function f, the partial sums S_n *always overshoot the mark by about 9%*. This fact was pointed out by Gibbs in a letter to *Nature* [1899]. (Actually Gibbs' phenomenon was first described by the British mathematician Wilbraham [1848]; see Carslaw [1925] for the history.) The function Gibbs considered was a sawtooth as in Fig. 1. Gibbs was replying to a letter by the physicist Michaelson to *Nature* [1898], in which the latter expressed himself doubtful as to the "idea that a real discontinuity [in f] can replace a sum of continuous curves [S_n]." Michaelson was reportedly put out with mathematics because the output of his machine for computing the first 80 terms of a Fourier series was not "close enough" to f at the jumps.

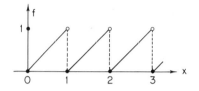

FIGURE 1

To investigate Gibbs' phenomenon, let us look at the function

$$f(x) = \begin{cases} -1 & \text{for} \quad -\tfrac{1}{2} \leqslant x < 0, \\ +1 & \text{for} \quad 0 \leqslant x < \tfrac{1}{2}. \end{cases}$$

The aim is to show that for large n and small $|x|$, S_n behaves as in Fig. 2: It approximates f well, except in a small neighborhood of $x = 0$, where it overshoots the levels by ± 1. A natural conjecture would be that the overshoot tends to 0 as $n \uparrow \infty$, but surprisingly, this does *not* take place; instead,

$$\lim_{n \uparrow \infty} \max S_n = 1.089490+,$$

i.e., the overshoot is ultimately about 9%. This is Gibbs' phenomenon; for better pictures: See Carslaw [1930, Chapter 9].

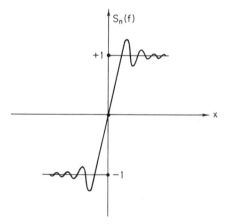

FIGURE 2

EXERCISE 1. Check that Gibbs' phenomenon must present itself in the vicinity of any jump of a piecewise smooth function. At such a jump $\lim_{n \uparrow \infty} S_n(x) = \frac{1}{2}f(x-) + \frac{1}{2}f(x+)$. *Hint:* The problem may be localized by Theorem 1.5.3. The other ingredient is an easy extension of Theorem 1.4.2 to jump-free piecewise smooth functions.

PROOF OF GIBBS' PHENOMENON. The first step is to express S_n in a convenient way, starting from the Dirichlet formula $S_n = f \circ D_n$ and taking advantage of the fact that D_n is even:

$$S_n = f \circ D_n = D_n \circ f = \int_{-\frac{1}{2}}^{\frac{1}{2}} D_n(x-y)f(y)\,dy$$

$$= -\int_{-\frac{1}{2}}^{0} D_n(y-x)\,dy + \int_{0}^{\frac{1}{2}} D_n(y-x)\,dy$$

$$= -\int_{-x-\frac{1}{2}}^{-x} D_n(y)\,dy + \int_{-x}^{-x+\frac{1}{2}} D_n(y)\,dy$$

$$= -\int_{x}^{x+\frac{1}{2}} D_n + \int_{x}^{-x+\frac{1}{2}} D_n + \int_{-x}^{x} D_n$$

$$= \int_{-x}^{x} D_n - \int_{-x+\frac{1}{2}}^{x+\frac{1}{2}} D_n.$$

The second integral is bounded by a constant multiple of $1/n$ if, for example, $|x| \leqslant \frac{1}{4}$: namely, by partial integration,

$$\int_{-x+\frac{1}{2}}^{x+\frac{1}{2}} D_n = \int_{-x+\frac{1}{2}}^{x+\frac{1}{2}} \frac{\sin \pi(2n+1)\,y}{\sin \pi y}\,dy$$

$$= -\frac{1}{\pi(2n+1)} \int_{-x+\frac{1}{2}}^{x+\frac{1}{2}} [\cos \pi(2n+1)\,y]' \frac{dy}{\sin \pi y}$$

$$= -\frac{1}{\pi(2n+1)} \frac{\cos \pi(2n+1)\,y}{\sin \pi y} \Big|_{-x+\frac{1}{2}}^{x+\frac{1}{2}}$$

$$-\frac{1}{2n+1} \int_{-x+\frac{1}{2}}^{x+\frac{1}{2}} \cos \pi(2n+1)\,y\, \frac{\cos \pi y}{\sin^2 \pi y}\,dy$$

so that

$$\left| S_n - \int_{-x}^{x} D_n \right| = \left| \int_{-x+\frac{1}{2}}^{x+\frac{1}{2}} D_n \right| \leqslant \text{constant} \times n^{-1}.$$

This permits you to replace S_n by $\int_{-x}^{x} D_n$ in estimating the overshoot. A second simplification is made by noticing that

$$\frac{1}{\sin \pi y} - \frac{1}{\pi y} = \frac{\pi y}{6} + \frac{7\pi^3 y^3}{360} + \frac{31\pi^5 y^5}{15,120} + \cdots$$

is of class $C^1[-\frac{1}{2},\frac{1}{2}]$. This permits you to integrate

$$\int_{-x}^{x} D_n(y)\, dy - \int_{-x}^{x} \frac{\sin \pi(2n+1)\, y}{\pi y} = \int_{-x}^{x} \left[\frac{1}{\sin \pi y} - \frac{1}{\pi y} \right] \sin \pi(2n+1)\, y\, dy$$

once by parts, and so to check that this expression is likewise bounded by a constant multiple of $1/n$ for $|x| \leqslant \frac{1}{4}$. The net result is that for $n \uparrow \infty$ and $|x| \leqslant \frac{1}{4}$, $S_n(x)$ is well approximated by

$$\int_{-x}^{x} \frac{\sin \pi(2n+1)\, y}{\pi y}\, dy = \frac{2}{\pi} \int_0^{\pi(2n+1)x} \frac{\sin y}{y}\, dy.$$

This integral achieves its maximum at the first root of $\sin \pi(2n+1)x = 0$, i.e., at $x = 1/(2n+1)$, and the corresponding peak value is

$$\frac{2}{\pi} \int_0^{\pi} \frac{\sin y}{y}\, dy = 1.089490+ .$$

This substantiates Gibbs' phenomenon.

EXERCISE 2.   Check Gibbs' phenomenon by hand for the sawtooth function of Fig. 1. *Hint:*   Use the standard recipe for the Riemann integral to verify that for $n \uparrow \infty$ and $|x| \leqslant \frac{1}{2}$,

$$S_n(x/n) = \frac{1}{2} - \sum_{k=1}^{n} (\pi k)^{-1} \sin(2\pi x k/n)$$

is well approximated by the integral

$$\frac{1}{2} - \frac{1}{\pi} \int_0^{2\pi x} \frac{\sin y}{y}\, dy.$$

1.7   MISCELLANEOUS APPLICATIONS

It is now time to see what you can *do* with Fourier series. A number of more or less elementary applications are explained in the present section. More elaborate (and important) applications to (one-dimensional) mathematical physics occupy Section 1.8.

1. Evaluation of Sums

The Plancherel identity can be used to evaluate a number of nontrivial infinite sums. The example

$$\sum_{n=1}^{\infty} n^{-2} = \pi^2/6$$

will illustrate this trick.

PROOF. The nth Fourier coefficient of the sawtooth function f of Fig. 1.6.1 is

$$\int_0^1 x \exp(-2\pi inx)\, dx = \begin{cases} \frac{1}{2} & \text{for } n = 0, \\ -(2\pi in)^{-1} & \text{for } n \neq 0. \end{cases}$$

Therefore, by the Plancherel identity,

$$\tfrac{1}{3} = \int_0^1 x^2\, dx = \|f\|^2 = \|\hat{f}\|^2 = \tfrac{1}{4} + (2\pi^2)^{-1}\sum_{n=1}^{\infty} n^{-2}.$$

Now solve for the sum.

EXERCISE 1. Check that $\sum_{n=1}^{\infty} n^{-4} = \pi^4/90$.

EXERCISE 2. Prove the binomial identity

$$\sum_{k=0}^{n} \binom{n}{k}^2 = \binom{2n}{n}.$$

Hint: $\sum_{k=0}^{n}\binom{n}{k}e_k = (1+e_1)^n$; also look at $\sum_{k=0}^{2n}\binom{2n}{k}e_{k-n} = 2^{2n}\cos^{2n}\pi x$.

2. Wirtinger's Inequality

This inequality states that *for any bounded interval $Q = [a,b]$ and any $f \in C^1(Q)$ with $f(a) = f(b) = 0$,*

$$\int_Q |f|^2 \leqslant [(b-a)^2/\pi^2]\int_Q |f'|^2;$$

the constant $(b-a)^2\pi^{-2}$ cannot be improved.

PROOF. The substitution $x \to y = \frac{1}{2}[(x-a)/(b-a)]$ reduces the general statement to the case $a = 0$, $b = \frac{1}{2}$: namely, if $f(x) = g(y)$, and if the inequality holds for g, then

$$\int_a^b |f(x)|^2\, dx = 2(b-a) \int_0^{\frac{1}{2}} |g(y)|^2\, dy$$

$$\leqslant \frac{2(b-a)}{4\pi^2} \int_0^{\frac{1}{2}} |g'(y)|^2\, dy$$

$$= \frac{(b-a)^2}{\pi^2} \int_a^b |f'(x)|^2\, dx .$$

Therefore you can take $Q = [0, \frac{1}{2}]$ to start. The function f is now extended over $-\frac{1}{2} \leqslant x < 0$ so as to be an odd periodic function of class $\mathbf{C}^1(S^1)$ as in Fig. 1. By the oddness of f, $\hat{f}(0) = \int_{-\frac{1}{2}}^{\frac{1}{2}} f = 0$, and by the Plancherel identity,

$$\int_{-\frac{1}{2}}^{\frac{1}{2}} |f'|^2 = \sum_{n \neq 0} |(f')^\wedge(n)|^2$$

$$= \sum_{n \neq 0} |2\pi i n \hat{f}(n)|^2$$

$$\geqslant 4\pi^2 \sum_{n \neq 0} |\hat{f}(n)|^2$$

$$= 4\pi^2 \int_{-\frac{1}{2}}^{\frac{1}{2}} |f|^2 .$$

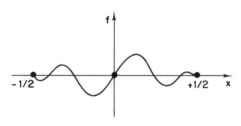

FIGURE 1

A second application of oddness permits you to replace $\int_{-\frac{1}{2}}^{\frac{1}{2}}$ by $2\int_0^{\frac{1}{2}}$ on both sides:

$$\int_0^{\frac{1}{2}} |f|^2 \leqslant (4\pi^2)^{-1} \int_0^{\frac{1}{2}} |f'|^2 .$$

The proof is finished by noticing that the bound is achieved by constant multiples of

$$f = 2i \sin 2\pi x = e^{2\pi i x} - e^{-2\pi i x} .$$

3. Weierstrass' Approximation Theorem

This theorem states that *on a bounded interval Q, any function f ∈ C(Q) can be uniformly approximated by a polynomial.*

PROOF FOR $a = -\frac{1}{4}$ *AND* $b = \frac{1}{4}$. Extend f from $|x| \leqslant \frac{1}{4}$ to $|x| \leqslant \frac{1}{2}$ so as to be continuous and periodic with period 1. The extended f can be uniformly approximated by a trigonometric polynomial: namely, by Fejér's theorem 1.4.3, the arithmetic mean $n^{-1}(S_0 + \cdots + S_{n-1})$ is uniformly close to f for large n. But also each of the $2n-1$ exponentials that figure in this sum may be uniformly well approximated for $|x| \leqslant \frac{1}{2}$ by sufficiently many terms of its power series about $x = 0$; and for $|x| \leqslant \frac{1}{4}$, the superposition of these cutoff power series is a uniformly close polynomial approximation to the original function f. This finishes the proof for $Q = [-\frac{1}{4}, \frac{1}{4}]$; the general case is left to the reader. As a by-product, you see that the powers $x^n : n \geqslant 0$ span $L^2(Q)$, as advertised for $Q = [-1, +1]$ in Exercise 1.3.7.

4. The Isoperimetric Problem

The isoperimetric inequality states that *of all* (*simple*) *closed plane curves of unit length, the circle encloses the biggest area.* To express this in a more precise way, let C be a plane curve swept out by a moving point $[x(t), y(t)]$ as t (time) runs from 0 to 1, and let this curve be

(a) *closed:* $x(0) = x(1)$ and $y(0) = y(1)$.

(b) *smooth:* x and y belong to $C^1(S^1)$.

(c) *simple:* $[x(t_1), y(t_1)] \neq [x(t_2), y(t_2)]$ for $t_1 \neq t_2$ unless $t_1 = 0$ and $t_2 = 1$, or $t_1 = 1$ and $t_2 = 0$,

(d) *of unit length:* $l = \int_0^1 [\dot{x}^2(t) + \dot{y}^2(t)]^{1/2} dt = 1$,
in which the overdot stands for differentiation with regard to time.

Denote by A the area enclosed by C. *The isoperimetric inequality states that*

$$A < 1/4\pi,$$

unless C is a circle, in which case $l = 1 = 2\pi R$, $R = 1/2\pi$, *and* $A = \pi R^2 = 1/4\pi$. The following proof, due to Hurwitz, is adapted from Courant and Hilbert [1953]; it is important for this proof that the presentation of the curve by means of x and y can be chosen so as to make

$$\dot{x}^2(t) + \dot{y}^2(t) = 1,$$

which is the same as to say that t may be taken as arc length:

$$\int_0^t [\dot{x}^2(s) + \dot{y}^2(s)]^{1/2} ds.$$

The only other fact you need is contained in

EXERCISE 3.   Deduce the area formula

$$A = \tfrac{1}{2}\int_0^1 [x(t)\,\dot{y}(t) - \dot{x}(t)\,y(t)]\,dt$$

either by hand or by Stokes' formula for a smooth plane vector field $\mathfrak{f} = (f_1, f_2)$:

$$\int_C (f_1\,dx + f_2\,dy) = \iint_D \frac{\partial f_2}{\partial x} \quad \frac{\partial f_1}{\partial y}\,dx\,dy,$$

D stands for the region enclosed by C and the left-hand integral is taken in the counterclockwise sense. *Hint for the Stokes proof:* $\mathfrak{f} = (-y, x)$.

PROOF.   The first step is to expand x, y, $\dot{x}$, and $\dot{y}$ into Fourier series

$$x(t) = \sum \hat{x}(n)\,e^{2\pi int}$$

with coefficients

$$\hat{x}(n) = \int_0^1 x(t)\,e^{-2\pi int}\,dt\,,$$

and so on, and to translate the length condition

$$l = \int_0^1 [\dot{x}^2(t) + \dot{y}^2(t)]\,dt = 1$$

and the area formula

$$A = \tfrac{1}{2}\int_0^1 [x(t)\,\dot{y}(t) - \dot{x}(t)\,y(t)]\,dt$$

into Fourier language. This gives

$$1 = \sum [|\dot{\hat{x}}(n)|^2 + |\dot{\hat{y}}(n)|^2] = \sum 4\pi^2 n^2 [|\hat{x}(n)|^2 + |\hat{y}(n)|^2]$$

and

$$A = \tfrac{1}{2}\sum [\hat{x}(n)\,\hat{y}(n)^* - \hat{x}(n)^*\,\hat{y}(n)]$$
$$= \sum \pi in [\hat{x}(n)^*\,\hat{y}(n) - \hat{x}(n)\,\hat{y}(n)^*]\,,$$

so that

$$\pi^{-1}[(4\pi)^{-1} - A] = \sum (n^2 [|\hat{x}(n)|^2 + |\hat{y}(n)|^2] - in [\hat{x}(n)^*\,\hat{y}(n) - \hat{x}(n)\,\hat{y}(n)^*])\,.$$

Put $\hat{x}(n) = \alpha(n) + i\beta(n)$ and $\hat{y}(n) = \gamma(n) + i\delta(n)$ with real $\alpha, \beta, \gamma, \delta$ and substitute into the last sum:

$$\sum_{n \neq 0} [n^2(\alpha^2 + \beta^2 + \gamma^2 + \delta^2) - 2n(\beta\gamma - \alpha\delta)]$$

$$= \sum_{n \neq 0} [(n\alpha + \delta)^2 + (n\beta - \gamma)^2 + (n^2 - 1)(\delta^2 + \gamma^2)].$$

Clearly, the sum is nonnegative [i.e. $(4\pi)^{-1} \geqslant A$] and vanishes only if

(a) $\alpha = \beta = \gamma = \delta = 0$ for $|n| \geqslant 2$,

(b) $\alpha(\pm 1) = \mp\delta(\pm 1)$,

(c) $\beta(\pm 1) = \pm\gamma(\pm 1)$.

But also

$$\alpha(+1) = \operatorname{Re} \int_0^1 x(t) e^{-2\pi it}\, dt = \int_0^1 x(t) \cos 2\pi t\, dt = \alpha(-1)$$

since x is a real function, and similarly

$$\beta(+1) = \int_0^1 x(t) \sin 2\pi t\, dt = -\beta(-1),$$

$$\gamma(+1) = \int_0^1 y(t) \cos 2\pi t\, dt = \gamma(-1),$$

$$\delta(+1) = \int_0^1 y(t) \sin 2\pi t\, dt = -\delta(-1),$$

so you can express the curve corresponding to $A = (4\pi)^{-1}$ by means of the formulas

$$x(t) = \hat{x}(0) + \hat{x}(-1)e^{-2\pi it} + \hat{x}(+1)e^{2\pi it}$$

$$= \hat{x}(0) + 2\alpha(1) \cos 2\pi t - 2\beta(1) \sin 2\pi t,$$

$$y(t) = \hat{y}(0) + \hat{y}(-1)e^{-2\pi it} + \hat{y}(+1)e^{2\pi it}$$

$$= \hat{y}(0) + 2\beta(1) \cos 2\pi t + 2\alpha(1) \sin 2\pi t.$$

But this is just a complicated way of saying that C is a circle:

$$[x(t) - \hat{x}(0)]^2 + [y(t) - \hat{y}(0)]^2 = 4[\alpha(1)^2 + \beta(1)^2] = 1/4\pi^2.$$

The last evaluation comes from the fact that

$$1 = (\dot{x})^2 + (\dot{y})^2 = 4[\alpha(1)^2 + \beta(1)^2] \times 4\pi^2.$$

The proof is finished.

5.   Jacobi's Identity for the Theta Function

The "theta function" is defined by the sum

$$\vartheta(t) = \sum_{n=-\infty}^{\infty} \exp(-\pi n^2 t), \qquad t > 0,$$

and Jacobi's formula states that

$$\vartheta(t) = t^{-\frac{1}{2}}\vartheta(1/t).$$

The function ϑ is an important transcendental function arising in a number of very different fields, including number theory, heat flow, elliptic and automorphic functions, and statistical mechanics. The reader will find other proofs of Jacobi's identity later [see Subsection 1.8.3 and Subsection 2.7.5]. The identity is very useful for computing ϑ for small t. For example, if $t = 0.01$ and if π is known, you must take about 21 terms ($|n| \leqslant 10$) of the sum on the left-hand side to compute to one significant figure, while the very first term ($n = 0$) of the right-hand sum gives the correct value to over 130 significant figures!

EXERCISE 4.   Convince yourself that these numbers are reasonable.

PROOF OF JACOBI'S IDENTITY.   Fix $t > 0$ and look at the (periodic) function

$$f(x) = \sum_{k=-\infty}^{\infty} \exp[-(x-k)^2/2t].$$

The sum converges uniformly for $0 \leqslant x \leqslant 1$ since

$$0 \leqslant \exp[-(x-n)^2/2t] \leqslant \exp(-n^2/4t) \qquad \text{for} \quad |n| \geqslant 2 \quad \text{and} \quad 0 \leqslant x \leqslant 1,$$

so you can compute the Fourier coefficients of f as follows:

$$\hat{f}(n) = \int_0^1 f e_n{}^* = \sum_{k=-\infty}^{\infty} \int_0^1 \exp[-(x-k)^2/2t]\, e^{-2\pi i n x}\, dx$$

$$= \sum_{k=-\infty}^{\infty} \int_{-k}^{-k+1} \exp(-x^2/2t)\, e^{-2\pi i n x}\, dx$$

$$= \int_{-\infty}^{\infty} \exp(-x^2/2t)\, e^{-2\pi i n x}\, dx$$

$$= \sqrt{2\pi t}\, \exp(-2\pi^2 n^2 t).$$

The last evaluation will be verified shortly. Granting this, it is clear from the rapid decrease of $\hat{f}$ that the formal sum $\sum \hat{f}(n) e_n$ converges uniformly to f:

$$f(x) = \sum_{n=-\infty}^{\infty} \exp[-(x-n)^2/2t] = \sqrt{2\pi t} \sum_{n=-\infty}^{\infty} \exp(-2\pi^2 n^2 t) e_n(x).$$

This identity was also known to Jacobi. It specializes to the identity for the theta function upon putting $x = 0$ and replacing t by $t/2\pi$. The actual evaluation of

$$\hat{f}(n) = \int_{-\infty}^{\infty} \exp(-x^2/2t) \, e^{-2\pi inx} \, dx = \sqrt{2\pi t} \exp(-2\pi^2 n^2 t)$$

is carried out by a couple of elegant tricks, the first of which is due to Feller [1966, Vol. 2, p. 476]. Bring in the function

$$\hat{f}(\gamma) = \int_{-\infty}^{\infty} \exp(-x^2/2t) \, e^{-2\pi i\gamma x} \, dx$$

defined for all $\gamma \in R^1$, and notice that

$$\hat{f}' = -2\pi i \int_{-\infty}^{\infty} x \exp(-x^2/2t) \, e^{-2\pi i\gamma x} \, dx$$

$$= 2\pi i t \int_{-\infty}^{\infty} [\exp(-x^2/2t)]' \, e^{-2\pi i\gamma x} \, dx$$

$$= -2\pi i t \int_{-\infty}^{\infty} \exp(-x^2/2t)(e^{-2\pi i\gamma x})' \, dx$$

$$= -4\pi^2 \gamma t \int_{-\infty}^{\infty} \exp(-x^2/2t) \, e^{-2\pi i\gamma x} \, dx$$

$$= -4\pi^2 \gamma t \hat{f}$$

by a self-evident partial integration. This may be solved for $\hat{f}$:

$$\hat{f}(\gamma) = \hat{f}(0) \exp(-2\pi^2 \gamma^2 t),$$

and to finish the proof you have only to evaluate

$$\hat{f}(0) = \int_{-\infty}^{\infty} \exp(-x^2/2t) \, dx = \sqrt{t} \int_{-\infty}^{\infty} \exp(-x^2/2) \, dx \equiv \sqrt{t} \, I$$

as $(2\pi t)^{1/2}$. The trick for doing that is an old standby:

$$I^2 = \int_{-\infty}^{\infty} \exp(-x^2/2) \, dx \int_{-\infty}^{\infty} \exp(-y^2/2) \, dy$$

$$= \int_{R^2} \exp[-(x^2+y^2)/2] \, dx \, dy$$

$$= \int_0^{2\pi} d\theta \int_0^\infty r \, dr \exp(-r^2/2)$$

$$= 2\pi \int_0^\infty \exp(-r^2/2) \, d(r^2/2)$$

$$= 2\pi .$$

Additional information about theta functions may be found in Bellman [1961].

6. Equidistribution of Arithmetic Sequences

A basic postulate of statistical mechanics is the so-called "ergodic principle" of Boltzmann and Gibbs, which states that the time average of a mechanical quantity should be the same as its phase average; see Ford and Uhlenbeck [1963, pp. 9–13] for a nice discussion of such matters. A simple instance of this phenomenon can be seen in the following model due to Weyl [1916].

As "phase space," bring in the circle $0 \leqslant x < 1$, pick a number $0 < \gamma < 1$, and look at the "rotation"

$$x \rightarrow x_1 = x + \gamma$$

with addition modulo 1. The "trajectory" of the phase point $x_0 = x$ is the arithmetic series

$$x_0 = x, \qquad x_1 = x + \gamma, \qquad x_2 = x + 2\gamma, \qquad x_3 = x + 3\gamma, \qquad \text{etc.,}$$

considered modulo 1. A "mechanical quantity" is a (periodic) phase function $f \in C(S^1)$. Its "time average" is

$$\lim_{n \uparrow \infty} n^{-1} \sum_{k=0}^{n-1} f(x_k),$$

assuming this limit to exist, while its "phase average" is just the arithmetic mean:

$$\int_0^1 f.$$

Weyl proved that *the two averages are the same if γ is an irrational number.*

EXERCISE 5. Check that the two averages cannot always be the same if γ is rational.

PROOF OF WEYL'S THEOREM. Begin with the function $f = e_m$ for fixed integral m: for $m = 0, f \equiv 1$ and both averages are unity, while for

$m \neq 0$, $e_m(\gamma) \neq 0$, so that

$$n^{-1} \sum_{k=0}^{n-1} e_m(x_k) = n^{-1} \sum_{k=0}^{n-1} e^{2\pi i m (x+k\gamma)}$$

$$= \frac{e_m(x)}{n} \frac{e_m(n\gamma)-1}{e_m(\gamma)-1}$$

tends to

$$0 = \int_0^1 e_m$$

when $n \uparrow \infty$, as it should. For the general phase function $f \in C(S^1)$, you have only to use Fejér's theorem 1.4.3 to approximate f uniformly by a trigonometric polynomial $f^0 = n^{-1}(S_0 + \cdots + S_{n-1})$ and to check the bound

$$\limsup_{n \uparrow \infty} \left| n^{-1} \sum_{k=0}^{n-1} f(x_k) - \int_0^1 f \right|$$

$$\leqslant \limsup_{n \uparrow \infty} \left| n^{-1} \sum_{k=0}^{n-1} f^0(x_k) - \int_0^1 f^0 \right| + 2 \| f^0 - f \|_\infty$$

$$= 2 \| f^0 - f \|_\infty .$$

The proof is finished.

An immediate conclusion is that

$$\lim_{n \uparrow \infty} n^{-1} \#(k < n : a \leqslant x_k \leqslant b) = b - a$$

for any circular arc ab. The symbol $\#(k < n : \cdots)$ is read "the number of integers $k < n$ for which the stated event takes place," so that the left-hand side is the empirical frequency with which the trajectory visits the arc ab. This explains the word "equidistribution." The proof is simple: Just approximate the indicator function 1_{ab} of the arc above and below by continuous functions f^+ and f^-, as in Fig. 2, so as to make

$$\int_0^1 (f^+ - f^-)$$

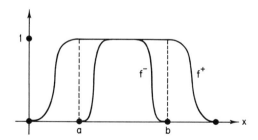

FIGURE 2

small, and observe that

$$\int_0^1 f^- = \lim_{n\uparrow\infty} n^{-1} \sum_{k=0}^{n-1} f^-(x_k)$$

$$\leqslant \liminf_{n\uparrow\infty} n^{-1} \sum_{k=0}^{n-1} 1_{ab}(x_k)$$

$$\leqslant \limsup_{n\uparrow\infty} n^{-1} \sum_{k=0}^{n-1} 1_{ab}(x_k)$$

$$\leqslant \lim_{n\uparrow\infty} n^{-1} \sum_{k=0}^{n-1} f^+(x_k)$$

$$= \int_0^1 f^+ .$$

Both $\int_0^1 f^-$ and $\int_0^1 f^+$ are very close to $\int_0^1 1_{ab} = b - a$, so

$$\lim_{n\uparrow\infty} n^{-1} \sum_{k=0}^{n-1} 1_{ab}(x_k) = b - a ,$$

and the proof is finished.

7.   An Eigenfunction Expansion

The purpose of this subsection is to study the map $K: f \to f''$ acting in $L^2[0,1]$ from a geometrical viewpoint. You will see that much of what you know about linear maps (matrices) of the finite-dimensional space C^n carries over to such maps of the infinite-dimensional space C^∞.

The map K is *not* defined on the whole of $L^2[0,1]$; it acts only on the smaller domain $D(K) \subset C^1[0,1]$ of functions f with

$$f'(x) = f'(0) + \int_0^x \tilde{f}(y)\, dy$$

for some function $\tilde{f} \in L^2[0,1]$ and

$$f(0) = f(1) = 0 .$$

$K: f \to \tilde{f}$ is defined thereby as a *linear* map of $D(K)$ *onto* $L^2[0,1]$.

At a first reading, the reader may choose to impose the more restrictive condition that $f \in C^2[0,1]$. This defines a smaller domain for K, and although it may seem more natural, it is not the best choice from the standpoint of Hilbert space. With this change in domain, for instance, K no longer maps *onto* $L^2[0,1]$ (nor is it a closed operator). For more information on $D(K)$ and a seemingly more natural definition, see Exercise 9. The function $Kf = \tilde{f}$ is sometimes called the L^2 derivative of f; it is uniquely determined by

$f \in \mathsf{D}(K)$ and coincides with f'' when $f \in \mathsf{C}^2[0,1]$. Adopting the terminology of finite-dimensional spaces, you can speak of K as *symmetric*:

$$(Kf,g) = \int_0^1 \tilde{f}g^* = -\int_0^1 f'g'^* = \int_0^1 f\tilde{g}^* = (f, Kg)$$

for functions from $\mathsf{D}(K)$, and *negative*:

$$(Kf,f) = -\int |f'|^2 < 0$$

if $f \in \mathsf{D}(K)$ is not identically 0.

As in the finite-dimensional case, you may expect that the eigenvalues of K are real and negative: If $f \in \mathsf{D}(K)$ is an eigenfunction [i.e., if $Kf = \gamma f$, and $f \not\equiv 0$], then

$$\gamma \|f\|^2 = (Kf,f) < 0.$$

To compute the actual eigenfunctions and eigenvalues is a simple task: If $f'' = \gamma f$ for some $\gamma < 0$ and if $f(0) = 0$, then

$$f = \sqrt{2} \sin \sqrt{-\gamma}\, x$$

up to a multiplicative constant, and to make $f(1) = 0$, you have to pick $\gamma = -n^2\pi^2$ for some integer $n \geqslant 1$. The factor $\sqrt{2}$ figuring in f is thrown in to make f of unit length.

EXERCISE 6.   Check that $\int_0^1 \sin^2 n\pi x\, dx = \frac{1}{2}$ for $n \geqslant 1$ by pure thought. *Hint:* $\sin^2 x + \cos^2 x = 1$.

EXERCISE 7.   Show that $f_n = \sqrt{2} \sin n\pi y : n \geqslant 1$ is a unit-perpendicular family. *Hint:* $-n^2\pi^2 \int_0^1 f_n f_m = (Kf_n, f_m)$ is symmetrical in n and m.

EXERCISE 8.   Check that $f_n : n \geqslant 1$ spans $\mathsf{L}^2[0,1]$. The resulting expansion

$$f = \sum_{n=1}^\infty (f, f_n) f_n = \sum_{n=1}^\infty \left(\int_0^1 f(y) \sqrt{2} \sin n\pi y\, dy \right) \sqrt{2} \sin n\pi x$$

is a "Fourier sine series" of the kind introduced in Exercise 1.4.1. *Hint:* f can be extended to $-1 \leqslant x \leqslant +1$ so as to be odd and periodic with period 2, and the extended function can be expanded into a series

$$\sum \left[(\sqrt{2})^{-1} \int_{-1}^{+1} f(y) \exp(-\pi i n y)\, dy \right] (\sqrt{2})^{-1} e_n(x/2).$$

EXERCISE 9.   Check that $f \in \mathsf{D}(K)$ iff

$$\sum n^4 |(f, f_n)|^2 < \infty,$$

and that K is "diagonal" in the coordinates f_n, $n \geqslant 1$:

$$K : f = \sum (f, f_n) f_n \to \sum (-n^2 \pi^2)(f, f_n) f_n .$$

Because $(Kf, f) = - \|f'\|^2 = 0$ only if $f = 0$, K is a 1:1 map from $D(K)$ onto $L^2[0, 1]$, and you may hope that the map $G : L^2[0, 1] \to D(K)$ inverse to $-K$ can be expressed as

$$Gf = \sum_{n=1}^{\infty} (n^2 \pi^2)^{-1} (f, f_n) f_n$$

$$= \sum_{n=1}^{\infty} (n^2 \pi^2)^{-1} f_n(x) \int_0^1 f(y) f_n(y)\, dy$$

$$= \int_0^1 \left[\sum_{n=1}^{\infty} (2/n^2 \pi^2) \sin n\pi x \sin n\pi y \right] f(y)\, dy$$

$$= \int_0^1 G(x, y) f(y)\, dy$$

with a self-evident notation. This is easy to verify: The sum for the so-called *Green function* $G(x, y)$ converges uniformly for $0 \leqslant x \leqslant 1$ and $0 \leqslant y \leqslant 1$, and both the composite maps

$$GK : D(K) \xrightarrow{K} L^2[0, 1] \xrightarrow{G} D(K)$$

$$KG : L^2[0, 1] \xrightarrow{G} D(K) \xrightarrow{K} L^2[0, 1]$$

act as the -1 times the identity on their respective domains.

EXERCISE 10.   Check $GK = KG = -1$ in detail.

The Green function G can be computed in a second way: For $f \in L^2[0, 1]$, the function $u = Gf$, is a solution of $Ku = u'' = -f$ subject to $u(0) = u(1) = 0$, and this is easily solved by hand:

$$u'(x) = u'(0) + \int_0^x u''(y)\, dy = u'(0) - \int_0^x f(y)\, dy,$$

so

$$u(x) = u'(0) x - \int_0^x dy \int_0^y f(z)\, dz$$

$$= u'(0) x - \int_0^x (x - y) f(y)\, dy$$

by interchanging the integrals in the next to last line. But also $u(1) = 0$. This determines

$$u'(0) = \int_0^1 (1-y) f(y) \, dy,$$

and substituting back into the formula for u, you find that

$$u(x) = \int_0^1 x(1-y) f(y) \, dy - \int_0^x (x-y) f(y) \, dy$$

$$= \int_0^x [x(1-y) - (x-y)] f(y) \, dy + \int_x^1 x(1-y) f(y) \, dy$$

$$= \int_0^x (1-x) y f(y) \, dy + \int_x^1 x(1-y) f(y) \, dy.$$

This expression may be compared with the previous sine series for $u = Gf$ to prove the following nontrivial identity:

$$G(x,y) = \sum_{n=1}^{\infty} (2/n^2 \pi^2) \sin n\pi x \sin n\pi y$$

$$= \begin{cases} (1-x) y & \text{for } y \leqslant x, \\ x(1-y) & \text{for } y \geqslant x. \end{cases}$$

EXERCISE 11.   Check this by expanding the function defined by lines 2 and 3 of the above formula into a sine series, calculating the coefficients

$$\sqrt{2}(1-x) \int_0^x y \sin n\pi y \, dy + \sqrt{2} x \int_x^1 (1-y) \sin n\pi y \, dy$$

by hand.

EXERCISE 12.   Use the sine series for G to check that $(Gf, f) \geqslant \pi^2 \|Gf\|^2$, and use this to give a new proof of the Wirtinger inequality of Subsection 2. *Hint:* $(Kf, f) = -\|f'\|^2$.

EXERCISE 13.   The spur (or trace) of G is the sum of its eigenvalues:

$$\sum_{n=1}^{\infty} (n^2 \pi^2)^{-1} = \tfrac{1}{6}.$$

Check that this is the same as the "sum" of G "down the diagonal":

$$\int_0^1 G(x,x) \, dx,$$

as computed either from the sine series for G or from $G(x, x) = x(1-x)$. Applications of the Green function G will be found in Exercises 1.8.4 and 1.8.8.

EXERCISE 14. Check that the functions $f_0 = 1$ and $f_n = \sqrt{2}\cos n\pi x$: $n \geqslant 1$ also form a unit-perpendicular basis for $L^2[0, 1]$. The corresponding expansion is like the "cosine series" of Exercise 1.4.1 but of period 2. Check that $f_n: n \geqslant 0$ is a complete list of the eigenfunctions of $K: f \to f''$ acting on the new domain $D(K)$ defined by $f'(0) = f'(1) = 0$ instead of by $f(0) = f(1) = 0$. K maps the new $D(K)$ onto

$$L^2[0, 1] \cap \left(f: \int_0^1 f = 0 \right) = D(G),$$

but is not 1:1. Why? Check that $-K$ *is* 1:1 if you cut it down to $D'(K) = D(K) \cap (f: \int_0^1 f = 0)$ and compute the inverse map $G: D(G) \to D'(K)$ in the form of a (symmetric) Green function. The condition $\int_0^1 f = 0$ is an instance of the "Fredholm alternative"; in the present case, this states that you cannot solve $u'' = -f$ for $u \in D(K)$ unless f is perpendicular to the solutions of class $D(K)$ of $u'' = 0$, namely, the constants.

Answer:

$$G(x, y) = \sum_{n=1}^{\infty} \frac{2}{n^2 \pi^2} \cos n\pi x \cos n\pi y$$

$$= \frac{x^2 - x + y^2 - y - |x-y|}{2} + \frac{1}{3}$$

is a Green function. It is unique up to an additive constant.

1.8   APPLICATIONS TO THE PARTIAL DIFFERENTIAL EQUATIONS OF ONE-DIMENSIONAL MATHEMATICAL PHYSICS

The two basic partial differential equations of one-dimensional mathematical physics are the "heat equation"

(a) $\partial u / \partial t = \frac{1}{2}(\partial^2 u / \partial x^2)$,

and the "wave equation"

(b) $\partial^2 u / \partial t^2 = \partial^2 u / \partial x^2$.

In the first, $u = u(t, x)$ is the temperature in a one-dimensional conductor
(e.g., a wire) as a function of the time $t \geqslant 0$ and the place x, while in the
second, it is the displacement of a string from its resting position. The purpose of this section is to explain how (a) and (b) reflect the natural phenomena
of heat flow and wave motion, and to indicate how Fourier series can be
used to solve them. In so doing, it is to be emphasized that a basic step in
understanding mathematics is to understand its relation to nature, if only
for the reason that pictures from nature can provide a feeling for deep
mathematical facts (and also the other way around). For example, if you
know that (a) has to do with heat flow, then you will believe at once that the
solution u (the temperature) must be zero or more throughout the future
if it is so at time $t = 0$. Naturally, you must give a mathematical proof of
this fact, but the point is that you would not be likely to suspect it coming
to the problem cold. A second obvious conjecture is that the solution of
(a) becomes flatter as time passes, while the solution of (b) may oscillate
indefinitely. For additional applications of Fourier series to similar problems,
see Sommerfeld [1949]. (Sommerfeld is one of the great masters of this
subject.) A fascinating philosophical piece on nature and mathematics will
be found in Wigner [1967, pp. 222–237].

1*. Derivation of the Heat Equation

The derivation of (a) is based upon Newton's law of cooling, which states
that the flux of heat *across* a point x is proportional to the gradient of the
temperature *at* x. This means that the amount of heat that flows past x from
left to right in a short time T is approximately

$$-\text{constant} \times \frac{\partial u}{\partial x} \times T$$

with a positive constant depending upon the material (water, copper, or
what have you). The minus sign is present because heat flows from the hotter
place to the cooler; see Flanders and Swan [1963]. This is an *experimental*
fact that is well verified for moderate temperature gradients; the full experimental relationship between flux and gradient is very complicated. To
proceed, the *net* amount of heat flowing out of a small interval $a \leqslant x \leqslant b$
in a short time T is therefore

$$-\text{constant} \times \frac{\partial u}{\partial x}\bigg|_a^b \times T,$$

or approximately so. This can also be computed in a second way: It is, in
fact, proportional to the product of the length of the interval and the

(average) decrease of the temperature inside. The constant of proportionality is the "specific heat" of the conducting material. Therefore,

$$\text{constant} \times (b-a)\,(-\partial u/\partial t)\,T = -\text{constant} \times \left.\frac{\partial u}{\partial x}\right|_a^b \times T,$$

or approximately so, and if you divide both sides by the product of $b-a$ and T and make them both small, you find

$$\frac{\partial u}{\partial t} = \text{constant} \times \frac{\partial^2 u}{\partial x^2}$$

with a new constant which may be put equal to $\frac{1}{2}$ for simplicity.

2*.   Derivation of the Wave Equation

The derivation of (b) is similar. The starting point is Newton's second law:

$$\text{force} = \text{mass} \times \text{acceleration}.$$

The small piece of string between the points a and $b > a$ has a mass proportional to the length $b-a$, and its (vertical) acceleration is approximately $\partial^2 u/\partial t^2$ evaluated at some place $a \leqslant x \leqslant b$. The force acting at the place x comes from the internal tension of the string which is nearly constant for small oscillations. This force acts *along* the string, so its vertical part is proportional to the sine of the angle of inclination:

$$\left[1 + \left(\frac{\partial u}{\partial x}\right)^2\right]^{-\frac{1}{2}} \frac{\partial u}{\partial x}.$$

Therefore, if $\partial u/\partial x$ is small, the *net* force acting on the piece ab is approximately

$$f = \text{constant} \times \left.\frac{\partial u}{\partial x}\right|_a^b,$$

and Newton's second law tells you that this is the same as

$$\text{mass} \times \text{acceleration} = \text{constant} \times (b-a) \times \frac{\partial^2 u}{\partial t^2},$$

that is,

$$\frac{\partial^2 u}{\partial t^2} = \text{constant} \times (b-a)^{-1} \times \left.\frac{\partial u}{\partial x}\right|_a^b = \text{constant} \times \frac{\partial^2 u}{\partial x^2},$$

or approximately so, and the constant may be put equal to one for simplicity. The derivation is finished.

3.   Heat Flow on the Circle

Think of the temperature in a circular wire (of length 1) or of periodic temperatures (of period 1) in a wire extending indefinitely in both directions (R^1). The flow of heat cannot destroy such periodicity, and in either case you end up trying to solve

$$\partial u/\partial t = \tfrac{1}{2}(\partial^2 u/\partial x^2)$$

on the circle $0 \leqslant x < 1$. Because u is supposed to be the temperature, it is natural to conjecture that the whole solution is determined by the temperature $u(0, \cdot) = f$ at time $t = 0$. A formal solution may be obtained by expanding u into a Fourier series

$$u(t, x) = \sum c_n(t) e_n(x)$$

with coefficients

$$c_n(t) = \int_0^1 u(t, x) e_n{}^*(x) \, dx$$

for each $t \geqslant 0$ separately, and then noticing that

$$\dot{c}_n = \frac{\partial}{\partial t} \int_0^1 u e_n{}^* = \int_0^1 \frac{\partial u}{\partial t} e_n{}^* = \int_0^1 \frac{1}{2} \frac{\partial^2 u}{\partial x^2} e_n{}^*$$

$$= \frac{1}{2} \int_0^1 u \frac{\partial^2 e_n{}^*}{\partial x^2} = -2\pi^2 n^2 \int_0^1 u e_n{}^* = -2\pi^2 n^2 c_n.$$

This can be solved for c_n:

$$c_n(t) = c_n(0) \exp(-2\pi^2 n^2 t),$$

and the constant $c_n(0)$ can be computed from the (known) function $f = u(0, \cdot)$:

$$c_n(0) = \int_0^1 u(0, x) e_n{}^*(x) \, dx = \hat{f}(n).$$

The proposed solution is therefore

$$u(t, x) = \sum \hat{f}(n) \exp(-2\pi^2 n^2 t) e_n(x)$$

$$= \sum \exp(-2\pi^2 n^2 t) e_n(x) \int_0^1 f(y) e_n{}^*(y) \, dy$$

$$= \int_0^1 \left[\sum \exp(-2\pi^2 n^2 t) e_n(x-y) \right] f(y) \, dy$$

$$= p \circ f.$$

in which p is the so-called fundamental solution or Green function:

$$p = p_t(x) = \sum_{n=-\infty}^{\infty} \exp(-2\pi^2 n^2 t)\, e_n(x).$$

To be more precise, the following claim is made for initial temperatures $f \in C(S^1)$: *the only solution to the problem*

(a) $\dfrac{\partial u}{\partial t} = \dfrac{1}{2}\dfrac{\partial^2 u}{\partial x^2}$, $t > 0$, $0 \leqslant x < 1$

(b) $u \in C^\infty[(0, \infty) \times S^1]$

(c) $\lim_{t \downarrow 0} \|u - f\|_\infty = \lim_{t \downarrow 0} \max_{0 \leqslant x < 1} |u(t, x) - f(x)| = 0$

is $u = p \circ f$.

PROOF. The proof that $u = p \circ f$ is the *only* possible solution is plain sailing over the course laid out above, so you only have to check that this function actually *is* a solution. To begin with, for any positive time $t > 0$, $\exp(-2\pi^2 n^2 t)$ is a rapidly decreasing function of n. Using this fact, it is easy to verify that p (and u) belong to $C^\infty[(0, \infty) \times S^1]$ and that for $t > 0$, the defining sum for p can be differentiated under the summation sign as often as you please; see Exercise 1.4.4 for help with this. Because each summand $\exp(-2\pi^2 n^2 t)\, e_n(x)$ is a solution of the problem of heat flow, so is p (and also u), and to finish the proof, you only have to check that $\lim_{t \downarrow 0} \|u - f\|_\infty = 0$. This is plain for $f \in C^2(S^1)$, since in that case

$$\|u(t, \cdot) - f\|_\infty \leqslant \sum [1 - \exp(-2\pi^2 n^2 t)]\,|\hat{f}(n)|$$

and

$$|\hat{f}(n)| = |(2\pi i n)^{-2}(f'')^\wedge(n)| \leqslant (2\pi n)^{-2}\|f''\|_\infty.$$

For the general initial temperature $f \in C(S^1)$, you have to do a little more. A key point is to check that p is nonnegative. To do this, it is enough to verify that $u \geqslant 0$ if $f \geqslant 0$ is of class $C^2(S^1)$. This is easy: $\lim_{t \downarrow 0} \|u - f\|_\infty = 0$ for such f, so that if $u(t_0, x_0) < 0$ for some $t_0 > 0$ and $0 \leqslant x_0 < 1$, then $v = e^{\beta t} u$ has a negative minimum $\alpha < 0$ in the band $[0, t_0] \times S^1$ at some place $0 < t_1 \leqslant t_0$ and $0 \leqslant x_1 < 1$. But at this place,

$$0 \geqslant \frac{\partial v}{\partial t} = \beta v + e^{\beta t}\frac{\partial u}{\partial t} = \alpha\beta + e^{\beta t}\frac{1}{2}\frac{\partial^2 u}{\partial x^2}$$

$$= \alpha\beta + \frac{1}{2}\frac{\partial^2 v}{\partial x^2} \geqslant \alpha\beta,$$

and that is impossible for $\beta < 0$. A moment's reflection will now convince you that p is nonnegative, and since

$$\int_0^1 p = \sum \exp(-2\pi^2 n^2 t) \int_0^1 e_n = 1,$$

you obtain, as a fringe benefit, the important bound ("maximum principle")

$$\|u\|_\infty = \|p \circ f\|_\infty \leqslant \|f\|_\infty \int_0^1 p = \|f\|_\infty$$

for every $f \in C(S^1)$. The rest of the proof that $\lim_{t \downarrow 0} \|u - f\|_\infty = 0$ for $f \in C(S^1)$ is easy: You have only to pick $f_n \in C^2(S^1)$ so as to make $\lim_{n \uparrow \infty} \|f_n - f\|_\infty = 0$, to compare $u_n = p \circ f_n$ and $u = p \circ f$ as in

$$\|u(t, \cdot) - f\|_\infty \leqslant \|u_n(t, \cdot) - f_n\|_\infty + 2\|f_n - f\|_\infty,$$

and then to make $t \downarrow 0$ and $n \uparrow \infty$, in that order. The fact that $\lim_{t \downarrow 0} u = f$ may be thought of as an analog of Theorem 1.4.3. The statement of the latter is that

$$n^{-1}(S_0 + \cdots + S_{n-1}) = \sum_{|k| < n} [1 - |k|/n] \hat{f}(n) e_n$$

is a uniform approximation to f for large n, while the present statement is that

$$p \circ f = \sum \exp(-2\pi^2 n^2 t) \hat{f}(n) e_n$$

is such an approximation for small $t > 0$; see Exercise 1.4.9 for another variant.

A second proof that $u = p \circ f$ solves the problem of heat flow may be based upon the Jacobi identity of Subsection 1.7.5:

$$p_t(x) = \sum \exp(-2\pi^2 n^2 t) e_n(x) = (2\pi t)^{-\frac{1}{2}} \sum \exp[-(x-n)^2/2t].$$

Think of f as a periodic function on the line of period 1. Then

$$u = p \circ f = (2\pi t)^{-\frac{1}{2}} \int_0^1 \sum \exp[-(x-y-n)^2/2t] f(y) \, dy$$

$$= (2\pi t)^{-\frac{1}{2}} \sum \int_n^{n+1} \exp[-(x-y)^2/2t] f(y) \, dy$$

$$= \int_{-\infty}^\infty \{\exp[-(x-y)^2/2t]/\sqrt{2\pi t}\} f(y) \, dy,$$

and all the desired properties of u can be read off from this much more transparent formula. For example, the evaluation

$$\int_{-\infty}^\infty \frac{\exp(-x^2/2t)}{\sqrt{2\pi t}} \, dx = 1,$$

from Subsection 1.7.5 permits you to express $u - f$ as

$$u - f = \int_{-\infty}^{\infty} \frac{\exp(-y^2/2t)}{\sqrt{2\pi t}} [f(x+y) - f(x)] \, dy,$$

and now it is easy to see that

$$\|u - f\|_{\infty} \leqslant \int_{-\infty}^{\infty} \frac{\exp(-y^2/2t)}{\sqrt{2\pi t}} \|f_y - f\|_{\infty} \, dy$$

$$= \int_{-\infty}^{\infty} \frac{\exp(-y^2/2)}{\sqrt{2\pi}} \|f_{y\sqrt{t}} - f\|_{\infty} \, dy$$

tends to 0 as $t \downarrow 0$, by bounded convergence.

This circle of ideas can also be used to give a new proof of Jacobi's identity. Bring in the sum

$$q = q_t(x) = (2\pi t)^{-\frac{1}{2}} \sum_{n=-\infty}^{\infty} \exp[-(x-n)^2/2t].$$

Then you may check that $q \circ f$ solves the problem of heat flow and tends to f as $t \downarrow 0$. But this means that $q \circ f = p \circ f$, since there is only one such solution, and you conclude that $q = p$. This is Jacobi's identity.

EXERCISE 1.   Check all the technical details of this proof of Jacobi's identity. The formula $p = q$ is an instance of the "method of images" of Lord Kelvin, in which the integers figuring in the sum for q are thought of as fictitious heat sources on the covering surface R^1 of the circle; see Sommerfeld [1949] for more information on this topic.

4.   Heat Flow on an Interval

The temperature in a rod of length 1 with ends held at temperature 0 is described by

(a) $\partial u / \partial t = \frac{1}{2}(\partial^2 u / \partial x^2)$, $t > 0$, $0 < x < 1$,

subject to

(b) $u(t, x) = 0$, $t > 0$, $x = 0, 1$.

To solve this problem, think of u as an odd function of $x \in R^1$ of period 2. This suggests that you expand it into a sine series of period 2 as in Exercise 1.7.8, and you find that *the* nice solution reducing to $f \in C[0, 1]$ at time $t = 0$ is

$$u(t, x) = \sum_{n=1}^{\infty} \exp(-n^2 \pi^2 t/2) \left[\int_0^1 f(y) \sqrt{2} \sin n\pi y \, dy \right] \sqrt{2} \sin n\pi x.$$

Naturally, you must insist upon $f(0) = f(1) = 0$ if you hope to make

$$\lim_{t\downarrow 0} \|u-f\|_\infty = 0.$$

EXERCISE 2.   Check that

$u(t,x)$

$$= \int_0^1 \sum_{n=-\infty}^{\infty} \left[\frac{\exp[-(x-y-2n)^2/2t]}{\sqrt{2\pi t}} - \frac{\exp[-(x+y-2n)^2/2t]}{\sqrt{2\pi t}} \right] f(y)\, dy$$

is also a solution of the preceding problem. This is a second instance of Kelvin's method of images; see Exercise 1.

EXERCISE 3.   Use Exercise 2 to verify the Jacobi-type identity

$$\sum_{n=-\infty}^{\infty} \left[\frac{\exp[-(x-y-2n)^2/2t]}{\sqrt{2\pi t}} - \frac{\exp[-(x+y-2n)^2/2t]}{\sqrt{2\pi t}} \right]$$

$$= \sum_{n=1}^{\infty} \exp(-n^2\pi^2 t/2)\, 2 \sin n\pi x \sin n\pi y.$$

Give a second proof by computing by hand the sine coefficients of the left-hand side; see Subsection 1.7.5.

A related problem is to take $\lim_{t\downarrow 0} u = 0$, but to keep supplying heat at the place $0 \leqslant x \leqslant 1$ at the rate $f(x)$. Now the problem is to solve

(a) $\partial u/\partial t = \frac{1}{2}(\partial^2 u/\partial x^2) + f, \qquad t > 0, \quad 0 < x < 1$

   subject to

(b) $u(t,x) = 0, \qquad t > 0, \quad x = 0, 1$

(c) $\lim_{t\downarrow 0} u = 0,$

and you can easily verify that

$$u(t,x) = \sum_{n=1}^{\infty} \frac{1 - \exp(-\pi^2 n^2 t/2)}{\pi^2 n^2/2} \left(\int_0^1 f(y)\, \sqrt{2} \sin n\pi y\, dy \right) \sqrt{2} \sin n\pi x.$$

EXERCISE 4.   Check that $u(\infty, x) = 2Gf$ in which G is the Green function of Subsection 1.7.7. This provides you with a physical interpretation of G: It describes the ultimate balancing between the cooling at the ends of the rod and the heating due to f.

A third problem is to find the temperature in a rod with insulated ends with initial temperature f and no supply of heat after time $t = 0$. The adjective "insulated" means that the outward flux of heat is 0 at the ends:

$$(\partial u / \partial x)\,(t, x) = 0, \qquad t > 0, \quad x = 0, 1 ;$$

see the Newtonian law of cooling in Subsection 1.

EXERCISE 5. Check that the solution of the insulated problem is

$u(t, x)$

$$= \int_0^1 f(y)\,dy + \sum_{n=1}^{\infty} \exp(-n^2\pi^2 t/2)$$

$$\times \left[\int_0^1 f(y)\,\sqrt{2}\cos n\pi y\,dy \right] \sqrt{2}\cos n\pi x$$

$$= \int_0^1 \sum_{n=-\infty}^{\infty} \left[\frac{\exp[-(x-y-2n)^2/2t]}{\sqrt{2\pi t}} + \frac{\exp[-(x+y-2n)^2/2t]}{\sqrt{2\pi t}} \right] f(y)\,dy$$

and deduce the Jacobi-type identity

$$\sum_{n=-\infty}^{\infty} \left[\frac{\exp[-(x-y-2n)^2/2t]}{\sqrt{2\pi t}} + \frac{\exp[-(x+y-2n)^2/2t]}{\sqrt{2\pi t}} \right]$$

$$= 1 + \sum_{n=1}^{\infty} \exp(-n^2\pi^2 t/2)\,2\cos n\pi x\,\cos n\pi y .$$

This is a third instance of the method of images.

5. Temperature of the Earth[1]

A simple but interesting problem is to determine the temperature deep in the Earth from the temperature on the surface. Think of the surface temperature as a known periodic function f of the time $t \geq 0$ of period 1 (1 year). The temperature $u(t, x)$ at time $t \geq 0$ and depth $x \geq 0$ is also periodic (in t), and it is natural to assume that $|u| \leq \|f\|_\infty < \infty$. Under these circumstances, $u(t, x)$ can be expanded into a Fourier series for each fixed $0 \leq x < \infty$:

$$u(t, x) = \sum c_n(x)\,e^{2\pi i n t} .$$

Because

$$\partial u / \partial t = \tfrac{1}{2}(\partial^2 u / \partial x^2),$$

---

[1] This section has been adapted from Sommerfeld [1949].

the coefficient

$$c_n(x) = \int_0^1 u(t,x)\, e^{-2\pi int}\, dt$$

is a solution of

$$c_n'' = \int_0^1 (\partial^2 u/\partial x^2)\, e^{-2\pi int}\, dt = 2\int_0^1 (\partial u/\partial t)\, e^{-2\pi int}\, dt = 4\pi inc_n$$

$$= [(2\pi|n|)^{\frac{1}{2}}(1\pm i)]^2\, c_n$$

with a plus or a minus sign according as $n>0$ or not. Besides,

$$c_n(0+) = \int_0^1 f(t)\, e^{-2\pi int}\, dt = \hat{f}(n),$$

and

$$|c_n| \leqslant \|u\|_\infty \leqslant \|f\|_\infty < \infty,$$

so the only possibility is

$$c_n = \hat{f}(n)\exp[-(2\pi|n|)^{\frac{1}{2}}(1\pm i)x],$$

and putting this back in, you see that

$$u(t,x) = \sum \hat{f}(n)\exp[-(2\pi|n|)^{\frac{1}{2}}x]\exp[2\pi int \mp (2\pi|n|)^{\frac{1}{2}}ix].$$

Especially, the temperature response at depth x to the simple harmonic $e_n = e^{2\pi int}$ is *damped* by the factor $\exp[-(2\pi|n|)^{\frac{1}{2}}x]$ and suffers a *phase shift* in the amount $(2\pi|n|)^{\frac{1}{2}}x$. To get a feeling for this, let the surface tem-

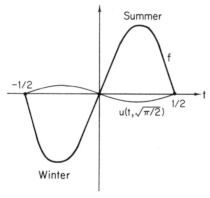

FIGURE 1

perature f be a simple sinusoid $\sin 2\pi t$ [the heavy curve of Fig. 1], so that the mean annual temperature $\hat{f}(0) = \int_0^1 f$ is 0 and

$$u(t,x) = \exp(-\sqrt{2\pi}\,x)\sin(2\pi t - \sqrt{2\pi}\,x).$$

Pick $\sqrt{2\pi}\,x = \pi$. The formula says that the temperature at this depth is damped by the factor

$$e^{-\pi} = 0.04 + \approx \tfrac{1}{25}$$

and *completely out of phase with the seasons*, as indicated by the light curve in Fig. 1. This is the reason for vegetable cellars: At the proper depth, not only is the "local climate" much more nearly constant (damping) but it is cooler in summer and warmer in winter (phase shift). In fact, if you put in the proper constants, you find that a good depth for such a cellar is about 4 meters or 13 feet, which is not far off!

6.   Oscillations of a String

The treatment of this problem is on a more formal level. The industrious reader can supply the missing technical points, but it will be perfectly satisfactory just to assume that every function in sight is infinitely differentiable.

Think of the string as stretched between $x = 0$ and $x = 1$ and *tied* at the ends, so that the displacement $u = u(t, x)$ from the resting position is a solution of

(a) $\partial^2 u/\partial t^2 = \partial^2 u/\partial x^2$, $t > 0$, $0 < x < 1$,

(b) $u(t, x) = 0$, $t > 0$, $x = 0, 1$.

The solution may be expanded into a sine series of period 2:

$$u(t, x) = \sum_{n=1}^{\infty} c_n(t)\,\sqrt{2}\sin n\pi x$$

with coefficients

$$c_n(t) = \int_0^1 u(t, x)\,\sqrt{2}\sin n\pi x\,dx = \int_0^1 u f_n.$$

The latter may be obtained by solving

$$\ddot{c}_n = \int_0^1 (\partial^2 u/\partial t^2) f_n = \int_0^1 (\partial^2 u/\partial x^2) f_n$$

$$= (u'f_n - u f_n')\Big|_0^1 + \int_0^1 u f_u''$$

$$= -n^2\pi^2 \int_0^{1.} u f_n$$

$$= -n^2\pi^2 c_n,$$

the answer being

$$c_n(t) = c_n(0+) \cos n\pi t + \dot{c}_n(0+) \left[(\sin n\pi t)/n\pi\right].$$

This indicates that to determine the motion of the string, you need to know not only the initial displacement $u(0+, \cdot) = f$ but also the initial speed $\partial u(0+, \cdot)/\partial t = g$, as is only natural, since in Newtonian mechanics the "phase" is always a pair: position *and* velocity. Going back to the sine series for u, you see that

$$u(t, x) = \sum_{n=1}^{\infty} \cos n\pi t \left[\int_0^1 f(y) \sqrt{2} \sin n\pi y \, dy\right] \sqrt{2} \sin n\pi x$$

$$+ \sum_{n=1}^{\infty} \left[(\sin n\pi t)/n\pi\right] \left[\int_0^1 g(y) \sqrt{2} \sin n\pi y \, dy\right] \sqrt{2} \sin n\pi x,$$

which expresses a fundamental fact about the vibrating string: *The general oscillation is a superposition of "fundamental modes"*

$$\sin n\pi t \sin n\pi x \quad \text{and} \quad \cos n\pi t \cos n\pi x.$$

at the natural frequencies or so-called fundamental tones

$$n\pi/2\pi = \tfrac{1}{2}, 1, \tfrac{3}{2}, 2, \tfrac{5}{2}, \quad \text{etc.}$$

The solution still looks very complicated, but it can be expressed in a much more transparent way by thinking of f and g as odd periodic functions on the line of period 2 and using the standard trigonometrical identity

$$\cos n\pi t \sin n\pi x = \tfrac{1}{2}[\sin n\pi (x+t) + \sin n\pi (x-t)].$$

The result is that u can be expressed by the classical formula of d'Alembert:

$$u(t, x) = \tfrac{1}{2}[f(x+t) + f(x-t)] + \tfrac{1}{2} \int_0^t [g(x+s) + g(x-s)] \, ds$$

$$= \tfrac{1}{2}[f(x+t) + f(x-t)] + \tfrac{1}{2} \int_{x-t}^{x+t} g(y) \, dy,$$

which says that u can be pictured as the superposition $u^+ + u^-$ of a wave traveling to the left at speed 1:

$$u^+(t, x) = \tfrac{1}{2}f(x+t) + \tfrac{1}{2} \int_0^{x+t} g(y) \, dy$$

and a wave traveling to the right at speed 1:

$$u^-(t, x) = \tfrac{1}{2}f(x-t) + \tfrac{1}{2} \int_{x-t}^0 g(y) \, dy.$$

EXERCISE 6. The sum of the kinetic and potential energies of the string is

$$H = K + U = \int_0^1 (\partial u/\partial t)^2 \, dx + \int_0^1 (\partial u/\partial x)^2 \, dx.$$

Check, both by manual differentiation and by the Plancherel identity for sine series, that H is preserved in time.

EXERCISE 7. A steady force density f is applied to the tied string, i.e., the force applied to the part of the string between $x = a$ and $x = b$ is $\int_a^b f$. Convince yourself that, with a proper choice of units, the displacement is a solution of

$$\partial^2 u/\partial t^2 = (\partial^2 u/\partial x^2) + f$$

and solve this problem if $u = \partial u/\partial t = 0$ at time $t = 0$. *Answer:*

$$u(t, x) = \sum_{n=1}^{\infty} [(1 - \cos n\pi t)/n^2\pi^2]\left[\int_0^1 f(y) \sqrt{2} \sin n\pi y \, dy\right] \sqrt{2} \sin n\pi x.$$

EXERCISE 8. The "mean position"

$$v(x) = \lim_{T \uparrow \infty} T^{-1} \int_0^T u(t, x) \, dt$$

of the string of Exercise 7 is to be expressed by means of the Green function G of Subsection 1.7.7. *Answer:* $v = Gf$. Check that v can also be viewed as the "resting position" of the string, that is, $u = v$ is the only solution of $\partial^2 u/\partial t^2 = \partial^2 u/\partial x^2 + f$, which is constant in time.

1.9*   MORE GENERAL EIGENFUNCTION EXPANSIONS

The success of Fourier series in solving the problems of heat flow and wave motion of Section 1.8 stems from the fact that the trigonometric functions are eigenfunctions of $K: f \to f''$. The purpose of this section is to study the eigenfunctions of more complicated differential operators K. First an application:

Think of a weighted string, tied at $x = a$ and at $x = b$, with a (nonconstant) mass density Δ; this means that the mass of the part of the string between $x = a'$ and $x = b'$ is

$$m = \int_{a'}^{b'} \Delta(x) \, dx \qquad \text{for any} \quad a \leqslant a' < b' \leqslant b.$$

The density is to be positive on the closed interval $a \leqslant x \leqslant b$ and of class

$C^2[a,b]$. By a simple adaptation of Subsection 1.8.2, you see that the displacement of the string is a solution of

(a) $\qquad \dfrac{\partial^2 u}{\partial t^2} = \Delta^{-1}\dfrac{\partial^2 u}{\partial x^2}, \qquad t > 0, \quad a < x < b$

(b) $\qquad u(t,x) = 0, \qquad t > 0, \quad x = a,b.$

To solve this problem, bring in the eigenfunctions f and eigenvalues γ of the differential operator

$$K: f \to \Delta^{-1}f''.$$

A nontrivial function f is an eigenfunction of K iff $f'' = \gamma\,\Delta f$ and $f(a) = f(b) = 0$. The admissible eigenvalues form a discrete series

$$0 > \gamma_1 > \gamma_2 > \gamma_3 > \cdots \downarrow -\infty,$$

and the corresponding eigenfunctions $f_n: n \geqslant 1$, properly scaled, form a unit-perpendicular basis of the space $L^2(Q, \Delta(x)\,dx)$ of measurable functions f on $Q = [a,b]$ with

$$\|f\|_\Delta^2 = \int_Q |f|^2\Delta = \int_a^b |f(x)|^2\Delta(x)\,dx < \infty,$$

see Exercise 1.2.12. The point is that the eigenfunctions provide you with a kind of Fourier series specially adapted to $Kf = \Delta^{-1}f''$ in the same way as the standard sine (and/or cosine) series is specially adapted to $Kf = f''$. The displacement of the string is expressed by the sum

$$u(t,x) = \sum_{n=1}^{\infty}\left[a_n\cos(-\gamma_n)^{1/2}t + b_n\,\frac{\sin(-\gamma_n)^{1/2}t}{(-\gamma_n)^{1/2}}\right]f_n(x)$$

with coefficients,

$$a_n = (f,f_n)_\Delta = \int_Q f\!f_n\,\Delta$$

and

$$b_n = (g,f_n)_\Delta = \int_Q g f_n\,\Delta,$$

computed from the initial data $u(0+,\cdot) = f$ and $\partial u(0+,\cdot)/\partial t = g$. The sum for the solution exhibits the fact that the general oscillation is a superposition of the "fundamental modes"

$$[\sin(-\gamma_n)^{1/2}t]\,f_n(x) \qquad \text{and} \qquad [\cos(-\gamma_n)^{1/2}t]\,f_n(x)$$

at the "natural" frequencies $(2\pi)^{-1}(-\gamma_n)^{1/2}$ ("fundamental tones").

The proof that there are "enough" eigenfunctions is outlined below in a series of simple steps. For more details, see Birkhoff and Rota [1962, pp. 286–310] and Courant and Hilbert [1953, pp. 291–295, 331–339, 414].

Amplification: The present discussion is of wider scope than may be apparent. It applies to any differential operator $Kf = Af'' + Bf'$ with smooth coefficients $A > 0$ and B: in fact, if

$$C(x) = \int_a^x \frac{B(t)}{A(t)} \, dt,$$

then

$$Af'' + Bf' = Ae^{-C}(e^C f')',$$

and this looks like

$$Kf = \frac{1}{\Delta(y)} \frac{d^2 f}{dy^2}$$

in the new scale

$$y = \int_a^x e^{-C(t)} \, dt$$

if you put

$$\Delta(y) = [1/A(x)] e^{2C(x)}.$$

Much the same discussion applies if you impose the general boundary conditions

$$\cos \alpha f(a) + \sin \alpha f'(a) = 0, \qquad \cos \beta f(b) + \sin \beta f'(b) = 0$$

in place of $f(a) = f(b) = 0$. You can even treat "singular" operators such as the Legendre operator $(\sin x)^{-1} (\sin x f')'$ of Section 4.12, but all that is left aside.

Step 1:   Take $a = 0$ and $b = 1$, for simplicity, and think of K as acting on the domain $\mathbf{D}(K)$ of Subsection 1.7.7, but with the new space $L^2(Q, \Delta \, dx)$ in place of the customary $L^2(Q) = L^2(Q, dx)$. Then K is symmetric and negative: namely, for functions from $\mathbf{D}(K)$

$$(Kf, g)_\Delta = \int_0^1 \Delta^{-1} f'' g^* \Delta = \int_0^1 f'' g^*$$

$$= \int_0^1 f(g'')^* = \int_0^1 f(\Delta^{-1} g'')^* \Delta$$

$$= (f, Kg)_\Delta$$

and

$$(Kf,f)_\Delta = \int_0^1 f''f^* = -\int_0^1 |f'|^2 < 0 \qquad \text{unless} \quad f = 0.$$

Just as in Subsection 1.7.7, this causes the eigenvalues of K to be real and negative (if such exist).

Step 2: Now extend Δ to $x \geqslant 1$, keep it positive, and make it ultimately identically 1. Fix $\gamma < 0$, and look for a solution f of

$$f'' = \gamma \Delta f, \qquad x > 0 \qquad \text{with} \quad f(0) = 0 \quad \text{and} \quad f'(0) = 1,$$

or what is the same, a solution of

$$f = x + \gamma I(\Delta f), \qquad x > 0,$$

in which I stands for the integral operator

$$I: f \to \int_0^x d\xi \int_0^\xi f(\eta) \, d\eta.$$

To this end, define $f_{n+1} = I(\Delta f_n)$, for $n \geqslant 0$, beginning with $f_0(x) = x$, and let

$$L = L(x) = \max(\Delta(y): 0 \leqslant y \leqslant x).$$

Then, since $f_0(x) = x$, you see that

$$0 \leqslant f_1 = I(\Delta f_0) \leqslant LI(f_0) = L\frac{x^3}{3!},$$

$$0 \leqslant f_2 \leqslant LI(f_1) \leqslant L^2 I\left(\frac{x^3}{3!}\right) = L^2 \frac{x^5}{5!},$$

$$\vdots$$

$$0 \leqslant f_n \leqslant LI(f_{n-1}) \leqslant L^n I\left[\frac{x^{2n-1}}{(2n-1)!}\right] = L^n \frac{x^{2n+1}}{(2n+1)!},$$

by induction. Therefore, the sum

$$f = \sum_{n=0}^\infty \gamma^n f_n$$

converges uniformly on compact sets, and

$$x + \gamma I(\Delta f) = f_0 + \gamma \sum_{n=0}^\infty \gamma^n I(\Delta f_n)$$

$$= f_0 + \sum_{n=0}^\infty \gamma^{n+1} f_{n+1}$$

$$= f_0 + \sum_{n=1}^\infty \gamma^n f_n$$

$$= f,$$

i.e., f is *a* solution. The only moot point is whether it is the *only* solution. But that is easily settled in the affirmative. You have only to notice that the difference $h = f - g$ of two solutions, f and g, satisfies $h = \gamma I(\Delta h)$, and use this to estimate

$$M(x) = \max(|h(y)| : 0 \leqslant y \leqslant x)$$

in an iterative way. The point is that

$$M \leqslant |\gamma| \, I(\Delta M) \leqslant |\gamma| \, LMI(1) = |\gamma| \, LM(x^2/2!)$$

implies

$$M \leqslant |\gamma| \, I\left(\Delta |\gamma| \, LM \frac{x^2}{2!}\right) \leqslant |\gamma|^2 \, L^2 MI\left(\frac{x^2}{2!}\right) = |\gamma|^2 \, L^2 M \frac{x^4}{4!}$$

which implies

$$M \leqslant |\gamma| \, I\left(\Delta |\gamma|^{n-1} \, L^{n-1} M \frac{x^{2n-2}}{(2n-2)!}\right) \leqslant |\gamma|^n \, L^n M \frac{x^{2n}}{(2n)!} \, .$$

The proof is finished by observing that this inequality cannot be maintained for all n unless $M = 0$.

EXERCISE 1.   Show that if f is any solution of $f'' = \gamma \Delta f$ such that $f(x_0) = f'(x_0) = 0$ at some point $x_0 \geqslant 0$, then $f \equiv 0$. *Hint:* The key is uniqueness.

The letter f is now reserved for the solution just constructed; for additional clarity, the dependence on γ is sometimes indicated explicitly by writing $f(\gamma, x)$ in place of f or $f(x)$. The formula $f = \sum \gamma^n f_n$, together with the estimates on f_n guarantee that f is a smooth (in fact, analytic) function of γ. A sketch of f is seen in Fig. 1.

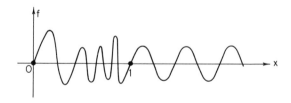

FIGURE 1

The general sinusoidal shape of f reflects the fact that f and $f'' = \gamma f \Delta$ are of opposite sign [$\gamma < 0$], causing f to bend back toward the level $f = 0$ after every upward or downward crossing. Indeed, f is ultimately a sinusoid, since $\Delta = 1$ far out. As indicated in the figure, the roots of $f(x) = 0$ are simple, i.e., $f' \neq 0$ at such a root. This is because you cannot have $f(x_0) = f'(x_0) = 0$ at the same place x_0 unless f vanishes identically; see Exercise 1. The roots of f are isolated for the same reason, the point being that x_0 can only be a limit of roots $x_n: n \geqslant 1$, if both

$$f(x_0) = \lim_{n\uparrow\infty} f(x_n) = 0$$

and

$$f'(x_0) = \lim_{n\uparrow\infty} \frac{f(x_0)-f(x_n)}{x_0 - x_n} = 0.$$

Because f is a sinusoid far out, its roots ultimately fall into an arithmetical series, and you can label them

$$0 < x_1(\gamma) < x_2(\gamma) < \cdots \uparrow \infty .$$

Step 3:   Check that K has infinitely many eigenfunctions f_n with eigenvalues $0 > \gamma_1 > \gamma_2 \cdots \downarrow -\infty$. The plan of attack is to check first that for small $\gamma < 0$, the first root x_1 of $f(x) = 0$ lies above $x = 1$ and then to verify that as $\gamma \downarrow -\infty$, all the roots x_n move continuously to the left, ultimately crossing $x = 1$ and approaching $x = 0$. The effect is to compress the oscillation of Fig. 1 by pushing everything to the left, as in (the highly schematic) Fig. 2. The point is that if $x_n = 1$ for $\gamma = \gamma_n$, then $f_n = f(\gamma_n, x)$ is an eigenfunction of K, because it satisfies *both* $f(0) = 0$ and $f(1) = 0$.

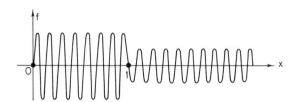

FIGURE 2

PROOF.   By the MacLaurin expansion of f about $x = 0$,

$$f(x) = f(0) + xf'(0) + \tfrac{1}{2}x^2 f''(y) = x + \tfrac{1}{2}x^2\gamma\,\Delta(y)\,f(y)$$

for some $0 \leqslant y \leqslant x$. This shows that as $\gamma \uparrow 0$, the first root $x_1(\gamma)$ moves out to ∞. To see that the roots x_n slide to the left as $\gamma \downarrow -\infty$, think of $f = f(\gamma, x)$ and $x_n = x_n(\gamma)$ as functions of $\gamma < 0$ and differentiate the identity $f(\gamma, x_n) = 0$ with respect to γ:

$$0 = \frac{\partial f}{\partial \gamma}(\gamma, x_n) + f'(\gamma, x_n)\frac{dx_n}{d\gamma} .$$

Because $f'(x_n) \neq 0$, the *existence* of $dx_n/d\gamma$ is guaranteed by the implicit function theorem. The only moot point is whether it is positive, which is the same as saying that f' and $\partial f/\partial \gamma$ are of opposite sign. To see this, differentiate $f'' = \gamma \Delta f$ by γ:

$$\left(\frac{\partial f}{\partial \gamma}\right)'' = \Delta f + \gamma \Delta \frac{\partial f}{\partial \gamma}.$$

Then

$$\left(f \frac{\partial f'}{\partial \gamma} = f' \frac{\partial f}{\partial \gamma}\right)' = f \frac{\partial f''}{\partial \gamma} - f'' \frac{\partial f}{\partial \gamma}$$

$$= f \left(\Delta f + \gamma \Delta \frac{\partial f}{\partial \gamma}\right) - \gamma \Delta f \frac{\partial f}{\partial \gamma}$$

$$= \Delta f^2,$$

and integrating back, you find that at x_n

$$-f' \frac{\partial f}{\partial \gamma} = f \frac{\partial f'}{\partial \gamma} - f' \frac{\partial f}{\partial \gamma} = \int_0^{x_n} \Delta f^2 \, dx > 0,$$

as advertised.

To see that each root ultimately crosses to the left of $x = 1$ as $\gamma \downarrow -\infty$ look first to the right of some root $x_n < 1$, for example, $x_0 = 0$, and observe that f and f'' have opposite but fixed signs in the interval $x_n < x < x_{n+1}$. For ease of discussion, take $f'' < 0 < f$ and let $\delta = \min(\Delta(x): x \geqslant 0)$. Then $\delta > 0$, and since

$$f'' = \gamma \Delta f \leqslant \gamma \, \delta f,$$

it follows that f must bend back toward $f = 0$ at least as rapidly as $g(x) = \sin(-\delta\gamma)^{\frac{1}{2}}(x - x_n)$. The point is that g is a solution of the equation

$$g'' = \gamma \, \delta g$$

which is positive on the open interval $x_n < x < x_n + [\pi/(-\delta\gamma)^{\frac{1}{2}}]$ and vanishes at the endpoints.

Now, if $x_{n+1} > \alpha_n = x_n + [\pi/(-\delta\gamma)^{\frac{1}{2}}]$, then f is positive for $x_n < x \leqslant \alpha_n$, and so

$$0 < (f'g - fg')\Big|_{x_n}^{\alpha_n}$$

$$= \int_{x_n}^{\alpha_n} (f'g - fg')'$$

$$= \int_{x_n}^{\alpha_n} f''g - fg''$$

$$= \int_{x_n}^{\alpha_n} (\Delta - \delta) fg,$$

which is impossible. But this means that

$$x_{n+1} \leqslant x_n + \frac{\pi}{(-\delta\gamma)^{1/2}},$$

and this is less than 1 for large γ, inasmuch as x_n can only move to the left. The proof is finished, and you now have at your disposal an infinite number of eigenfunctions $f_n = f(\gamma_n, x)$, the associated eigenvalues

$$0 > \gamma_1 > \gamma_2 > \gamma_3 \cdots \downarrow -\infty$$

being the roots γ_n of $x_n(\gamma) = 1$.

EXERCISE 2.[1]   Show that if the slope of g is adjusted so that $g'(x_n) = f'(x_n)$ then $f \leqslant g$ on the whole open interval $x_n < x < x_{n+1}$. Hint: Check that $f'/f \leqslant g'/g$ for $x_n < x < x_{n+1}$. Then integrate from $x_n + \varepsilon$ to x and let $\varepsilon \downarrow 0$.

Because K is symmetric, the f's are automatically perpendicular:

$$(\gamma_n - \gamma_m)\,(f_n, f_m)_\Delta = (Kf_n, f_m)_\Delta - (f_n, Kf_m)_\Delta = 0,$$

and you can scale them in order to make a unit-perpendicular family. Henceforth, it is assumed that this has been done. The final task is to check that this family actually spans $L^2([0, 1], \Delta\, dx)$.

Step 4:   This step prepares the way with a sharp estimate of f_n for $n \uparrow \infty$. To make the actual estimate plausible, look at the function $g = \Delta^{1/2} f$ in the new scale $y = \int_0^x \sqrt{\Delta}$. A simple computation shows that

$$d^2g/dy^2 = [\gamma + \tfrac{1}{4}\Delta^{-2}\Delta'' - \tfrac{5}{16}\Delta^{-3}(\Delta')^2]g.$$

Because $\Delta \in C^2[0, 1]$, the expression $\tfrac{1}{4}\Delta^{-2}\Delta''$, etc. is bounded for $0 \leqslant x \leqslant 1$, and you can expect g to be well approximated by

$$k \sin\left(\sqrt{-\gamma}\int_0^x \sqrt{\Delta}\right)$$

for a suitable amplitude $k(\gamma)$ and $\gamma \downarrow -\infty$. Detailed estimates bear this out: In fact, you find that γ_n and f_n are well approximated by

$$-n^2\pi^2\bigg/\left(\int_0^1 \sqrt{\Delta}\right)^2$$

and

$$e_n = \Delta^{-1/4}\left(\int_0^1 \sqrt{\Delta}\right)^{-1/2}\sqrt{2}\,\sin\left(n\pi \int_0^x \sqrt{\Delta}\bigg/\int_0^1 \sqrt{\Delta}\right),$$

---

[1] H. E. Rauch, private communication.

respectively, up to errors not exceeding a constant multiple of $1/n$. Now the family $e_n: n \geqslant 1$ is a unit-perpendicular basis of $L^2([0,1], \Delta\, dx)$, as you can easily check with the help of Exercise 1.7.8. At the same time, the unit-perpendicular family $f_n: n \geqslant 1$ is close to $e_n: n \geqslant 1$ in the sense that

$$\sum_{n=1}^{\infty} \|f_n - e_n\|_\Delta^2 \leqslant \text{constant} \times \sum_{n=1}^{\infty} n^{-2} < \infty.$$

Step 5:   This forces the family $f_n: n \geqslant 1$ to span also. The proof is adapted from Birkhoff and Rota [1962, pp. 307–309]. Pick a positive integer d so large that

$$\sum_{n>d} \|f_n - e_n\|_\Delta^2 < 1.$$

Then the only function $f \in L^2([0,1], \Delta(x)\, dx)$ that is perpendicular to e_n for $n \leqslant d$, and also to f_n for $n > d$, is the trivial function $f \equiv 0$. Otherwise, you could obtain a contradiction: namely,

$$\|f\|_\Delta^2 = \sum_{n>d} |(f, e_n)_\Delta|^2 = \sum_{n>d} |(f, e_n - f_n)_\Delta|^2$$

$$\leqslant \sum_{n>d} \|f\|_\Delta^2 \|f_n - e_n\|_\Delta^2$$

$$< \|f\|_\Delta^2.$$

However, now you see that the matrix $(f_i, e_j)_\Delta: i,j \leqslant d$ is nonsingular: If the vector $(b_1, ..., b_d)$ is perpendicular to this matrix, i.e., if

$$\sum_{i=1}^{d} b_i (f_i, e_j)_\Delta = 0 \qquad \text{for every } j \leqslant d,$$

then $\sum_{i=1}^{d} b_i f_i$ is perpendicular to e_j for $j \leqslant d$ as well as to f_n for $n > d$. But this means that $\sum_{i=1}^{d} b_i f_i = 0$, by the above, and so too that $b_i = 0$ for every $i \leqslant d$. Hence, the matrix is nonsingular as was claimed. Therefore, if f is perpendicular to $f_n: n \geqslant 1$, you can pick complex numbers $c_1, ..., c_d$ so as to make

$$f^0 = f - \sum_{i=1}^{d} c_i f_i$$

perpendicular not only to $f_n: n > d$ but also to $e_j: j \leqslant d$, with the result that $f^0 = 0$. Hence $\sum_{i=1}^{d} c_i f_i$ must also vanish, since it was chosen perpendicular to all the $f_n: n \geqslant 1$ to begin with. But this is just another way of saying that $f_n: n \geqslant 1$ spans. The proof is finished. For other proofs, see Akhiezer [1962, pp. 141–143, 219–233] and Yosida [1960, Chapter 2]. The latter studies the integral operator based on the Green function for the given differential operator. A special instance of the techniques used to handle such integral operators will be found in Section 2.8.

1.10   SEVERAL-DIMENSIONAL FOURIER SERIES

Consider the standard n-dimensional real number space R^n and the "lattice" $Z^n \subset R^n$ of points with integral coordinates. A function f on R^n is said to have "periods from Z^n" if it is periodic (of period 1) in each of its n variables:

$$f(x) = f(x+k) \qquad \text{for every} \quad k \in Z^n.$$

If you like, you can think of such a function as living on the standard n-dimensional torus

$$T^n \colon x = (x_1, \ldots, x_n), \qquad 0 \leqslant x_i < 1, \quad 1 \leqslant i \leqslant n.$$

The relation between R^n, Z^n, and T^n can be expressed as $T^n = R^n / Z^n$; for $n = 1$, this is simply the statement that the circle can be pictured as the real numbers mod 1.

1.   Fourier Series on a Standard Torus

The space $L^2(T^n)$ is the set of measurable functions f on the standard torus T^n with

$$\|f\|^2 = \int_{T^n} |f(x)|^2 \, d^n x = \int_0^1 \cdots \int_0^1 |f(x_1, \ldots, x_n)|^2 \, dx_1 \ldots dx_n < \infty.$$

EXERCISE 1.   Check that finite sums of products $f_1(x_1) \ldots f_n(x_n)$ of functions from $L^2(S^1)$ span $L^2(T^n)$.

By Exercise 1, the exponentials

$$e_k(x) = e^{2\pi i k \cdot x} = \exp[2\pi i (k_1 x_1 + \cdots + k_n x_n)] = e_{k_1}(x_1) \ldots e_{k_n}(x_n)$$

form a unit-perpendicular basis of $L^2(T^n)$ as $k = (k_1, \ldots, k_n)$ runs over the lattice of integral points Z^n. Any function $f \in L^2(T^n)$ can be expanded into an n-dimensional Fourier series

$$f = \sum_{k \in Z^n} \hat{f}(k) \, e_k$$

with coefficients

$$\hat{f}(k) = (f, e_k) = \int_{T^n} f e_k{}^* \, d^n x,$$

and there is a Plancherel formula

$$\|f\|^2 = \int_{T^n} |f|^2 = \|\hat{f}\|^2 = \sum_{Z^n} |\hat{f}(k)|^2.$$

The extension of most of the one-dimensional results developed in Sections 1.4–1.8 is easy, although some small technical changes must be made. The following sample will suffice.

EXERCISE 2.   The problem of heat flow on T^n is $\partial u/\partial t = \Delta u/2$ in which $\Delta = \partial^2/\partial x_1^2 + \cdots + \partial^2/\partial x_n^2$. Compute the solution and check that it tends to $\int f(x) d^n x$ as $t \uparrow \infty$ for any nice initial temperature f. *Hint:*   Imitate Subsection 1.7.3, first for $n = 2$ and then for general n.

2*.   Application to Random Walks

Pólya [1921] discovered a very beautiful application of several-dimensional Fourier series to "random walks." Think of a particle moving on the d-dimensional lattice Z^d according to the following rule. The particle starts at time 0 at the origin and moves at time $n \geqslant 1$ by a unit step e_n to a neighboring lattice point; for example, if $d = 3$, the possible steps are

$$\mathbf{e} = (\pm 1, 0, 0), \quad (0, \pm 1, 0), \quad \text{and} \quad (0, 0, \pm 1).$$

The position of the particle at time $n \geqslant 1$ is the sum of the individual steps: $\mathbf{s}_n = \mathbf{e}_1 + \cdots + \mathbf{e}_n$. The step $\mathbf{e}_n$ is statistically independent of the preceding steps $\mathbf{e}_j : j < n$, and the possible steps are equally likely at each stage. This means that

$$P(\mathbf{e}_1 = e_1, \ldots, \mathbf{e}_n = e_n) = P(\mathbf{e}_1 = e_1) \times \cdots \times P(\mathbf{e}_n = e_n)$$
$$= (2d)^{-n}$$

for any fixed unit steps $e_1, \ldots, e_n$, in which $P(E)$ means "the probability of the event E." The problem is to compute $P(\mathbf{s}_n = k)$ and to study the behavior of $\mathbf{s}_n$ for $n \uparrow \infty$. Pólya's idea is to think of $P(\mathbf{s}_n = k)$ as the Fourier coefficient $\hat{f}(k)$ of a function $f \in L^2(T^d)$:

$$f(x) = \sum_{k \in Z^d} P(\mathbf{s}_n = k) e^{2\pi i k \cdot x}.$$

This sum is just the "expectation" or "mean value" of $\exp(2\pi i \mathbf{s}_n \cdot x)$ and is easily computed using the independence of the individual steps:

$$f(x) = \sum_{e_1} \cdots \sum_{e_n} (2d)^{-n} \exp(2\pi i e_1 \cdot x) \cdots \exp(2\pi i e_n \cdot x)$$
$$= \left[(2d)^{-1} \sum_{e_1} \exp(2\pi i e_1 \cdot x)\right]^n$$
$$= \left[(\cos 2\pi x_1 + \cdots + \cos 2\pi x_d)/d\right]^n$$
$$= [f_d(x)]^n.$$

Pólya's formula is immediate from this;

$$P(\mathbf{s}_n = k) = \hat{f}(k) = (f_d^n)^{\wedge}(k) = \int_{T^d} f_d^n e^{-2\pi i k \cdot x}.$$

In particular,

$$P(\mathbf{s}_n = 0) = \int_{T^d} f_d^n,$$

and since $|f_d| \leqslant 1$, the expected number of times the particle visits the origin can be expressed as

$$\sum_{n=0}^{\infty} P(\mathbf{s}_n = 0) = \lim_{\varepsilon \uparrow 1} \sum_{n=0}^{\infty} \varepsilon^n P(\mathbf{s}_n = 0)$$

$$= \lim_{\varepsilon \uparrow 1} \int_{T^d} \sum_{n=0}^{\infty} \varepsilon^n f_d^n$$

$$= \lim_{\varepsilon \uparrow 1} \int_{T^d} (1 - \varepsilon f_d)^{-1}$$

$$= \int_{T^d} (1 - f_d)^{-1}$$

by an application of monotone convergence to the region where $0 \leqslant f_d \leqslant 1$. Because

$$\frac{1}{2} \times \frac{4\pi^2 |x|^2}{2d} \leqslant 1 - f_d \leqslant \frac{4\pi^2 |x|^2}{2d}$$

for small $|x|$, the integral diverges for $d \leqslant 2$ and converges for $d \geqslant 3$. Pólya used this to prove a very striking fact about the ultimate behavior of the walk:

$$P(\mathbf{s}_n = 0, \text{ i.o.}) = 1 \qquad \text{for} \quad d \leqslant 2,$$

$$P(\lim_{n \uparrow \infty} |\mathbf{s}_n| = \infty) = 1 \qquad \text{for} \quad d \geqslant 3.$$

PROOF FOR $d \geqslant 3$. The integral $\int (1 - f_d)^{-1} < \infty$ says that the expected number of times that the particle visits the origin is less than ∞. This can happen only if the actual number of visits is less than ∞ with probability 1, and since the origin is not special in any way, the same must be true of every lattice point in Z^d. But this means that for any $R < \infty$, the particle ultimately stops visiting the ball $|k| < R$, and that is the same as to say

$$P(\lim_{n \uparrow \infty} |\mathbf{s}_n| = \infty) = 1.$$

PROOF FOR $d \leqslant 2$. At time $n = 1$, the particle steps to one of the $2d$ nearest neighbors of the origin. The problem is to check that the probability p of ultimately returning to the origin is 1. But that is self-evident as soon as you reflect that the probability of visiting the origin m or more times (including the visit at time $n = 0$) is p^{m-1}, for then the probability of precisely m visits is

$$p^{m-1} - p^m = p^{m-1}(1-p),$$

and if p were less than 1, the expected number of visits would be

$$\sum_{m=1}^{\infty} m p^{m-1}(1-p) = (1-p)^{-1} < \infty,$$

contradicting the evaluation $\int (1-f_d)^{-1} = \infty$ $(d \leqslant 2)$. The proof is finished; for additional information on the subject, see Feller [1968, Vol. 1, pp. 342–371].

3*. Fourier Series on a Nonstandard Torus

A variant of the several-dimensional Fourier series of Subsection 1 arises by looking at functions on a nonstandard torus. For simplicity, only dimension $n = 2$ is discussed. Pick numbers $-\infty < n < \infty$ and $b > 0$ and introduce the [nonstandard] lattice $Z \subset R^2$ of all plane points of the form

$$\omega = j(1,0) + k(a,b) \qquad \text{with integral } j \text{ and } k,$$

as in Fig. 1. As for the standard lattice of article 1, a function with "periods from Z" can be thought of as living on the torus $T = R^2/Z$ obtained by identifying opposite sides of the "fundamental cell" shaded in the figure. Define Z' to be the "dual lattice" of points $\omega' \in R^2$ such that the inner product $\omega' \cdot \omega$ is integral for every $\omega \in Z$.

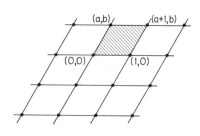

FIGURE 1

EXERCISE 3. Z' is the lattice of points

$$\omega' = j(1, -a/b) + k(0, 1/b) \qquad \text{with integral } j \text{ and } k;$$

in particular, the standard lattice $(a = 0, b = 1)$ is its own dual. Check this.

EXERCISE 4.   Check that

$$e_\gamma(x) = e^{2\pi i \gamma \cdot x}$$

can be regarded as a function on the torus T iff $\gamma \in Z'$.

EXERCISE 5.   Check that any function f of class $\mathsf{L}^2(T)$ can . be expanded into a Fourier series

$$f = \sum_{\omega' \in Z'} \hat{f}(\omega') e_{\omega'}$$

with coefficients

$$\hat{f}(\omega') = [\text{area}(T)]^{-1} \int_T f e_{\omega'}^* \, d^2 x$$

and verify the Plancherel identity:

$$\|f\|^2 = \int_T |f(x)|^2 \, d^2 x = \text{area}(T)\|\hat{f}\|^2 = \text{area}(T)\sum_{Z'} |\hat{f}(\omega')|^2 .$$

Hint: Think of f and $e_{\omega'}$ as functions of $y_1 = x_1 - (a/b)x_2$ and $y_2 = x_2/b$. This will bring you back to the standard torus of Subsection 1. An application of such Fourier series to a new Jacobi-type theta-function identity will be presented in Subsection 2.11.3.

EXERCISE 6.   Develop similar Fourier series for nonstandard tori in dimensions $n \geqslant 3$.

Chapter 2 *Fourier Integrals*

2.1 FOURIER INTEGRALS

The standard Fourier series

$$f = \sum \hat{f}(n) e_n = \sum \hat{f}(n) e^{2\pi i n x}$$

on the circle $0 \leqslant x < 1$ may be thought of as an expansion of a periodic function (of period 1) into simple harmonics of the same period. Clearly, the choice of period is just a matter of convenience: For functions of period $T > 0$, the appropriate harmonics are

$$e_n(x) = (\sqrt{T})^{-1} e^{2\pi i n x / T}, \qquad n \in Z^1,$$

and any nice function f of period T can be expanded as

$$f(x) = \sum_{n=-\infty}^{\infty} \left[T^{-1} \int_{-T/2}^{T/2} f(y) \, e^{-2\pi i n y / T} \, dy \right] e^{2\pi i n x / T}.$$

A small change in viewpoint leads at once to the Fourier integral: The idea is that the right-hand side is really a Riemann sum over a subdivision with spacing $1/T$, and with any luck, it should approximate the integral

$$f(x) = \int_{-\infty}^{\infty} \left[\int_{-\infty}^{\infty} f(y) \, e^{-2\pi i \gamma y} \, dy \right] e^{2\pi i \gamma x} \, d\gamma$$

as $T \uparrow \infty$. This does not make too much sense for a periodic function f (the integral cannot converge well), but it does suggest that something can be done to recover a nice function f from its *Fourier integral (or transform)*:

$$\hat{f}(\gamma) = \int_{-\infty}^{\infty} f(y) \, e^{-2\pi i \gamma y} \, dy$$

via the *inverse Fourier integral*

$$(\hat{f})^{\vee}(x) = \int_{-\infty}^{\infty} \hat{f}(\gamma) \, e^{2\pi i \gamma x} \, d\gamma \, .$$

The purpose of the next few sections is to put this formal discussion on a solid mathematical foundation, chiefly for functions from $C_{\downarrow}^{\infty}(R^1)$, $L^2(R^1)$, and $L^1(R^1)$. $C_{\downarrow}^{\infty}(R^1)$ is the class of infinitely differentiable rapidly decreasing functions on R^1: "Infinitely differentiable" means that $f \in C^{\infty}(R^1)$; "rapidly decreasing" applies to $f, f', f'', \ldots$, and means that $x^p D^q f$ tends to 0 as $|x| \uparrow \infty$ for every nonnegative integral p and q, separately. Here, D stands for differentiation: $Df = f'$.

   EXERCISE 1. $\exp(-x^2)$ belongs to $C_{\downarrow}^{\infty}(R^1)$, but $(1+x^2)^{-1}$ does not; nor does $\exp(-|x|)$, but for a different reason. Verify this statement.

   Usage: $\int f$ stands for $\int_{-\infty}^{\infty} f(x) \, dx$ throughout this chapter; especially, the limits of integration are always $\pm \infty$ if nothing is said to the contrary. $L^2(R^1)$ is the familiar space of measurable functions f with

$$\|f\|_2 = \left(\int |f|^2 \right)^{\frac{1}{2}} < \infty \, .$$

$L^1(R^1)$ is the space of summable functions with

$$\|f\|_1 = \int |f| < \infty \, ;$$

it is an algebra under the "convolution" product

$$f \circ g = \int_{-\infty}^{\infty} f(x-y) g(y) \, dy \, .$$

as you might expect by comparison with the case of the circle.

EXERCISE 2.   Check that the product ∘ is associative and commutative, and verify the bound $\|f \circ g\|_1 \leqslant \|f\|_1 \|g\|_1$.

$C_{\downarrow}^{\infty}(R^1)$ is dense in both $L^1(R^1)$ and $L^2(R^1)$; see Exercise 1.2.11.

EXERCISE 3.   Check that translation is continuous both in $L^1(R^1)$ and in $L^2(R^1)$, i.e., verify that in either space $f_y(x) = f(x+y)$ is close to f for small $|y|$; see Exercise 1.5.7.

2.2   FOURIER INTEGRALS FOR $C_{\downarrow}^{\infty}(R^1)$

To define the Fourier transform for functions f of class $L^2(R^1)$, it is not enough merely to write down the formal integral for $\hat{f}$ since the integrand is not necessarily summable. This trouble does not arise for $f \in L^1(R^1)$, but then $\hat{f}$ may not be summable, so you may have trouble with the inversion $\hat{f}^{\vee}$; for example, the indicator function of $-1 \leqslant x \leqslant 1$ is summable, but

$$(\pi \gamma)^{-1} \sin 2\pi \gamma = \int_{-1}^{+1} e^{-2\pi i \gamma x} \, dx$$

is *not*. $C_{\downarrow}^{\infty}(R^1)$ does not suffer from any trouble of this kind, so it will be a good place to begin. The statement is threefold:

(a)   *The Fourier transform ∧ maps $C_{\downarrow}^{\infty}(R^1)$ onto itself.*
(b)   *For such functions the inverse transform does what it should:*

$$(\hat{f})^{\vee} = f;$$

(c)   *There is a Plancherel identity:*

$$\|f\|_2 = \|\hat{f}\|_2 .$$

The proof consists in justifying the formal development of Section 2.1. A second very similar approach, via the "Poisson summation formula," will be found in Subsection 2.7.5.

PROOF.   Pick a function $f \in C_{\downarrow}^{\infty}(R^1)$. Then, by partial integration

$$(f')^{\wedge} = \int f'(x) \, e^{-2\pi i \gamma x} \, dx = -\int f(x) \, (e^{-2\pi i \gamma x})' \, dx = 2\pi i \gamma \hat{f};$$

also, by the rapid decrease of f,

$$(-2\pi i x f)^{\wedge} = (\hat{f})',$$

and so, by induction,

$$(2\pi i \gamma)^p D^q \hat{f} = [D^p(-2\pi i x)^q f]^{\wedge},$$

for any nonnegative integral p and q. Therefore,

$$|\gamma|^p |D^q \hat{f}| \leqslant (2\pi)^{q-p} \|D^p x^q f\|_1 < \infty,$$

so that $\hat{f}$ also belongs to $\mathbf{C}_\downarrow^\infty(R^1)$. Now let f be compact and regard it as an infinitely differentiable function on the circle $-T/2 \leqslant x \leqslant T/2$, as is permissible if $f = 0$ for $|x| \geqslant T/3$, say. Then you can express f for $|x| < T/2$ as a rapidly convergent Fourier series of period T:

$$f(x) = \sum_{n=-\infty}^{\infty} e^{2\pi inx/T} T^{-1} \int_{-T/2}^{T/2} f(y) e^{-2\pi iny/T} dy$$

$$= \sum_{n=-\infty}^{\infty} e^{2\pi inx/T} T^{-1} \hat{f}(n/T).$$

But this is just a Riemann sum approximating to the integral

$$\hat{f}^\vee(x) = \int_{-\infty}^{\infty} \hat{f}(\gamma) e^{2\pi i\gamma x} d\gamma,$$

and so in order to prove that

$$(\hat{f})^\vee = f$$

for compact functions f, you have only to check that the sum converges to the integral as $T \uparrow \infty$. The same line of reasoning leads from the formula

$$\|f\|_2^2 = \int_{-T/2}^{T/2} |f|^2 = \sum_{n=-\infty}^{\infty} T^{-1} |\hat{f}(n/T)|^2$$

to the Plancherel identity:

$$\|f\|_2 = \|\hat{f}\|_2 = \left(\int_{-\infty}^{\infty} |\hat{f}(\gamma)|^2 d\gamma \right)^{1/2}.$$

EXERCISE 1. Complete the proof of the two arguments sketched above, putting in all the necessary estimates. *Hint:* Use the fact that $\hat{f} \in \mathbf{C}_\downarrow^\infty(R^1)$ to bound $|\hat{f}|$ by a constant multiple of $(1+\gamma^2)^{-1}$.

To deal with noncompact functions $f \in \mathbf{C}_\downarrow^\infty(R^1)$, pick a compact infinitely differentiable function e as in Fig. 1 and put $f_n = f \times e(x/n)$.

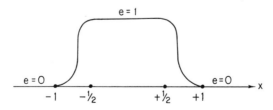

FIGURE 1

Then

$$\|\hat{f} - \hat{f}_n\|_\infty \leqslant \|f - f_n\|_1 \leqslant \int_{|x| > n/2} |f(x)| \, dx$$

tends to 0 as $n \uparrow \infty$. Besides, $\hat{f}_n$, and so too $\hat{f}$, is dominated by a function from $L^1(R^1) \cap L^2(R^1)$: namely,

$$|\hat{f}_n| \leqslant \|f_n\|_1 \leqslant \|f\|_1 < \infty$$

for $|\gamma| \leqslant 1$, while for $|\gamma| \geqslant 1$,

$$|\hat{f}_n| = (4\pi^2 \gamma^2)^{-1} |(f''_n)^\wedge|$$

$$\leqslant (4\pi^2 \gamma^2)^{-1} \|f''_n\|_1$$

$$\leqslant (4\pi^2 \gamma^2)^{-1} \|f''(x) e(x/n) + (2/n) f'(x) e'(x/n) + (n^2)^{-1} f(x) e''(x/n)\|_1$$

$$\leqslant (4\pi^2 \gamma^2)^{-1} [\|f''\|_1 \|e\|_\infty + (2/n) \|f'\|_1 \|e'\|_\infty + (n^2)^{-1} \|f\|_1 \|e''\|_\infty]$$

is bounded by a constant multiple of γ^{-2}, independently of $n \geqslant 1$. The domination from $L^1(R^1)$ permits you to verify the Fourier inversion formula:

$$f(x) = \lim_{n \uparrow \infty} f_n(x) = \lim_{n \uparrow \infty} \int \hat{f}_n(\gamma) e^{2\pi i \gamma x} \, d\gamma$$

$$= \int \hat{f}(\gamma) e^{2\pi i \gamma x} \, d\gamma$$

$$= \hat{f}^{\,\vee}(x),$$

while the domination from $L^2(R^1)$ justifies the Plancherel identity:

$$\|f\|_2 = \lim_{n \uparrow \infty} \|f_n\|_2 = \lim_{n \uparrow \infty} \|\hat{f}_n\|_2 = \|\hat{f}\|_2.$$

The rest is easy: As you already know, $\wedge$ maps $C_\downarrow^\infty(R^1)$ *into* itself and to see that the map is *onto*, you have only to use the identity

$$f = \hat{f}^{\,\vee} = \int \hat{f}(-\gamma) e^{-2\pi i \gamma x} \, d\gamma,$$

which displays the general function f of class $C_\downarrow^\infty(R^1)$ as the Fourier transform of

$$\hat{f}(-\cdot)$$

which is likewise of class $C_\downarrow^\infty(R^1)$.

   Warning: The practice of defining Fourier integrals by the recipe

$$\hat{f}(\gamma) = \int f(x) e^{-2\pi i \gamma x} \, dx, \qquad f(x) = \int \hat{f}(\gamma) e^{2\pi i \gamma x} \, d\gamma$$

is not universal (nor even very common). It simplifies the theoretical discussion, but for applications, the alternative pair

$$\hat{f}(\gamma) = (2\pi)^{-\frac{1}{2}} \int f(x)\, e^{i\gamma x}\, dx, \qquad f(x) = (2\pi)^{-\frac{1}{2}} \int \hat{f}(\gamma)\, e^{-i\gamma x}\, d\gamma$$

is often better. Another pair in common use is

$$\hat{f}(\gamma) = \int f(x)\, e^{i\gamma x}\, dx, \qquad f(x) = (2\pi)^{-1} \int \hat{f}(\gamma)\, e^{i\gamma x}\, d\gamma,$$

but be careful: The first alternative pair still obeys the Plancherel identity, but for the second, a constant factor is needed:

$$\|f\|_2 = (2\pi)^{-\frac{1}{2}} \|\hat{f}\|_2!$$

2.3   FOURIER INTEGRALS FOR $L^2(R^1)$: FIRST METHOD

A nice Fourier integral for functions of class $L^2(R^1)$ can now be obtained with hardly any extra effort.

Given $f \in L^2(R^1)$, pick $f_n \in C_\downarrow^\infty(R^1)$ so as to make $\lim_{n\uparrow\infty} \|f_n - f\|_2 = 0$. By the Plancherel identity for $C_\downarrow^\infty(R^1)$,

$$\|\hat{f}_n - \hat{f}_m\|_2 = \|(f_n - f_m)^\wedge\|_2 = \|f_n - f_m\|_2$$
$$\leqslant \|f_n - f\|_2 + \|f_m - f\|_2$$

so that you can define $\hat{f} = \lim_{n\uparrow\infty} \hat{f}_n$ and be assured that the limit really exists, in $L^2(R^1)$.

EXERCISE 1.   Show that $\hat{f}$ is well defined, i.e., that it depends only on f and not upon the particular approximating functions employed. Check also that $\wedge$ is a linear map.

The Plancherel identity is immediate:

$$\|f\|_2 = \lim_{n\uparrow\infty} \|f_n\|_2 = \lim_{n\uparrow\infty} \|\hat{f}_n\|_2 = \|\hat{f}\|_2,$$

and so to this point, you know that $\wedge$ is a 1:1 *length-preserving map of* $L^2(R^1)$ *into itself.* The map $\vee$ is extended from $C_\downarrow^\infty(R^1)$ to $L^2(R^1)$ in the same way, and you obtain the inversion formula with no extra effort:

$$\hat{f}^\vee = (\lim_{n\uparrow\infty} f_n)^{\wedge\vee} = (\lim_{n\uparrow\infty} \hat{f}_n)^\vee = \lim_{n\uparrow\infty} (\hat{f}_n{}^\vee) = \lim_{n\uparrow\infty} f_n = f = f^{\vee\wedge}.$$

To sum up, $\wedge [\vee]$ *is an isomorphism of* $L^2(R^1)$ *onto itself, and* $\wedge \vee = \vee \wedge =$ the identity map.

The transforms $\wedge$ and $\vee$ can be expressed more concretely as

$$\hat{f} = \lim_{\substack{b \to \infty \\ a \to -\infty}} \int_a^b f(x)\, e^{-2\pi i \gamma x}\, dx,$$

$$f = \hat{f}^{\vee} = \lim_{\substack{b \to \infty \\ a \to -\infty}} \int_a^b \hat{f}(\gamma)\, e^{2\pi i \gamma x}\, d\gamma,$$

the limits being understood in the sense of $L^2(R^1)$. To see this, let f_{ab} be the product of f and the indicator function of a bounded interval $a \leqslant x \leqslant b$. Then

$$\|\hat{f} - \hat{f}_{ab}\|_2 = \|(f - f_{ab})^{\wedge}\|_2 = \|f - f_{ab}\|_2$$

tends to 0 as $a \downarrow -\infty$ and $b \uparrow \infty$. To finish the proof, you have only to check that

$$\hat{f}_{ab} = \int_a^b f(x)\, e^{-2\pi i \gamma x}\, dx.$$

This is not as transparent as it looks, but with this warning, can be left as the next exercise.

EXERCISE 2.   Check the formula for $\hat{f}_{ab}$. Hint: $\hat{f}_{ab}$ is *defined* as a limit in $L^2(R^1)$, but you can also make the convergence take place *pointwise*. How? Now see Exercise 1.2.8.

EXERCISE 3.   Check that $(f,g) = (\hat{f}, \hat{g})$ for functions from $L^2(R^1)$; this is *Parseval's identity*. Hint: See Exercise 1.3.8.

EXERCISE 4.   Check that $(fg)^{\wedge} = \hat{f} \circ \hat{g}$ for functions from $L^2(R^1)$. Hint: Use Exercise 3.

EXERCISE 5.   Check the Fourier cosine and sine integral formulas for functions f from $L^2[0, \infty)$:

$$\hat{f}_{\text{even}}(\gamma) = 2 \int_0^\infty f(x) \cos 2\pi \gamma x\, dx,$$

$$f(x) = 2 \int_0^\infty \hat{f}_{\text{even}}(\gamma) \cos 2\pi \gamma x\, d\gamma,$$

and

$$\hat{f}_{\text{odd}}(\gamma) = 2 \int_0^\infty f(x) \sin 2\pi \gamma x\, dx,$$

$$f(x) = 2 \int_0^\infty \hat{f}_{\text{odd}}(\gamma) \sin 2\pi \gamma x\, d\gamma.$$

Hint: Extend $f \in L^2[0, \infty)$ to the whole line, first in an odd, and then in an even way.

EXERCISE 6.   Check that if $\hat{f} \in L^2(R^1)$ then $\hat{f}$ is rapidly decreasing in the sense of $\gamma^n \hat{f}(\gamma) \in L^2(R^1)$ for every $n \geqslant 0$ iff $f \in C^\infty$ and all its derivative $D^n f$ belong to $L^2(R^1)$. Also check the formula $(D^n f)^\wedge = (2\pi i \gamma)^n \hat{f}$ for such functions. *Hint:*   One half of the proof is a little tricky if f is not compact: You need to show, e.g., that $f(x) \to 0$ as $x \uparrow \infty$ if f and f' belong to $L^2(R^1)$. To do this, write

$$f^2(x) = f^2(0) + \int_0^x (f^2)' = f^2(0) + 2 \int_0^x ff'$$

in order to convince yourself that the limit exists, and then argue separately that it had better be zero.

EXERCISE 7.   Check that the "Hilbert transform"

$$T: f \in L^2(R^1) \to [\hat{f} \operatorname{sign}(\gamma)]^\vee$$

can be expressed as

$$Tf = -\frac{1}{\pi i} \int \frac{f(y)}{x-y} \, dy,$$

interpreting the integral as a "principal value"

$$\lim_{\varepsilon \downarrow 0} \int_{|x-y| > \varepsilon} \frac{f(y)}{x-y} \, dy$$

in $L^2(R^1)$. *Hint:*   By Exercise 3,

$$\int_{|y| > \varepsilon} \frac{f(x-y)}{y} \, dy = f \circ g_\varepsilon = (\hat{f} \hat{g}_\varepsilon)^\vee,$$

in which

$$g_\varepsilon(y) = \begin{cases} y^{-1} & \text{for } |y| > \varepsilon, \\ 0 & \text{elsewhere.} \end{cases}$$

Now make $\varepsilon \downarrow 0$, and use the evaluation

$$\lim_{R \uparrow \infty} \int_0^R \frac{\sin x}{x} \, dx = \frac{\pi}{2}.$$

Amplification: An amusing aspect of the inversion formula $(\hat{f})^\vee = f$ appears if you try to write it out like a Fourier series. Bring in a (formal) differential

$$e_\gamma (d\gamma)^{1/2} = e^{2\pi i \gamma x} (d\gamma)^{1/2}.$$

In this language, $\hat{f}(\gamma)\,(d\gamma)^{\frac{1}{2}}$ appears as the (formal) Fourier coefficient

$$\hat{f}(\gamma)\,(d\gamma)^{\frac{1}{2}} = (f, e_\gamma(d\gamma)^{\frac{1}{2}}) = \int f(x)\,e_\gamma^*(x)\,dx\,(d\gamma)^{\frac{1}{2}}$$

and f is expressed as a (formal) sum, or better, as an integral:

$$f(x) = \int (f, e_\gamma(d\gamma)^{\frac{1}{2}})\,e_\gamma(d\gamma)^{\frac{1}{2}} = \int \hat{f}(\gamma)\,e^{2\pi i \gamma x}\,d\gamma.$$

The suggestion is that the family of (formal) differentials $e_\gamma(d\gamma)^{\frac{1}{2}} \colon \gamma \in R^1$ fills the office of a unit-perpendicular basis of $L^2(R^1)$, and while this is not perfectly correct, it may be partly justified by the fact that for any bounded intervals I and J,

$$\left(\int_I e_\gamma\,d\gamma, \int_J e_\gamma\,d\gamma \right) = (1_I^\vee, 1_J^\vee) = (1_I, 1_J) = m(I \cap J),$$

whence

$$\left(\int_I e_\gamma\,d\gamma, \int_J e_\gamma\,d\gamma \right) = 0 \qquad \text{if } I \text{ and } J \text{ are disjoint,}$$

and

$$\left\| \frac{\int_I e_\gamma\,d\gamma}{[m(I)]^{\frac{1}{2}}} \right\|_2 = 1.$$

This is the point of departure for the next section.

2.4* FOURIER INTEGRALS FOR $L^2(R^1)$: SECOND METHOD

A second way of establishing the Fourier integral for functions of class $L^2(R^1)$ is to begin by proving

$$(\hat{1}_I)^\vee = 1_I,$$

and

$$(\hat{1}_I, \hat{1}_J) = (1_I, 1_J) = m(I \cap J)$$

for bounded intervals I and J, by hand. This leads at once to the inversion formula and the Plancherel identity for the class K of compact piecewise constant functions. The extension to $L^2(R^1)$ is plain sailing over the course laid down in Section 2.3: You have only to repeat the whole thing with K in place of $C_\downarrow^\infty(R^1)$. [Why?]

Step 1: Check the inversion formula $(\hat{1}_I)^\vee = 1_I$: A little scaling will convince you that it is enough to deal with the interval I: $-\frac{1}{2} \leqslant x \leqslant \frac{1}{2}$. Put $1_I = f$. Then

$$\hat{f}(\gamma) = \int_{-\frac{1}{2}}^{\frac{1}{2}} e^{-2\pi i \gamma x} \, dx = \frac{\sin \pi \gamma}{\pi \gamma},$$

and the only moot point is whether

$$f_{ab}(x) = \int_a^b \frac{\sin \pi \gamma}{\pi \gamma} e^{2\pi i \gamma x} \, d\gamma$$

tends to f in $L^2(R^1)$ as $a \downarrow -\infty$ and $b \uparrow \infty$. By Exercise 1 of this section,

$$\|f_{ab} - f\|_2^2 = \int_{-\infty}^{\infty} \left| \int_a^b \frac{\sin \pi \gamma}{\pi \gamma} e^{2\pi i \gamma x} \, d\gamma \right|^2 dx$$

$$- 2 \int_{-\frac{1}{2}}^{\frac{1}{2}} \left[\int_a^b \frac{\sin \pi \gamma}{\pi \gamma} \cos 2\pi \gamma x \, d\gamma \right] dx + 1$$

$$= \int_a^b \left(\frac{\sin \pi \gamma}{\pi \gamma} \right)^2 d\gamma - 2 \int_a^b \left(\frac{\sin \pi \gamma}{\pi \gamma} \right)^2 d\gamma + 1,$$

so it is enough to check the evaluation

$$\int_{-\infty}^{\infty} \left(\frac{\sin \pi \gamma}{\pi \gamma} \right)^2 d\gamma = \frac{1}{\pi} \int_{-\infty}^{\infty} \left(\frac{\sin \gamma}{\gamma} \right)^2 d\gamma = 1.$$

This is a standard formula, which you can find in any integral table; a proof is also indicated in Exercise 3.1.12. Another method is to adapt a trick used in Section 1.6. By partial integration and a self-evident substitution,

$$\int_0^{\infty} \frac{\sin^2 \gamma}{\gamma^2} \, d\gamma = -\left. \frac{\sin^2 \gamma}{\gamma} \right|_0^{\infty} + \int_0^{\infty} \frac{2 \sin \gamma \cos \gamma}{\gamma} \, d\gamma$$

$$= \int_0^{\infty} \frac{\sin \gamma}{\gamma} \, d\gamma.$$

The latter is a standard *improper* integral, which means that it should be interpreted as

$$\lim_{B \uparrow \infty} \int_0^B \frac{\sin \gamma}{\gamma} \, d\gamma = I.$$

The existence of this limit is assured by the preceding development, although it can also be checked directly by elementary estimates. [Sketch a picture of $\gamma^{-1} \sin \gamma$ and try it!]. To evaluate the limit let $B = (2n+1)\pi/2$ so that

$$I = \lim_{n \uparrow \infty} \int_0^{(2n+1)\pi/2} \frac{\sin \gamma}{\gamma} \, d\gamma = \lim_{n \uparrow \infty} \int_0^{\frac{1}{2}} \frac{\sin(2n+1)\pi\gamma}{\gamma} \, d\gamma.$$

Now recall from Section 1.4 the formula

$$\int_0^{\frac{1}{2}} \frac{\sin(2n+1)\pi\gamma}{\sin \pi\gamma}\, d\gamma = \frac{1}{2}.$$

This shows that

$$I - \frac{\pi}{2} = \lim_{n\uparrow\infty} \int_0^{\frac{1}{2}} \left(\frac{1}{\gamma} - \frac{\pi}{\sin \pi\gamma}\right) \sin(2n+1)\pi\gamma\, d\gamma,$$

and since the first factor of the integrand is summable, the limit is 0 by the Riemann–Lebesgue lemma.

EXERCISE 1.   Show that

$$\int_A^B \left| \int_a^b \frac{\sin \pi\gamma}{\pi\gamma} e^{2\pi i\gamma x}\, d\gamma \right|^2 dx$$

$$= \int_a^b \frac{\sin \pi\alpha}{\pi\alpha} \left[\int_a^b \frac{\sin \pi\beta}{\pi\beta} \frac{\sin 2\pi B(\alpha-\beta) - \sin 2\pi A(\alpha-\beta)}{2\pi(\alpha-\beta)}\, d\beta \right] d\alpha$$

tends to $\int_a^b (\sin \pi\gamma/\pi\gamma)^2\, d\gamma$ as $A\downarrow-\infty$ and $B\uparrow\infty$. *Hint:* The key point is to check that

$$\lim_{B\uparrow\infty} \int_a^b \frac{\sin \pi\beta}{\pi\beta} \frac{\sin 2\pi B(\alpha-\beta)}{\pi(\alpha-\beta)}\, d\beta = \frac{\sin \pi\alpha}{\pi\alpha}$$

if $a < \alpha < b$. Use Exercise 1.5.8 as your guide, taking it for granted that the Riemann–Lebesgue lemma holds for summable functions on the line as well as on the circle. The latter is proved in Section 2.6.

Step 2:   Prove

$$(\hat{1}_I, \hat{1}_J) = (1_I, 1_J)$$

for bounded intervals I and J. Both $\hat{1}_I$ and $\hat{1}_J$ belong to $L^2(R^1)$, so $(\hat{1}_I, \hat{1}_J)$ exists, and can be evaluated as follows:

$$(\hat{1}_I, \hat{1}_J) = \int \hat{1}_I(\gamma)\, \hat{1}_J{}^*(\gamma)\, d\gamma$$

$$= \int \hat{1}_I(\gamma)\, d\gamma \int_J e^{2\pi i\gamma x}\, dx$$

$$= \lim_{n\uparrow\infty} \int_{-n}^n \hat{1}_I(\gamma)\, d\gamma \int_J e^{2\pi i\gamma x}\, dx$$

$$= \lim_{n\uparrow\infty} \int_J dx \int_{-n}^n e^{2\pi i\gamma x}\, \hat{1}_I(\gamma)\, d\gamma.$$

But the inside integral tends to 1_I in $L^2(R^1)$ as $n\uparrow\infty$, as you know from Step 1, so

$$(\hat{1}_I, \hat{1}_J) = \int_J 1_I = \int 1_I 1_J^* = (1_I, 1_J),$$

as advertised.

Step 3: A function $f \in K$ is a sum $\sum c_I 1_I$ of complex multiples of in-dicator functions of a finite number of bounded intervals I. For such functions, the inversion formula $(\hat{f})^\vee = f$ is now immediate, and the Plancherel identity takes only one more line:

$$\|f\|_2^2 = \sum c_I c_J^* (\hat{1}_I, \hat{1}_J) = \sum c_I c_J^* (1_I, 1_J) = \|f\|_2^2.$$

Step 4: Extend from K to the whole of $L^2(R^1)$. Since this is done just as in Section 2.3, the details are left to the reader.

EXERCISE 2. Show by direct calculation that the (improper) integral

$$\int_{-\infty}^{\infty} \frac{\sin \pi\gamma}{\pi\gamma} e^{2\pi i\gamma x} \, d\gamma = \begin{cases} 1 & \text{if } |x| < \tfrac{1}{2}, \\ 0 & \text{if } |x| > \tfrac{1}{2}. \end{cases}$$

Why is this result consistent with Step 1? *Hint:* Check first that

$$\int_{-\infty}^{\infty} \frac{\sin \pi\gamma x}{\pi\gamma} \, d\gamma = \pm 1$$

depending on whether x is positive or negative.

2.5* FOURIER INTEGRALS FOR $L^2(R^1)$: THIRD METHOD

Because the Fourier transform preserves inner products [see Exercise 2.3.3], you may think of it as an (infinite-dimensional) rotation. The pur-pose of this section is to investigate this suggestion more fully.

To begin with, *the fourth power of $\wedge$ is the identity map*: in fact, with the notation $f^- = f(-\cdot)$,

$$f^{\wedge\wedge}(x) = \int \hat{f}(\gamma)\, e^{-2\pi i\gamma x}\, d\gamma = \hat{f}^\vee(-x) = f(-x) = f^-(x),$$

and so

$$f^{\wedge\wedge\wedge\wedge} = f^{--} = f.$$

This means that if $f \in L^2(R^1)$ is an eigenfunction of $\wedge$ [$\hat{f} = \text{constant} \times f$], then the eigenvalue can only be a fourth root of unity, i.e., it can only be

i, -1, $-i$, or $+1$. The bulk of this section is devoted to proving that the eigenfunctions of $\wedge$ are precisely the "Hermite functions":

$$h_n(x) = \frac{(-1)^n}{n!} \exp(\pi x^2)\, D^n \exp(-2\pi x^2), \qquad n \geqslant 0.$$

The eigenvalue for h_n is $(-i)^n$, i.e.,

$$\hat{h}_n = (-i)^n h_n.$$

The scaled Hermite functions

$$e_n = \|h_n\|_2^{-1} h_n = [(4\pi)^{-n}\sqrt{2n!}]^{\frac{1}{2}} h_n$$

form a unit-perpendicular basis of $L^2(R^1)$, and you can use this fact to re-establish the Fourier transform on $L^2(R^1)$ via the formula

$$\hat{f} = \sum_{n=0}^{\infty} (f, e_n)\,(-i)^n e_n.$$

This approach goes back to Wiener [1933, pp. 51–71]. The geometrical content of this story is that $L^2(R^1)$ can be split into the sum

$$\mathsf{H}_0 \oplus \mathsf{H}_1 \oplus \mathsf{H}_2 \oplus \mathsf{H}_3$$

of the four perpendicular eigenspaces of $\wedge$:

$$\mathsf{H}_k = \left(f \in L^2(R^1) : f = \sum_{n=0}^{\infty} (f, e_{4n+k})\, e_{4n+k} \right), \qquad k = 0, 1, 2, 3,$$

in the sense that any $f \in L^2(R^1)$ can be split in precisely one way into a sum $f_0 + f_1 + f_2 + f_3$ of pieces, one from each eigenspace. The action of $\wedge$ upon H_k is just multiplication by $(-i)^k$, that is, a rotation through the angle $0°, 270°, 180°, 90°$, respectively. The proof is broken up into a series of simple exercises, with hints for their solutions.

EXERCISE 1. h_n is the product of $\exp(-\pi x^2)$ and a polynomial of precise degree n; especially, $h_n \in C_\downarrow^\infty(R^1)$.

EXERCISE 2. $h_n' - 2\pi x h_n = -(n+1)h_{n+1}$. Hint: Compute $h_n' - 2\pi x h_n$ from the definition.

EXERCISE 3. $h_n' + 2\pi x h_n = 4\pi h_{n-1}$, with the understanding that $h_{-1} \equiv 0$. Hint: Compute $h_n' + 2\pi x h_n$ from the definition, using Leibnitz' rule to check the identity $4\pi x D^n - D^n 4\pi x = -4\pi n D^{n-1}$.

EXERCISE 4. $\hat{h}_n = (-i)^n h_n$ and $\check{h}_n = (i)^n h_n$. Hint: $h_n \in C_\downarrow^\infty(R^1)$, so the

Fourier transform can be applied directly. Compute the transform of the identity of Exercise 2 using the rules

$$(-2\pi i x f)^\wedge = (\hat{f})', \qquad (f')^\wedge = 2\pi i \gamma \hat{f}$$

[see Section 2.2], to verify that $\hat{h}_n$ and $(-i)^n h_n$ obey the *same* recursion formula. The known transform $\hat{h}_0 = h_0$ [see Subsection 1.7.5] will finish the proof.

EXERCISE 5. h_n is an eigenfunction of the operator $K: f \to f'' - 4\pi^2 x^2 f$; the eigenvalue is $-4\pi(n+\frac{1}{2})$. *Hint:*   Combine Exercises 2 and 3.

EXERCISE 6. $(h_n, h_m) = 0$ for $n \neq m$. *Hint:* $(Kh_n, h_m) = (h_n, Kh_m)$, and $4\pi(n+\frac{1}{2}) \neq 4\pi(m+\frac{1}{2})$.

EXERCISE 7.   The family $h_n: n \geqslant 0$ spans $L^2(R^1)$. *Hint:*   For any $f \in L^2(R^1)$, the Fourier transform of $f \exp(-\pi x^2)$ can be expanded in a power series:

$$[f \exp(-\pi x^2)]^\wedge = \int f(x) \exp(-\pi x^2) e^{-2\pi i \gamma x} dx$$

$$= \int f(x) \exp(-\pi x^2) \sum_{n=0}^{\infty} (-2\pi i \gamma)^n (x^n/n!) dx$$

$$= \sum_{n=0}^{\infty} (-2\pi i \gamma^n/n!) \int f(x) \exp(-\pi x^2) x^n dx.$$

Check that this is legitimate, and infer that $f \equiv 0$ is the only function that is perpendicular to all the Hermite functions; see Exercise 11 for an alternative proof. *Amplification:*   To do things this way, you have to know that $[f(-\pi x^2)]^\wedge = 0$ only if $f = 0$. Prove this by hand. *Hint:*

$$([f \exp(-\pi x^2)]^\wedge, \hat{g}) = (f \exp(-\pi x^2), g)$$

may be proved by hand for any $g \in C_\downarrow^\infty(R^1)$. Do it!

EXERCISE 8.   Define $\hat{f}$ for the general $f \in L^2(R^1)$ by Wiener's recipe:

$$\hat{f} = \sum_{n=0}^{\infty} (f, e_n) (-i)^n e_n,$$

in which e_n is the nth Hermite function, rescaled so as to be of unit length [see Exercises 12 and 13 for the actual lengths $\|h_n\|_2$]. Check that $\wedge$ is a length-preserving map of $L^2(R^1)$ onto itself. Give a similar definition of $\vee$ and check $\hat{f}^\vee = \check{f}^\wedge = f$. Check that the present recipe is the same as that of sections 2.3 and 2.4. *Hint for the last part:* $\sum_{k=1}^n (f, e_k) e_k \in C_\downarrow^\infty(R^1)$.

This finishes the explanation of Wiener's approach to the Fourier transform for $L^2(R^1)$; the following exercises are supplementary.

EXERCISE 9. $f \in C_{\uparrow}^{\infty}(R^1)$ iff its Hermite coefficients (f, e_n) are rapidly
decreasing. Hint: By Exercise 5,

$$[-4\pi(n+\tfrac{1}{2})]^p(f, e_n) = (K^p f, e_n) \qquad \text{for any} \quad p \geqslant 1$$

if $f \in C_{\uparrow}^{\infty}(R^1)$; that takes care of one implication. The converse depends
upon Exercises 2 and 3 which may be used to check that if $f \in L^2(R^1)$ has
rapidly decreasing Hermite coefficients, then so does

$$[(D+2\pi x) - (D-2\pi x)]^p \, [(D+2\pi x) + (D-2\pi x)]^q f = (4\pi x)^p (2D)^q f$$

for any integral p and q, formally at any rate. Check that this is really legiti-
mate and infer (a) $f \in C^{\infty}(R^1)$, (b) $x^p D^q f \in L^2(R^1)$, and (c) $f \in C_{\downarrow}^{\infty}(R^1)$,
in that order.

EXERCISE 10.   Use Exercise 5 to verify that

$$\int x^2 |f(x)|^2 \, dx + \int \gamma^2 |\hat{f}(\gamma)|^2 \, d\gamma = \pi^{-1} \sum_{n=0}^{\infty} (n+\tfrac{1}{2}) |(f, e_n)|^2 \geqslant (2\pi)^{-1} \|f\|_2^2$$

and that the lower bound is achieved only for constant multiples of $\exp(-\pi x^2)$.
The content of this bound is that f and $\hat{f}$ cannot both be too sharply peaked
at the origin. A sharper bound of the same kind is Heisenberg's inequality:

$$\int x^2 |f(x)|^2 \, dx \times \int \gamma^2 |\hat{f}(\gamma)|^2 \, d\gamma \geqslant (16\pi^2)^{-1} \|f\|_2^4,$$

to be proved in Section 2.8; the extra sharpness is due to the fact that the
arithmetic mean $\tfrac{1}{2}(a+b)$ is always bigger than the geometrical mean $(ab)^{\frac{1}{2}}$.
Hint: For $f \in C_{\downarrow}^{\infty}(R^1)$,

$$4\pi \sum_{n=0}^{\infty} (n+\tfrac{1}{2}) |(f, e_n)|^2 = -(Kf, f)$$

$$= \int |f'(x)|^2 \, dx + 4\pi^2 \int x^2 |f(x)|^2 \, dx$$

$$= 4\pi^2 \left[\int x^2 |f(x)|^2 \, dx + \int \gamma^2 |\hat{f}(\gamma)|^2 \, d\gamma \right]$$

by partial integration and the rule $(f')^{\wedge} = 2\pi i \gamma \hat{f}$.

EXERCISE 11.   A second proof that the Hermite functions span $L^2(R^1)$

may be based upon the formula

$$\sum_{n=0}^{\infty} y^n h_n(x) = \exp(\pi x^2) \sum_{n=0}^{\infty} [(-y)^n/n!] \, D^n \exp(-2\pi x^2)$$

$$= \exp(\pi x^2) \exp[-2\pi(x-y)^2]$$

$$= \exp[-\pi(x-2y)^2] \exp(2\pi y^2).$$

The sum is a power series, as discussed in Subsection 3.1.4. Granting that it converges in the ordinary numerical sense use Bessel's inequality to verify that for each fixed y it also converges in $L^2(R^1)$ and infer that $\exp(-\pi x^2) \circ f = 0$ if $f \in L^2(R^1)$ is perpendicular to all the Hermite functions. Deduce that $f = 0$ using the Fourier integral of Section 2.3. *Hint:*

$$\exp(-\pi x^2) \circ f \in L^2(R^1).$$

Why? Justify $[\exp(-\pi x^2) \circ f]^\wedge = (\exp(-\pi x^2))^\wedge \hat{f}.$

EXERCISE 12.   Use the sum of Exercise 11 to evaluate

$$\|h_n\|_2^2 = (4\pi)^n (\sqrt{2} n!)^{-1}.$$

EXERCISE 13.   Evaluate $\|h_n\|_2$ a second way by use of the formula

$$(n+1)\|h_{n+1}\|_2^2 = 4\pi \|h_n\|_2^2 :$$

Hint: $(n+1)h_n = 4\pi h_{n-1} - 2h_n'$ by Exercises 2 and 3.

2.6   FOURIER INTEGRALS FOR $L^1(R^1)$

The Fourier integral is easily established for summable functions, too. The principal facts are summed up in the following theorem.

THEOREM 1.   *For any summable function f, the Fourier transform*

$$\hat{f}(\gamma) = \int f(x) \, e^{-2\pi i \gamma x} \, dx$$

exists as an ordinary Lebesgue integral, and enjoys the following properties:

(a) $\|\hat{f}\|_\infty \leqslant \|f\|_1$

(b) $\hat{f} \in C(R^1)$

(c) $\lim_{|\gamma| \uparrow \infty} \hat{f}(\gamma) = 0$

(d) $(f \circ g)^\wedge = \hat{f}\hat{g}$

(e) $\hat{f} = 0$ iff $f = 0$.

Amplification: Item (c) is the Riemann–Lebesgue lemma for $L^1(R^1)$. Items (a)–(e) state that $\wedge$ is a 1:1 mapping of $L^1(R^1)$ into the algebra of functions of class $C(R^1)$ which "vanish at ∞." The actual image is a very complicated subalgebra, which is not completely known to date. The same unfortunate circumstance was already encountered for the Fourier coefficients of summable functions on the circle [see Section 1.5].

PROOF. Item (a) is self-evident from the definition, and (b) is proved by noting that

$$|\hat{f}(\beta) - \hat{f}(\alpha)| \leqslant \int |e^{-2\pi i \beta x} - e^{-2\pi i \alpha x}| \, |f(x)| \, dx$$

tends to 0 as $|\beta - \alpha| \downarrow 0$ by dominated convergence. For (c), pick $f_n \in C_\downarrow^\infty(R^1)$ with $\|f_n - f\|_1 \leqslant 1/n$. Then $\hat{f}_n$ is rapidly decreasing, and

$$|\hat{f}_n(\gamma) - \hat{f}(\gamma)| \leqslant \|f_n - f\|_1 \leqslant n^{-1},$$

so

$$\limsup_{|\gamma| \uparrow \infty} |\hat{f}(\gamma)| = n^{-1} + \lim_{|\gamma| \uparrow \infty} |\hat{f}_n(\gamma)| = n^{-1}$$

for any $n \geqslant 1$.

EXERCISE 1. Give a second proof of (c) using the fact that

$$\hat{f}(\gamma) = -\int f(x) \, e^{-2\pi i \gamma (x-y)} \, dx = -\int f(x+y) \, e^{-2\pi i \gamma x} \, dx$$

for $y = 1/2\gamma$. *Hint:* See Theorem 1.5.2 for a model; Exercise 2.1.3 will also be helpful.

The proof of (d) is easy: Fubini's theorem justifies the interchange of integrals, and the formula drops out. The proof of (e) is only a little deeper: If $\hat{f} = 0$, then by Fubini's theorem,

$$\int f(x)g(x) \, dx = \int f(x) \left[\int g^\vee(\gamma) e^{-2\pi i \gamma x} \, d\gamma \right] dx$$

$$= \int \hat{f}(\gamma) g^\vee(\gamma) \, d\gamma = 0$$

for any $g \in C_\downarrow^\infty(R^1)$, and you see that $f = 0$ by making g approximate the indicator function of the interval $a \leqslant x \leqslant b$ in a bounded way. The details, which can be patterned on Exercise 1.1.13, are left to the reader.

EXERCISE 2. Check that $\|\hat{f}\|_\infty = \lim_{n \uparrow \infty} (\|f^n\|_1)^{1/n}$, where f^n is the

n-fold product $f \circ \cdots \circ f$. Confine yourself to the case $f \in L^1(R^1) \cap L^2(R^1)$.
Hint: Use the hint in Exercise 1.5.13.

EXERCISE 3.   Let $L^1(R^+, x^{-1} dx)$ denote the set of functions f defined
on the half-line $R^+ = (0, \infty)$, which are summable relative to the weight
$x^{-1} dx$:

$$\int_0^\infty |f(x)| x^{-1} dx < \infty.$$

Check that $L^1(R^+, x^{-1} dx)$ is an algebra under the convolution product

$$f_1 \circ f_2(x) = \int_0^\infty f_1(x/y) f_2(y) y^{-1} dy$$

and that

$$(f_1 \circ f_2)^\# = f_1^\# f_2^\#,$$

in which # stands for the "Mellin transform"

$$f^\#(\gamma) = \int_0^\infty f(x) x^{-2\pi i \gamma} x^{-1} dx.$$

Prove that the map

$$T: f \to f(e^x)$$

establishes a $1:1$, linear, product-preserving transformation from the algebra
$L^1(R^+, x^{-1} dx)$ *onto* the convolution algebra $L^1(R^1)$, and that the Mellin
transform $f^\#$ equals the Fourier transform $(Tf)^\wedge$.

Generally speaking, the inverse transform cannot be applied directly
to "undo" $\wedge$ since $\hat{f}$ may be nonsummable. For example,

$$\hat{f} = \int_0^1 x^{-\frac{1}{2}} e^{-2\pi i \gamma x} dx$$

is not summable, and even

$$\int |\hat{f}|^2 = \int_0^1 \frac{dx}{x} = \infty!$$

The difficulty is circumvented by the following Fejér-type recipe; see Subsection 1.8.3 for the analog for continuous functions on the circle.

THEOREM 2.   *A summable function f may be recovered from $\hat{f}$ by the
recipe:*

$$f = \lim_{t \downarrow 0} [\exp(-2\pi^2 \gamma^2 t) \hat{f}]^\vee,$$

the limit being taken in the sense of $L^1 R^1$):

$$\lim_{t\downarrow 0} \| [\exp(-2\pi^2\gamma^2 t)\hat{f}]^\vee - f \|_1 = 0.$$

PROOF.   Bring in the familiar "Gauss kernel"

$$p_t(x) = \frac{\exp(-x^2/2t)}{(2\pi t)^{1/2}}, \qquad t > 0, \quad x \in R^1$$

and recall the formula

$$\hat{p}_t(\gamma) = \exp(-2\pi^2\gamma^2 t)$$

of Subsection 1.7.5. Because $p_t \in C_\downarrow^\infty(R^1)$, this transform can be inverted in the naive way. Now look at $p_t \circ f$ and *its* transform:

$$(p_t \circ f)^\wedge = \hat{p}_t\hat{f} = \exp(-2\pi^2\gamma^2 t)\hat{f}.$$

This is a summable function, so you can apply $\vee$ to it directly, and with the help of Fubini's theorem, you find that

$$[\exp(-2\pi^2\gamma^2 t)\hat{f}]^\vee = \int e^{2\pi i\gamma x} \exp(-2\pi^2\gamma^2 t)\hat{f}(\gamma)\, d\gamma$$

$$= \int e^{2\pi i\gamma x} \exp(-2\pi^2\gamma^2 t) \left[\int f(y)\, e^{-2\pi i\gamma y}\, dy \right] d\gamma$$

$$= \int f(y) \left[\int e^{2\pi i\gamma(x-y)} \exp(-2\pi^2\gamma^2 t)\, d\gamma \right] dy$$

$$= p_t \circ f.$$

To finish the proof, you have only to check

$$\lim_{t\downarrow 0} \| p_t \circ f - f \|_1 = 0,$$

which is easy: If f_y is the translated function $f_y(x) = f(x+y)$, then

$$p_t \circ f - f = \int [f_y(x) - f(x)]\, p_t(y)\, dy,$$

since $\int p_t = 1$, and so

$$\| p_t \circ f - f \|_1 \leqslant \int \| f_{-y} - f \|_1\, p_t(y)\, dy$$

$$\leqslant \int_{-\delta}^{\delta} \| f_{-y} - f \|_1\, p_t(y)\, dy + 4\|f\|_1 \int_{\delta/\sqrt{t}}^\infty \frac{\exp(-y^2/2)\, dy}{(2\pi)^{1/2}}.$$

The first integral is small for small δ by Exercise 2.1.3, while the second tends to zero as $t\downarrow 0$ for fixed δ. The proof is finished.

EXERCISE 4. The formula $\lim_{t \downarrow 0} p_t \circ f = f$ states that p_t is an "approximate multiplicative identity" for $L^1(R^1)$. Check that no exact multiplicative identity exists. *Hint:*   See Exercise 1.5.12.

EXERCISE 5. Check that f may be recovered from $\hat{f}$ by simple application of $\vee$ if $\hat{f}$ is summable. *Hint:* $\lim_{t \downarrow 0} [\exp(-2\pi^2 \gamma^2 t)\hat{f}]^\vee = f$, pointwise.

EXERCISE 6. Check the formula
$$[(t/\pi)(x^2 + t^2)^{-1}]^\wedge = \exp(-2\pi|\gamma|t)$$
for $t > 0$. *Hint:* $[\exp(-2\pi|\gamma|t)]^\vee$ is an elementary integral.

EXERCISE 7. Use Exercise 6 to prove the following variant of Theorem 2.
$$\lim_{t \downarrow 0} \|[\exp(-2\pi|\gamma|t)\hat{f}]^\vee - f\|_1 = 0.$$

EXERCISE 8. Check that if f is summable and if
$$\int_{-1}^{1} |f(x+y) - f(x)| \frac{dy}{|y|} < \infty$$
for fixed x, then
$$f(x) = \lim_{\substack{b \to \infty \\ a \to -\infty}} \int_a^b \hat{f}(\gamma) e^{2\pi i \gamma x} \, d\gamma$$
in the usual numerical sense. This is the analogue of Dini's test for summable functions on the circle [see Exercise 1.5.8]. *Hint:*   The integral
$$\int_a^b \hat{f}(\gamma) e^{2\pi i \gamma x} \, d\gamma$$
can be expressed as $f \circ D$ with the Dirichlet-type kernel
$$D(x) = \frac{e^{2\pi i b x} - e^{2\pi i a x}}{2\pi i x}.$$
Now split the integral $f \circ D = \int f(x-y) D(y) \, dy$ into two pieces according as $|y| \leq 1$ or $|y| > 1$ and use the Riemann–Lebesgue lemma to get rid of the second one. The first may be dealt with much as in Exercise 1.5.8. The evaluation $\int_0^\infty x^{-1} \sin x \, dx = \pi/2$ from Section 2.4 is needed.

EXERCISE 9. Check that for summable f,
$$\lim_{b \uparrow \infty} b^{-1} \int_0^b da \int_{-a}^a \hat{f}(\gamma) e^{2\pi i \gamma x} \, d\gamma = f$$

in $L^1(R^1)$. This is the analogue of the fact that $\lim_{n\uparrow\infty} n^{-1}(S_0 + \cdots + S_{n-1}) = f$ for summable functions on the circle [see Theorem 1.5.1]. *Hint:* The left-hand side may be expressed as $f \circ F$ with the Fejér-type kernel $F = b^{-1}[(\pi x)^{-1} \sin \pi b x]^2$. Now follow the proof of Theorem 2. The evaluation $\int F = 1$ is needed. Obtain this from the Plancherel identity.

EXERCISE 10.[1]   Check that if f is a real, even, summable function and if $f(0+)$ and $f(0-)$ exist, then either $f(0-) = f(0+)$ *or* $\hat{f}$ changes sign infinitely often as $|\gamma| \uparrow \infty$. $\hat{f}$ is a *real* function, so its "sign" makes sense! The content of this result is that to build up a jump in f, you not only have to use very high frequencies but you must also have complicated cancellations in the tail of the integral $\int \hat{f}(\gamma)\, e^{2\pi i \gamma x}\, d\gamma$. Here you see another illustration of how the local behavior of f reflects the global behavior of $\hat{f}$. *Hint:* The function $\hat{f}$ is summable if it is of one sign far out, as you can see from

$$\tfrac{1}{2}[f(0-) + f(0+)] = \lim_{t\downarrow 0} p_t \circ f(0) = \lim_{t\downarrow 0} \int \exp(-2\pi^2 \gamma^2 t)\hat{f}(\gamma)\, d\gamma.$$

The basic facts about Fourier integrals for $C_\downarrow^\infty(R^1)$, $L^2(R^1)$, and $L^1(R^1)$ are now before you. The rest of this chapter is devoted to deeper points and to applications.

2.7   MISCELLANEOUS APPLICATIONS

The purpose of this section is to present a number of simple applications of Fourier integrals. For practical use, it is of the first importance to have a good table of transforms. Bateman [1954] is much the best one around.

1.   An Ordinary Differential Equation

Perhaps the most important formal feature of the Fourier integral is that it maps differential operators with constant coefficients into multiplication by polynomials, according to the rule:

$$(f')^\wedge = 2\pi i \gamma \hat{f}.$$

A simple example illustrating the usefulness of this fact is provided by the problem

$$u'' - u = -f,$$

in which f is known and you have to find u. A formal Fourier transform

---

[1] Adapted from Kac [1938].

on both sides gives

$$(4\pi^2 \gamma^2 + 1)\hat{u} = \hat{f},$$

or what is the same,

$$\hat{u} = (1 + 4\pi^2 \gamma^2)^{-1}\hat{f}.$$

But $(1 + 4\pi^2 \gamma^2)^{-1}$ is the Fourier transform of $\frac{1}{2}e^{-|x|}$ [see Exercise 2.6.6], so

$$u = [(1 + 4\pi^2 \gamma^2)^{-1}\hat{f}]^{\vee} = \frac{1}{2}(e^{-|x|} \circ f)^{\wedge\vee} = \frac{1}{2}e^{-|x|} \circ f$$

$$= \frac{1}{2}\int e^{-|x-y|}f(y)\, dy.$$

EXERCISE 1. Check by (formal) differentiation that $u = e^{-|x|} \circ f$ is indeed a solution.

EXERCISE 2. To make everything just said legitimate, you must subject u and f to appropriate technical conditions, for example, if $f \in C^{\infty}_{\downarrow}(R^1)$, then $u = \frac{1}{2}e^{-|x|} \circ f$ is the only solution of the same class. Give a justification if only $u \in C^2(R^1) \cap L^1(R^1)$ and $f \in C(R^1) \cap L^1(R^1)$. *Hint:* $u'' = u + f$ is summable, so $(u'')^{\wedge}$ makes sense. Check that $(u'')^{\wedge} = -4\pi^2 \gamma^2 \hat{u}$ by letting $a \downarrow -\infty$ and $b \uparrow \infty$ in the formula

$$\int_a^b u''(x)\, e^{-2\pi i \gamma x}\, dx = [u'(x) + 2\pi i \gamma u(x)]\, e^{-2\pi i \gamma x}\Big|_a^b$$

$$- 4\pi^2 \gamma^2 \int_a^b u(x)\, e^{-2\pi i \gamma x}\, dx.$$

To do this, check first that $\lim_{|x|\uparrow\infty} u'(x) = 0$, owing to the summability of u and u''. Then use the formula to infer that $\lim_{|x|\uparrow\infty} u(x)e^{-2\pi i \gamma x}$ also exists. What is its value?

2. Heat Flow

The problem of heat flow is an infinite rod:

(a) $\qquad \dfrac{\partial u}{\partial t} = \dfrac{1}{2}\dfrac{\partial^2 u}{\partial x^2}, \qquad t > 0, \quad x \in R^1,$

(b) $\qquad \lim_{t \downarrow 0} u = f$

is easily solved in much the same way. Do a (formal) Fourier transform of both sides of (a):

$$\partial \hat{u}/\partial t = -2\pi^2 \gamma^2 \hat{u},$$

solve for $\hat{u}$:

$$\hat{u} = \hat{f} \exp(-2\pi^2 \gamma^2 t),$$

and invert to obtain

$$u(t,x) = [\exp(-2\pi^2 \gamma^2 t)\hat{f}]^\vee = p_t \circ f = \int_{-\infty}^\infty \frac{\exp[-(x-y)^2/2t]}{(2\pi t)^{\frac{1}{2}}} f(y)\, dy,$$

in which $p_t(x)$ is the Gauss kernel $(2\pi t)^{-\frac{1}{2}}\exp(-x^2/2t)$ of Section 2.6.

EXERCISE 3.   Justify the result, if not the procedure. For example, prove that if f is summable, then $u = p_t \circ f$ is the only solution of class $C^\infty[(0,\infty)\times R^1]$ for which $\|u\|_1 \leqslant \|f\|_1$ and $\lim_{t\downarrow 0}\|u-f\|_1 = 0$.

EXERCISE 4.   Find the temperature in the "semi-infinite" rod $x \geqslant 0$ if the initial data $u(0+,\cdot) = f$ is known and the left end ($x = 0$) is held at temperature 0, imposing upon f whatever technical conditions you need. Answer:

$$u(t,x) = (2\pi t)^{-\frac{1}{2}} \int_0^\infty \{\exp[-(x-y)^2/2t] - \exp[-(x+y)^2/2t]\} f(y)\, dy.$$

This is another instance of Kelvin's method of images; see Exercises 1.8.1, 1.8.2, and 1.8.5.

A trickier problem for the semi-infinite rod is to keep the left end ($x = 0$) at a known temperature $u(t,0) = f(t)$ and to solve for the temperature inside, putting $\lim_{t\downarrow 0} u(t,x) = 0$ for simplicity's sake. Take $u = 0$ for $x < 0$. Then

$$\frac{\partial \hat{u}}{\partial t} = \frac{1}{2} \int_{-\infty}^\infty \frac{\partial^2 u}{\partial x^2} e^{-2\pi i \gamma x}\, dx$$

$$= -\frac{1}{2}\frac{\partial u}{\partial x}(t,0) - \pi i \gamma f - 2\pi^2 \gamma^2 \hat{u},$$

by a couple of partial integrations, so

$$\hat{u}(t,\gamma) = \int_0^t \exp[-2\pi^2 \gamma^2(t-s)]\left[-\frac{1}{2}\frac{\partial u}{\partial x}(s,0) - \pi i \gamma f(s)\right] ds,$$

and inverting, you find

$$u(t,x) = \int_0^t \left[-\frac{1}{2}\frac{\partial u}{\partial x}(s,0) - \frac{1}{2}f(s)\frac{\partial}{\partial x}\right] \frac{\exp[-x^2/2(t-s)]}{[2\pi(t-s)]^{\frac{1}{2}}}\, ds.$$

This expression must vanish for $x < 0$, and since the contribution from $\partial u/\partial x$ is an even function of x, whereas the contribution from $f\partial/\partial x$ is odd, the two contributions are the same for $x > 0$, and you can eliminate the former:

$$u(t,x) = 2 \times \int_0^t -\frac{1}{2} f(s) \frac{\partial}{\partial x} \frac{\exp[-x^2/2(t-s)]}{[2\pi(t-s)]^{1/2}} ds$$

$$= \int_0^t \frac{x \exp[-x^2/2(t-s)]}{[2\pi(t-s)^3]^{1/2}} f(s) ds, \qquad t > 0, \qquad x > 0.$$

EXERCISE 5. Check by hand that u solves the stated problem if $f \in C[0 \; \infty)$. *Hint:*

$$\int_{t-\varepsilon}^t \frac{x \exp[-x^2/2(t-s)]}{(2\pi)^{1/2}(t-s)^{3/2}} ds = 2 \int_{x\varepsilon^{-1/2}}^\infty \frac{\exp(-y^2/2)}{(2\pi)^{1/2}} dy .$$

3. Wave Motion

The problem of wave motion:

(a) $\qquad \dfrac{\partial^2 u}{\partial t^2} = \dfrac{\partial^2 u}{\partial x^2}, \qquad t > 0, \quad x \in R^1,$

(b) $\qquad \lim_{t \downarrow 0} u = f,$

(c) $\qquad \lim_{t \downarrow 0} \dfrac{\partial u}{\partial t} = g,$

is just as easy. Do the transform:

$$\partial^2 \hat{u}/\partial t^2 = -4\pi^2 \gamma^2 \hat{u},$$

solve for $\hat{u}$:

$$\hat{u}(t,\gamma) = \cos 2\pi\gamma t \, \hat{f}(\gamma) + \frac{\sin 2\pi\gamma t}{2\pi\gamma} \hat{g}(\gamma)$$

$$= \frac{1}{2} \left[e^{2\pi i \gamma t} + e^{-2\pi i \gamma t} \right] \hat{f}(\gamma) + \frac{1}{2} \int_{-t}^t e^{2\pi i \gamma y} dy \, \hat{g}(\gamma),$$

and invert:

$$u = u(t,x) = \frac{1}{2}[f(x+t) + f(x-t)] + \frac{1}{2} \int_{x-t}^{x+t} g(y) \, dy .$$

This is precisely the old formula of d'Alembert; see also Subsection 1.8.6.

EXERCISE 6. D'Alembert's formula is a bona fide solution of the stated problem if $f \in C^2(R^1)$ and $g \in C^1(R^1)$. Justify the formal procedure just presented under whatever extra technical conditions you need.

4*.   A Circuit Equation

A further application is to the problem

$$A\ddot{f} + B\dot{f} + Cf = \dot{e}, \qquad t > 0,$$

which describes the current f in the electrical circuit of Fig. 1, where A is

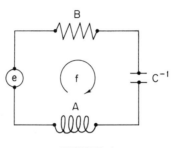

B

e   f C^{-1}

A

FIGURE 1

the inductance, B is the resistance, C^{-1} the capacitance, and e is the impressed voltage. Think of f as vanishing for $t < 0$ and let it tend to 0 together with $\dot{f}$ as $t \uparrow \infty$. Then

$$(\dot{f})^{\wedge} = \int_0^{\infty} \dot{f}(t)\, e^{-2\pi i \gamma t}\, dt = -f(0) + 2\pi i \gamma \hat{f},$$

$$(\ddot{f})^{\wedge} = -\dot{f}(0) + 2\pi i \gamma (\dot{f})^{\wedge} = -\dot{f}(0) - 2\pi i \gamma f(0) - 4\pi^2 \gamma^2 \hat{f},$$

so

$$[-A 4\pi^2 \gamma^2 + B 2\pi i \gamma + C]\hat{f} = [A 2\pi i \gamma + B]f(0) + A\dot{f}(0) + (\dot{e})^{\wedge},$$

and you can solve for f by inverting the transform

$$\hat{f} = \frac{A 2\pi i \gamma + B}{D} f(0) + \frac{A}{D}\dot{f}(0) + \frac{1}{D}(\dot{e})^{\wedge},$$

in which

$$D = D(\gamma) = -A 4\pi^2 \gamma^2 + B 2\pi i \gamma + C.$$

To make life simpler, put in some actual numbers: $A = 1$, $B = 2$, $C = 2$, $f(0) = \dot{f}(0) = 0$. Then

$$D = -4\pi^2 \gamma^2 + 4\pi i \gamma + 2$$

$$= (1 - i + 2\pi i \gamma)(1 + i + 2\pi i \gamma),$$

so

$$D^{-1} = \frac{1}{2i}\left[\frac{1}{1-i+2\pi i\gamma} - \frac{1}{1+i+2\pi i\gamma}\right]$$

$$= \frac{1}{2i}\int_0^\infty \left[e^{-(1-i)t} - e^{-(1+i)t}\right]e^{-2\pi i\gamma t}\,dt$$

$$= \int_0^\infty e^{-t}\sin t\, e^{-2\pi i\gamma t}\,dt,$$

and the solution is

$$f(t) = \hat{f}^{\,\vee} = \left[\int_0^\infty e^{-t}\sin t\, e^{-2\pi i\gamma t}\,dt\,(\dot{e})^{\wedge}\right]^{\vee}$$

$$= \int_0^t e^{-(t-s)}\sin(t-s)\,\dot{e}(s)\,ds.$$

The content of this formula is that the (current) response of the circuit to a differential voltage de, impressed t units of time in the past, is a damped sinusoid $e^{-t}\sin t \times de$. The total current $f = f(t)$ is the sum of the responses to all the voltage differentials arriving at times $0 \leqslant s < t$.

EXERCISE 7.   Check by hand that if $\lim_{t\uparrow\infty}\dot{e} = 0$, then f solves the stated problem, and f and $\dot{f}$ do indeed tend to 0 as $t\uparrow\infty$, as was assumed for the computation of $(\dot{f})^{\wedge}$ and $(\ddot{f})^{\wedge}$.

EXERCISE 8.   Relate the form of the circuit response f to the discriminant $(B^2 - 4AC)^{1/2}$, keeping in mind that $A, B, C \geqslant 0$. As a guide, look first at the special cases: $A = 1$, $B = 2$, $C = 1$, and $A = 2$, $B = 5$, $C = 2$.

5.   The Poisson Summation Formula

A fascinating aspect of Fourier integrals is the Poisson summation formula, which states that if $f \in C_{\downarrow}^{\infty}(R^1)$, or even if only $f \in C^2(R^1)$ and

$$|f(x)| + |f'(x)| + |f''(x)| \leqslant \text{constant} \times (1+x^2)^{-1},$$

then

$$\sum_{n=-\infty}^{\infty} f(n) = \sum_{n=-\infty}^{\infty} \hat{f}(n).$$

EXAMPLE 1.   Pick for f the familiar Gauss kernel

$$p_t(x) = (2\pi t)^{-1/2}\exp(-x^2/2t).$$

Then $\hat{p}_t = \exp(-2\pi^2\gamma^2 t)$, and the Poisson formula states that

$$(2\pi t)^{-\frac{1}{2}} \sum_{n=-\infty}^{\infty} \exp(-n^2/2t) = \sum_{n=-\infty}^{\infty} \exp(-2\pi^2 n^2 t).$$

This is just the Jacobi theta-function identity of Subsection 1.7.5 if you replace t by $t/2\pi$.

EXAMPLE 2. The Fourier transform of $(t/\pi)(x^2+t^2)^{-1}$ is $e^{-2\pi|\gamma|t}$ [see Exercise 2.6.6], so

$$\sum_{n=-\infty}^{\infty} (n^2+t^2)^{-1} = \frac{\pi}{t} \sum_{n=-\infty}^{\infty} e^{-2\pi|n|t} = \frac{\pi}{t} \frac{1+e^{-2\pi t}}{1-e^{-2\pi t}},$$

by the Poisson formula. The evaluation $\sum_{n\geqslant 1} n^{-2} = \pi^2/6$ of Subsection 1.7.1 comes out for free by making $t \downarrow 0$.

PROOF OF THE POISSON SUMMATION FORMULA. Pick $f \in C^2(R^1)$ with f, f', and f'' bounded as indicated. It is easy to see that the sum

$$f^0(x) = \sum_{k=-\infty}^{\infty} f(x+kT)$$

converges uniformly for $0 \leqslant x < T$ to a function f^0 of period T and class $C^2[0, T)$ [see Exercise 1.4.4 for help with this], and you can expand it into a uniformly convergent Fourier series

$$f^0(x) = \sum_{n=-\infty}^{\infty} T^{-1} \int_0^T f^0(y) e^{-2\pi iny/T} dy\, e^{2\pi inx/T}$$

with coefficients

$$T^{-1} \int_0^T f^0(x) e^{-2\pi inx/T} dx = T^{-1} \sum \int_{kT}^{(k+1)T} f(x) e^{-2\pi inx/T} dx$$

$$= T^{-1}\hat{f}(n/T).$$

The Poisson formula is just the special case of this series for $T = 1$ and $x = 0$.

EXERCISE 9. Give a new proof of the Fourier inversion formula for $f \in C_1^\infty(R^1)$ by making $T \uparrow \infty$ in the above Fourier series. Hint: The tail of the series can be controlled by the bound $|\hat{f}| \leqslant (4\pi^2\gamma^2)^{-1}\|f''\|_1$, as in Section 2.2.

6*.   The Euler–MacLaurin Summation Formula

A simple variant of the Poisson summation formula of considerable practical importance is the Euler–MacLaurin summation formula. Take an even function $f \in C_\downarrow^\infty (R^1)$. The Poisson formula tells you that

$$\sum_{n=0}^{\infty} f(n) = \tfrac{1}{2}f(0) + \tfrac{1}{2} \sum_{n=-\infty}^{\infty} f(n)$$

$$= \tfrac{1}{2}f(0) + \tfrac{1}{2} \sum_{n=-\infty}^{\infty} \hat{f}(n)$$

$$= \tfrac{1}{2}f(0) + \tfrac{1}{2}\hat{f}(0) + \sum_{n=1}^{\infty} 2 \int_0^{\infty} f(x) \cos 2\pi nx \, dx$$

$$= \tfrac{1}{2}f(0) + \int_0^{\infty} f(x) \, dx - \sum_{n=1}^{\infty} \int_0^{\infty} f'(x) \left[(\sin 2\pi nx)/\pi n \right] dx.$$

The final line is justified by a partial integration under the sum. A second partial integration under the sum produces the formula

$$\sum_{n=0}^{\infty} f(n) = \frac{1}{2}f(0) + \int_0^{\infty} f(x) \, dx - \sum_{n=1}^{\infty} \frac{f'(0)}{2\pi^2 n^2} - \sum_{n=1}^{\infty} \int_0^{\infty} f''(x) \frac{\cos 2\pi nx}{2\pi^2 n^2} \, dx$$

$$= \int_0^{\infty} f(x) \, dx + \frac{1}{2}f(0) - \frac{1}{12}f'(0) - \sum_{n=1}^{\infty} \int_0^{\infty} f''(x) \frac{\cos 2\pi nx}{2\pi^2 n^2} \, dx,$$

and you can go on this way, grinding out more and more elaborate formulas for the left-hand sum. The final formula looks like

$$\sum_{n=0}^{\infty} f(n) = \int_0^{\infty} f(x) \, dx + \frac{1}{2}f(0) - \frac{1}{12}f'(0) + \frac{1}{720}f'''(0) - \frac{1}{30240}f''''(0) + \cdots .$$

EXAMPLE 3.   The function $f = (1+x)^{-2}$ does not meet the stated conditions, but let us just see how accurate the formula is:

$$\sum_{n=0}^{\infty} f(n) = \sum_{n=1}^{\infty} n^{-2} = \frac{\pi^2}{6} = 1.645-,$$

$$\int_0^{\infty} f(x) \, dx = \int_1^{\infty} x^{-2} \, dx = 1,$$

$$\frac{1}{2}f(0) = 0.500,$$

$$-\frac{1}{12}f'(0) = 0.167,$$

$$\frac{1}{720}f'''(0) = -0.033,$$

$$-\frac{1}{30240}f''''(0) = 0.024,$$

and

$$\int_0^\infty f(x)\,dx + \frac{1}{2}f(0) - \frac{1}{12}f'(0) + \frac{1}{720}f'''(0) - \frac{1}{30240}f''''(0) = 1.658+,$$

so the error is about 0.6%.

7*. The Central Limit Theorem

The mathematical content of the "central limit theorem" of probability theory is that if f is a nonnegative summable function with $\int f = 1$, $\int xf = 0$, and $\int x^2 f = 1$, and if f^n is the n-fold product $f \circ \cdots \circ f$, then

$$\lim_{n \uparrow \infty} \int_{a\sqrt{n}}^{b\sqrt{n}} f^n(x)\,dx = \int_a^b \frac{\exp(-x^2/2)}{(2\pi)^{\frac{1}{2}}}\,dx$$

for any $-\infty < a < b < \infty$. The probabilistic content is as follows: Think of an infinite number of statistically independent copies $e_1, e_2, \ldots$ of a statistical quantity e distributed according to the rule

$$P(a \leqslant e < b) = \int_a^b f(x)\,dx,$$

in which the letter P stands for the probability of the indicated event. The adjective "independent" means that probabilities multiply:

$$P(a_1 \leqslant e_1 < b_1, a_2 \leqslant e_2 < b_2, \ldots) = \int_{a_1}^{b_1} f(x)\,dx \int_{a_2}^{b_2} f(x)\,dx \cdots$$

and you infer that the sum $s_n = e_1 + \cdots + e_n$ is distributed according to the rule:

$$P(a \leqslant s_n < b) = \int_{a \leqslant x_1 + \cdots + x_n < b} f(x_1)f(x_2) \cdots f(x_n)\,d^n x$$

$$= \int_a^b dx \int f(x - y_{n-1}) \cdots f(y_3 - y_2)f(y_2 - y_1)f(y_1)\,d^{n-1}y$$

$$= \int_a^b f^n(x)\,dx,$$

as may be seen by making the substitution $x_1 + \cdots + x_k = y_k$ ($k \leqslant n$) and putting $y_n = x$. The content of the central limit theorem is now seen to be that the scaled sum $n^{-\frac{1}{2}}s_n$ is nearly Gaussian distributed for large n:

$$P(a \leqslant n^{-\frac{1}{2}}s_n < b) = \sqrt{n} \int_a^b f^n(x\sqrt{n})\,dx$$

is approximately

$$(2\pi)^{-\frac{1}{2}} \int_a^b \exp(-x^2/2)\, dx.$$

The fact goes back to de Moivre and Laplace in the 18th century; for additional information, see Feller [1968, Vol. 1, pp. 174–195; 1966, Vol. 2, pp. 252–259].

PROOF OF THE CENTRAL LIMIT THEOREM. The key step is to use the fact that

$$\lim_{n \uparrow \infty} \left[\sqrt{n}\, f^n(x\sqrt{n})\right]^\wedge = \lim_{n \uparrow \infty} \left[\hat{f}\!\left(\frac{\gamma}{\sqrt{n}}\right)\right]^n = \exp(-2\pi^2\gamma^2) = \left[\frac{\exp(-x^2/2)}{(2\pi)^{\frac{1}{2}}}\right]^\wedge$$

to check that

$$\lim_{n \uparrow \infty} \int \sqrt{n}\, f^n(x\sqrt{n})\, k(x)\, dx = \int \frac{\exp(-x^2/2)}{(2\pi)^{\frac{1}{2}}}\, k(x)\, dx$$

for every function k belonging to $C_{\uparrow}^{\infty}(R^1)$. To do this, use the Fourier inversion formula to write

$$\int \sqrt{n}\, f^n(x\sqrt{n})\, k(x)\, dx = \int \sqrt{n}\, f^n(x\sqrt{n}) \left[\int k^\vee(\gamma)\, e^{-2\pi i \gamma x}\, d\gamma\right] dx$$

$$= \int \check{k}(\gamma) \left[\int \sqrt{n}\, f^n(x\sqrt{n}) e^{-2\pi i \gamma x}\, dx\right] d\gamma$$

$$= \int \check{k}(\gamma)\, [\hat{f}(\gamma/\sqrt{n})]^n\, d\gamma.$$

Now look at $[\hat{f}(\gamma/\sqrt{n})]^n$ for large n. This function is bounded by

$$\|\hat{f}\|_\infty^n \leqslant \|f\|_1^n = 1,$$

and for any fixed $\gamma \in R^1$,

$$\hat{f}\!\left(\frac{\gamma}{\sqrt{n}}\right) = \int \exp\!\left(\frac{-2\pi i \gamma x}{\sqrt{n}}\right) f(x)\, dx$$

$$= \int \left[1 - \frac{2\pi i \gamma x}{\sqrt{n}} - \frac{2\pi^2 \gamma^2 x^2}{n}\, [1 + \delta_n(x)]\right] f(x)\, dx,$$

by the MacLaurin expansion for the exponential, in which δ_n is bounded and approaches 0 pointwise as $n \uparrow \infty$. Because $\int f = 1$, $\int xf = 0$, and $\int x^2 f = 1$, you find

$$\hat{f}\left(\frac{\gamma}{\sqrt{n}}\right) = 1 - \frac{2\pi^2\gamma^2}{n}\left(1 + \int x^2\delta_n(x)f(x)\,dx\right)$$

$$= 1 - \frac{2\pi^2\gamma^2}{n}[1+o(1)],$$

with an error $o(1)$ which tends to 0 as $n\uparrow\infty$, since x^2f is summable. But then

$$\left[\hat{f}\left(\frac{\gamma}{\sqrt{n}}\right)\right]^n = \left[1 - \frac{2\pi^2\gamma^2}{n}[1+o(1)]\right]^n$$

is bounded (by 1) and tends to

$$\exp(-2\pi^2\gamma^2) = \left[\frac{\exp(-x^2/2)}{(2\pi)^{\frac{1}{2}}}\right]^{\wedge}$$

as $n\uparrow\infty$, so

$$\lim_{n\uparrow\infty}\int \sqrt{n}\,f^n(x\sqrt{n})\,k(x)\,dx = \lim_{n\uparrow\infty}\int \check{k}(\gamma)\,[\hat{f}(\gamma/\sqrt{n})]^n\,d\gamma$$

$$= \int \check{k}(\gamma)\,[(2\pi)^{-\frac{1}{2}}\exp(-x^2/2)]^{\wedge}\,d\gamma$$

$$= (2\pi)^{-\frac{1}{2}}\int \exp(-x^2/2)\,k(x)\,dx,$$

by dominated convergence. The last line comes from Parseval's formula and makes use of the fact that $\exp(-x^2/2)$ is real and even whence

$$[\exp(-x^2/2)]^{\wedge\wedge*} = \exp(-x^2/2).$$

The proof may be finished by approximating the indicator function of the interval $a \leqslant x \leqslant b$ above and below by functions k from $C^\infty(R^1)$; see Subsection 1.7.6 for a model for this type of proof.

2.8⋆   HEISENBERG'S INEQUALITY

An important theme of the subject of Fourier transforms is the local/global "duality" between f and $\hat{f}$. A simple instance is the formula

$$(2\pi i\gamma)^p\,D^q\hat{f} = [D^p(-2\pi ix)^q f]^{\wedge}$$

of 2.2 for functions of class $C_\downarrow^\infty(R^1)$ which relates the decay of f [or $\hat{f}$] to the smoothness of $\hat{f}$ [or f]. Another example is provided by Exercise 2.6.10.

This kind of duality was already important for Fourier series on the circle and is now present in a more symmetrical and beautiful form. A very striking example of it is *Heisenberg's inequality* which states that for $f \in L^2(R^1)$,

$$\int x^2 |f(x)|^2 \, dx \times \int \gamma^2 |\hat{f}(\gamma)|^2 \, d\gamma \geqslant (16\pi^2)^{-1} \|f\|_2^4.$$

The lower bound is attained only by constant multiples of $f = \exp(-kx^2)$ $(k > 0)$; see Exercise 2.5.10 for a less precise variant, Section 2.9 for precise lower bounds to the closely allied quantity

$$\cos^{-1} \left(\int_{-a}^a |f|^2 \right)^{1/2} + \cos^{-1} \left(\int_{-b}^b |\hat{f}|^2 \right)^{1/2},$$

and Section 3.2 for another instance of this kind of thing.

The proof of Heisenberg's inequality to be presented below is due to Weyl [1931]; it is very simple mathematically. The original proof by Heisenberg was entirely different, being based upon a quantum-mechanical picture. For the sake of general culture, it is important to understand what he had in mind. First, though, you need an outline of the formalism of quantum mechanics, for a one-dimensional particle.

A "state" of such a particle is a function $\psi \in L^2(R^1)$. The interpretation is that the probability of finding the particle in the interval $a \leqslant x \leqslant b$ is

$$\int_a^b \psi^* \psi = \int_a^b |\psi|^2 ;$$

naturally, the total probability should be 1, so you have to take $\|\psi\|_2 = 1$.

Usage: The placement of the conjugation in $\int \psi^* \psi$ follows quantum mechanical usage and is adopted only for the moment.

An "observable" is a symmetric operator A acting on an appropriate domain $D(A) \subset L^2(R^1)$. The "average" of A in the state ψ is declared to be

$$\text{average}(A) = \int \psi^* A \psi,$$

for ψ belonging to the domain of A. For example, the "position" of the particle is associated with the operation of "multiplication by x" [$A\psi = x\psi$], the domain of A is

$$D(A) = (\psi \in L^2(R^1): \|x\psi\|_2 < \infty),$$

and

$$\text{average}(A) = \int \psi^* \psi = \int x |\psi|^2.$$

The adjective "symmetric" means that

$$\int \psi^* A\psi = (A\psi, \psi) = (\psi, A\psi) = \int (A\psi)^*\psi = \left(\int \psi^* A\psi\right)^*,$$

so that the average is always a real number. A second important observable is the "momentum" which is associated with the operator

$$B = (2\pi i)^{-1} D$$

$[B\psi = (2\pi i)^{-1}\psi']$ acting on the domain

$$\mathsf{D}(B) = (\psi \in \mathsf{L}_2(R^1): \|\psi'\|_2 = \|2\pi i\gamma\hat{\psi}\|_2 < \infty).$$

The average of a power of the momentum is

$$\int \psi^* B^n\psi = \int \psi^* [(2\pi i)^{-1} D]^n\psi = \int \hat{\psi}^*\gamma^n\hat{\psi} = \int \gamma^n |\hat{\psi}|^2$$

so it is natural to interpret

$$\int_a^b \hat{\psi}^*\hat{\psi} = \int_a^b |\hat{\psi}|^2$$

as the probability that the momentum finds itself in the interval $a \leqslant \gamma \leqslant b$; fortunately, the total probability is 1 by the Plancherel identity: $\|\psi\|_2 = \|\hat{\psi}\|_2 = 1$. The position and momentum are noncommuting operators:

$$AB - BA = (xD - Dx)/2\pi i = i/2\pi,$$

and in quantum mechanics, this is taken to mean that they cannot be measured to an arbitrary degree of precision, *simultaneously*. Heisenberg's inequality reflects this fact by saying that in any state

$$\text{average}[A - \text{ave}(A)]^2 \times \text{average}[B - \text{ave}(B)]^2 \geqslant (16\pi^2)^{-1}.$$

To see that this is actually the same as the first form of Heisenberg's inequality stated above, notice that in the state ψ the left-hand side can be expressed as

$$\int [x - \text{ave}(A)]^2 |\psi(x)|^2 \, dx \times \int [\gamma - \text{ave}(B)]^2 |\hat{\psi}(\gamma)|^2 \, d\gamma$$

$$= \int x^2 |\psi_1(x)|^2 \, dx \times \int [\gamma - \text{ave}(B)]^2 |\hat{\psi}_1(\gamma)|^2 \, d\gamma,$$

in which $\psi_1(x) = \psi(x + \text{ave}(A))$, and you take advantage of the fact that $\hat{\psi}_1$ has the same modulus as $\hat{\psi}$. A second application of the same trick with $\hat{\psi}_2(\gamma) = \hat{\psi}_1(\gamma + \text{ave}(B))$ reduces this new expression to

$$\int x^2 |\psi_2(x)|^2 \, dx \times \int \gamma^2 |\hat{\psi}_2(\gamma)|^2 \, d\gamma,$$

and the latter is greater than or equal to $(16\pi^2)^{-1}$ since $\|\psi_1\|_2 = \|\psi_2\|_2 = \|\psi\|_2 = 1$.

EXERCISE 1.   Check that

$$\text{average}[A - \text{ave}(A)]^2 \times \text{average}[B - \text{ave}(B)]^2 \geqslant \tfrac{1}{4}|\text{average}(AB - BA)|^2$$

for any observables A and B. Do not worry about domains. A formal proof is all that is asked. This is the general form of Heisenberg's inequality. It reduces to the previous one if $A = $ position and $B = $ momentum. *Hint:* The average of $(AB - BA)$ is $2i \times$ the imaginary part of $(B\psi, A\psi)$. Use Schwarz's inequality to check the bound if $\text{ave}(A) = \text{ave}(B) = 0$. Reduce the general case to this one.

The little book by Heitler [1945] is recommended for additional information about quantum mechanics.

PROOF OF HEISENBERG'S INEQUALITY.   The proof is very simple if $f \in C_\downarrow^\infty(R^1)$:

$$4\pi^2 \int x^2 |f(x)|^2 \, dx \int \gamma^2 |\hat{f}(\gamma)|^2 \, d\gamma$$

$$= \int |xf(x)|^2 \, dx \int |2\pi i \gamma \hat{f}(\gamma)|^2 \, d\gamma$$

$$= \int |xf(x)|^2 \, dx \int |f'(x)|^2 \, dx \qquad \text{[by Plancherel's identity]}$$

$$\geqslant \left[\int |xf'f^*| \, dx \right]^2 \qquad \text{[by Schwarz's inequality]}$$

$$\geqslant \left[\int x\tfrac{1}{2}(f'f^* + f'^*f) \, dx \right]^2$$

$$= \frac{1}{4} \left[\int x(|f|^2)' \, dx \right]^2$$

$$= \frac{1}{4} \left[\int |f(x)|^2 \, dx \right]^2$$

$$= \tfrac{1}{4} \|f\|_2^4.$$

Now suppose that f is of class $L^2(R^1)$ only, and let us try to copy the proof. Because $\|xf\|_2 > 0$, you may as well suppose that $\|\gamma \hat{f}\|_2 < \infty$, because there is nothing to prove in the opposite case. This means that $f' = [2\pi i \gamma \hat{f}]^\vee$ is of class $L^2(R^1)$. The fact that f' is (or ought to be) the derivative of f is not

yet needed, and the prime is not intended to prejudge the issue. The proof
for $C_{\downarrow}^{\infty}(R^1)$ can now be copied, down to the place at which the integral

$$\int x(f'f^* + f'^*f)\, dx$$

makes its appearance. Now pick $f_n \in C_{\downarrow}^{\infty}(R^1)$ in order to approximate f
in the sense that

$$\lim_{n\uparrow\infty} \int (1 + 4\pi^2\gamma^2)|\hat{f}_n - \hat{f}|^2\, d\gamma = 0;$$

this may easily be done by a small amplification of exercise 1.2.11. Then

$$\|f_n - f\|_2^2 + \|f_n' - f'\|_2^2 = \int (1 + 4\pi^2\gamma^2)|\hat{f}_n - \hat{f}|^2\, d\gamma$$

tends to 0 as $n\uparrow\infty$, as does

$$|f_n(x) - f(x)| \leq \|\hat{f}_n - \hat{f}\|_1$$

$$\leq \left[\int (1 + 4\pi^2\gamma^2)^{-1}\, d\gamma\right]^{1/2}\left[\int (1 + 4\pi^2\gamma^2)|\hat{f}_n - \hat{f}|^2\, d\gamma\right]^{1/2}$$

for any fixed x, and you have

$$\int x(f'f^* + f'^*f)\, dx = \lim_{l\uparrow\infty}\lim_{n\uparrow\infty}\int_{|x|\leq l} x(f_n'f_n^* + f_n'^*f_n)\, dx$$

$$= \lim_{l\uparrow\infty}\lim_{n\uparrow\infty}\int_{|x|\leq l} x(|f_n|^2)'\, dx$$

$$= \lim_{l\uparrow\infty}\lim_{n\uparrow\infty}\left[x|f_n|^2\Big|_{-l}^{l} - \int_{|x|\leq l}|f_n|^2\, dx\right]$$

$$= \lim_{l\uparrow\infty} l[|f(l)|^2 + |f(-l)|^2] - \|f\|_2^2$$

$$= -\|f\|_2^2.$$

The last step employs the fact that $\lim_{l\uparrow\infty} l[|f(l)|^2 + |f(-l)|^2] = 0$; other-
wise, $|f(x)|^2 + |f(-x)|^2$ could not be summable.

The case in which the lower bound is attained is easily identified. The
key step in the proof of Heisenberg's inequality is the application of Schwarz's
inequality to the integral $\int xf'f^*$, so the bound is attained only if $f' = kxf^*$
a.e. with a complex constant k. Now you need to know that f' *really is the
derivative of f.* This is easily checked as follows:

$$\int_0^x f'(y)\, dy = \lim_{n\uparrow\infty}\int_0^x f_n'(y)\, dy$$

$$= \lim_{n\uparrow\infty}[f_n(x) - f_n(0)]$$

$$= f(x) - f(0),$$

and you see, upon replacing f' by kxf^* under the integral sign that f is actually differentiable in the ordinary sense and is a bona fide solution of $f' = kxf^*$! But then

$$x^{-1}(x^{-1}f')' = x^{-1}(kf^*)' = |k|^2 f,$$

and constant multiples of $f = \exp(-|k| x^2/2)$ are the only solutions that belong to $L^2(R^1)$. The proof is finished.

2.9*   BAND- AND TIME-LIMITED FUNCTIONS

A problem of considerable practical interest in communication engineering is that of making "signals" $f = f(t)$ of total "power" $\|f\|_2^2 = 1$ with both

$$\alpha^2 = \int_{-a}^{a} |f(t)|^2 \, dt$$

and

$$\beta^2 = \int_{-b}^{b} |\hat{f}(\gamma)|^2 \, d\gamma$$

as close to 1 as possible for fixed positive numbers a and b. $\alpha = 1$ means that the signal is confined to the period $|t| \leqslant a$ ("time-limited"); $\beta = 1$ means that its "power-spectrum" is confined to the band $|\gamma| \leqslant b$ ("band-limited"). Both $\alpha = 1$ and $\beta = 1$ cannot be achieved for the same signal $f \not\equiv 0$; in fact, for such f,

$$f(t) = \int_{-b}^{b} \hat{f}(\gamma) e^{2\pi i \gamma t} \, d\gamma = 0 \qquad \text{for} \quad |t| > a,$$

and by differentiation under the integral sign,

$$\int_{-b}^{b} \hat{f}(\gamma) e^{2\pi i \gamma a} \gamma^n \, d\gamma = 0 \qquad \text{for} \quad n \geqslant 0.$$

But then

$$f(t) = \int_{-b}^{b} \hat{f}(\gamma) e^{2\pi i \gamma (t-a)} e^{2\pi i \gamma a} \, d\gamma$$

$$= \sum_{n=0}^{\infty} \frac{[2\pi i (t-a)]^n}{n!} \int_{-b}^{b} \hat{f}(\gamma) e^{2\pi i \gamma a} \gamma^n \, d\gamma$$

$$= 0,$$

contradicting $f \not\equiv 0$. The same proof shows that a band-limited signal cannot vanish on *any* interval; see Exercise 3.1.16 for a variant of this proof.

You may ask: If $\alpha = \beta = 1$ cannot be achieved, what *can* be done? The answer is contained in a Heisenberg-like bound due to Landau and Pollack [1961] and Pollack and Slepian [1961]: the pairs $\alpha\beta$ corresponding to actual signals of f of unit power fill up the subregion of the unit square $[0, 1] \times [0, 1]$ demarcated by

$$\cos^{-1}\alpha + \cos^{-1}\beta \geqslant \cos^{-1}\sqrt{\gamma_1},$$

with the proviso that if α or $\beta = 0 \, [=1]$, then the other is $< 1 \, [> 0]$. The number γ_1 is the supremum of α^2, taken over the class of band limited signals of unit power $[\|f\|_2 = \beta = 1]$; it is a function of the product ab alone. A few samples of the curve $\cos^{-1}\alpha + \cos^{-1}\beta = \cos^{-1}\sqrt{\gamma_1}$ are shown in Fig. 1. Figure 2 depicts γ_1 as a function of ab.

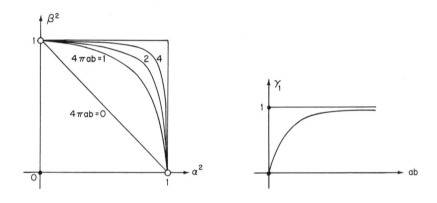

FIGURE 1 (left) and **FIGURE 2** (right)   Copyright, 1961, The American Telephone and Telegraph Co., reprinted by permission.

EXERCISE 1.   The inequality $\alpha^2 + \beta^2 \leqslant 1$ implies that

$$\cos^{-1}\alpha + \cos^{-1}\beta \geqslant \pi/2 \geqslant \cos^{-1}\sqrt{\gamma_1}.$$

The moral is that except for the excluded corners $(0, 1)$ and $(1, 0)$, no limitation is placed upon α and β unless the power sum $\alpha^2 + \beta^2$ exceeds the total power of the signal: $\alpha^2 + \beta^2 > \|f\|_2^2 = 1$.

PROOF OF $\cos^{-1}\alpha + \cos^{-1}\beta \geqslant \cos^{-1}\sqrt{\gamma_1}$.   The principal tools are the projection A onto the class of time limited functions:

$$Af = \begin{cases} f(t) & \text{if } |t| \leqslant a, \\ 0 & \text{otherwise} \end{cases}$$

and the projection B onto the class of band-limited functions:

$$Bf = \int_{-b}^{b} \hat{f}(\gamma) e^{2\pi i \gamma t} \, d\gamma .$$

Actually, the important thing is not the individual projections A and B but the product

$$BA : f \rightarrow \int_{-a}^{a} \frac{\sin 2\pi b(t-s)}{\pi (t-s)} f(s) \, ds .$$

What you need to know is that BA has an infinite number of eigenvalues

$$1 > \gamma_1 \geqslant \gamma_2 \geqslant \cdots \downarrow 0$$

and that the associated eigenfunctions e_n have the following properties:

(a) $e_n : n \geqslant 1$ is a unit-perpendicular basis of the class of band-limited functions.

(b) $\gamma_n^{-1/2} A e_n : n \geqslant 1$ is a unit-perpendicular basis of the class of time-limited functions.

(c) e_1 coincides, on the interval $|t| < a$, with the "top" eigenfunction of the differential operator

$$K : f \rightarrow ((\alpha^2 - t^2) f')' - 4\pi^2 b^2 t^2 f .$$

The proofs of (a) and (b) are presented below after some preparation. Item (c) is not really needed and will *not* be proved; the point of it is that the eigenfunctions of K are tabulated transcendental functions ("spheroidal wave functions"), so the number γ_1 can be computed numerically from the formula $BA e_1 = \gamma_1 e_1$. The best account of spheroidal wave functions is given by Robin [1959, Vol. 3, pp. 212–262].

PROOF OF THE EXISTENCE OF $e_n : n \geqslant 1$. The method is standard. Define γ_1 to be the supremum of $\alpha^2 = \|Af\|_2^2$, taken over the class of band-limited functions f of unit length, and pick a series of such functions f_n so as to make $(BAf_n, f_n) = \|Af_n\|_2^2 \uparrow \gamma_1$ as $n \uparrow \infty$. The plan is to use the functions f_n to make a band-limited function f_∞ with $\|Af_\infty\|_2^2 = \gamma_1$ and $\|f_\infty\|_2 = 1$. That $\gamma_1 < 1$ will be automatic, and f_∞ will turn out to be an eigenfunction of BA with eigenvalue $\gamma_1 : BAf_\infty = \gamma_1 f_\infty$. Here is the proof. The function f_n is band-limited, so

$$|f_n| = |Bf_n| \leqslant \int_{-b}^{b} |\hat{f}_n(\gamma)| \, d\gamma \leqslant (2b)^{1/2} \|f_n\|_2 = (2b)^{1/2},$$

and

$$|f_n(t) - f_n(s)| \leqslant \int_{-b}^{b} |\hat{f}_n(\gamma)| \, |e^{2\pi i \gamma t} - e^{2\pi i \gamma s}| \, d\gamma$$

$$\leqslant (2b)^{1/2} \|f_n\|_2 \times 2\pi b |t-s| .$$

By the first appraisal, you can weed out the f_n's so as to make $f_n(t)$ tend to some limit $f_\infty(t)$ at every rational place t, simultaneously. The second appraisal permits you to carry this convergence over from these special points to the whole line. The limit function f_∞ is of class $C(R^1)$ with $\|f_\infty\|_2 \leqslant \liminf_{n \uparrow \infty} \|f_n\|_2 = 1$, and $\|Af_\infty\|_2^2 = \gamma_1$ since $|f_n|$ is bounded independently of $n \geqslant 1$. The function f_∞ is even band-limited, as you can see by taking $k \in C_\downarrow^\infty(R^1)$, vanishing for $|\gamma| \leqslant b$, and computing

$$(\hat{k}, \hat{f}_\infty) = (k, f_\infty) = \lim_{n \uparrow \infty}(k, f_n) = \lim_{n \uparrow \infty}(\hat{k}, \hat{f}_n) = 0.$$

$\|f_\infty\|_2 = 1$ is now self-evident: Namely, if $\|f_\infty\|_2 < 1$, then $\|Af\|_2^2$ would be greater than γ_1 for $f = \|f\|_\infty^{-1} f_\infty$, contradicting the definition of γ_1. The final point is to verify that $BAf_\infty = \gamma_1 f_\infty$. To do this, introduce the band-limited function $k = (\gamma_1 - BA)f_\infty$, and look at

$$p(\varepsilon) = \gamma_1 \|f_\infty + \varepsilon k\|_2^2 - \|A(f_\infty + \varepsilon k)\|_2^2$$

as a quadratic polynomial in $-1 < \varepsilon < 1$. Because this expression is least ($=0$) for $\varepsilon = 0$, its slope has to vanish at that place:

$$0 = p'(0) = 2\,\mathrm{Re}\,(k, (\gamma_1 - A)f_\infty) = 2\|(\gamma_1 - BA)f_\infty\|_2^2$$

as advertised. To verify the identity, keep in mind that k and f are band limited and that B is a projection. This means that

$$(k, (\gamma_1 - A)f_\infty) = (Bk, (\gamma_1 - A)f_\infty)$$
$$= (k, B(\gamma_1 - A)f_\infty)$$
$$= (k, (\gamma_1 - BA)f_\infty)$$

Now call this eigenfunction e_1 instead of f_∞ and let γ_2 be the supremum of $\|Af\|_2^2$, taken over the class of band-limited functions f of unit length *perpendicular to* e_1. The same construction provides you with a new eigenfunction e_2 with eigenvalue γ_2, and you can go on for $n = 3, 4, \ldots$, successively, to make eigenfunctions e_n with eigenvalue $\gamma_n =$ the supremum of $\|Af\|_2^2$, taken over the class of band-limited functions f of unit length *perpendicular to e_k for $k < n$*. This process never stops, since the class of band-limited functions is of infinite dimension.

PROOF OF $\gamma_n \downarrow 0$. This is an important point for the proof of (a) and is checked as follows. Fix t and look at

$$f(s) = \frac{\sin 2\pi b(t-s)}{\pi(t-s)}.$$

Then

$$(Af, e_n) = \int_{-a}^{a} \frac{\sin 2\pi b(t-s)}{\pi(t-s)} e_n(s) \, ds$$

$$= BAe_n(t)$$

$$= \gamma_n e_n(t),$$

so

$$\sum_{n=1}^{\infty} \gamma_n^2 |e_n(t)|^2 = \sum_{n=1}^{\infty} |(Af, e_n)|^2 \leqslant \|Af\|_2^2 = \int_{-a}^{a} \left| \frac{\sin 2\pi b(t-s)}{\pi(t-s)} \right|^2 ds$$

by Bessel's inequality, and integrating over the line produces the bound

$$\sum_{n=1}^{\infty} \gamma_n^2 \leqslant 2a \int_{-\infty}^{\infty} \left| \frac{\sin 2\pi b t}{\pi t} \right|^2 dt = 4ab,$$

which is more than enough.

PROOF OF (a). A band-limited function f, which is perpendicular to all the eigenfunctions, e_n satisfies

$$\|Af\|_2^2 \leqslant \gamma_n \|f\|^2.$$

Because $\gamma_n \downarrow 0$, $f = 0$ for $|t| \leqslant a$, and as you know, the only band-limited function that vanishes on an interval is $f \equiv 0$.

PROOF OF (b). The functions $\gamma_n^{-\frac{1}{2}} Ae_n$ form a unit-perpendicular family, since

$$(Ae_m, Ae_k) = (Ae_m, e_k) = (Ae_m, Be_k) = (BAe_m, e_k) = \gamma_m(e_m, e_k).$$

Now pick a time-limited function f, which is perpendicular to every e_n. In view of (a)

$$\hat{f}(\gamma) = \int_{-a}^{a} f(t) e^{-2\pi i \gamma t} \, dt = 0$$

for $|\gamma| < b$, since $e^{-2\pi i \gamma t}$ is the bounded limit of the band-limited functions

$$n \int_{0}^{1/n} e^{-2\pi i (\gamma + \delta) t} \, d\delta$$

as $n \uparrow \infty$. The proof is finished as before: Such a function $\hat{f} \not\equiv 0$ cannot vanish on any interval.

LEMMA.   *Suppose $\alpha \beta$ is an admissible pair, i.e., suppose that $\alpha = \|Af\|_2$ and $\beta = \|Bf\|_2$ for an actual function f of unit length. Then any pair $\alpha' \beta'$ with $0 < \alpha' \leqslant \alpha$ and $0 < \beta' \leqslant \beta$ is also admissible.*

PROOF. Pick a function f of unit length with $\alpha = \|Af\|_2$ and $\beta = \|Bf\|_2$, and put

$$f_1 = fe^{2\pi i\delta t}.$$

Then $\|f_1\|_2 = 1$, $\|Af_1\|_2 = \alpha$, and

$$\beta' = \|Bf_1\|_2^2 = \int_{-b}^{b} |\hat{f}(\gamma - \delta)|^2 \, d\gamma \, ;$$

especially, $\beta' = \beta$ for $\delta = 0$, while β' tends to 0 as $\delta \uparrow \infty$, so any pair $\alpha\beta'$ with $0 < \beta' \leq \beta$ is admissible. The proof is finished by a self-evident symmetry between α and β.

At this point, you have all the machinery needed for the proof of $\cos^{-1}\alpha + \cos^{-1}\beta \geq \cos^{-1}\sqrt{\gamma_1}$. The number γ_1 is the top eigenvalue of BA, and the proof is as follows:

Case 1 $[\alpha = 0]$: The problem is to check that the only limitation on β is that $\beta \neq 1$. The value $\beta = 1$ is not possible, since a band-limited function $f \not\equiv 0$ cannot vanish on any interval. At the opposite extreme, $\beta = 0$ *is* possible, i.e., you can find functions perpendicular both to the class of time-limited functions and to the class of band-limited functions. If not, then functions of the form $f = Af_1 + Bf_2$ would be dense in $L^2(R^1)$. But such a function can be written as the sum of two perpendicular pieces:

$$f = (1 - A) Bf_2 + A(f_1 + Bf_2),$$

so

$$\|f\|_2 \geq \|(1 - A) Bf_2\|_2$$
$$\geq \|Bf_2\|_2 - \|ABf_2\|$$
$$\geq (1 - \sqrt{\gamma_1}) \|Bf_2\|_2 ,$$

or, what is the same in view of $\gamma_1 < 1$,

$$\|Bf_2\|_2 \leq (1 - \sqrt{\gamma_1})^{-1} \|f\|_2 .$$

Also,

$$\|Af_1\|_2 = \|f - Bf_2\|_2 \leq [1 + (1 - \sqrt{\gamma_1})^{-1}] \|f\|_2 ,$$

and now if you make

$$f_n = Af_{1n} + Bf_{2n}$$

tend to f in $L^2(R^1)$, then these bounds will ensure that the sequence Af_{1n} $[Bf_{2n}]$ converges to a time-limited [band-limited] function $Af_{1\infty}$ $[Bf_{2\infty}]$

in $L^2(R^1)$. In other words, the assumption that functions of the form $Af_1 + Bf_2$ are dense implies that *every* function $f \in L^2(R^1)$ can be expressed in this way. But this means that every $f \in L^2(R^1)$ coincides with a band-limited function far out, and that is certainly not the case. As to the intermediate values $0 < \beta < 1$

$$f = \frac{(1-A)e_n}{(1-\gamma_n)^{1/2}}$$

satisfies

$$\|f\|_2 = \frac{\|(1-A)e_n\|_2}{(1-\gamma_n)^{1/2}} = \frac{(\|e_n\|_2^2 - \|Ae_n\|_2^2)^{1/2}}{(1-\gamma_n)^{1/2}} = 1,$$

$$\|Af\|_2 = 0,$$

$$\|Bf\|_2 = \frac{\|(1-BA)e_n\|_2}{(1-\gamma_n)^{1/2}} = (1-\gamma_n)^{1/2} \uparrow 1 \qquad \text{as} \quad n \uparrow \infty,$$

and the lemma fills in the values $\beta < (1-\gamma_n)^{1/2}$.

Case 2 $\left[0 < \alpha < \sqrt{\gamma_1}\right]$: The statement is that all values $0 \leqslant \beta \leqslant 1$ are permitted. Pick $\gamma_n < \alpha^2$. The function f defined by

$$(\gamma_1 - \gamma_n)^{1/2} f = (\alpha^2 - \gamma_n)^{1/2} e_1 + (\gamma_1 - \alpha^2)^{1/2} e_n$$

satisfies

$$\|f\|_2 = 1,$$

$$\|Bf\|_2 = \|f\|_2 = 1,$$

$$\|Af\|_2 = \left(\frac{\alpha^2 - \gamma_n}{\gamma_1 - \gamma_n} \gamma_1 + \frac{\gamma_1 - \alpha^2}{\gamma_1 - \gamma_n} \gamma_n\right)^{1/2} = \alpha,$$

and the lemma fills in the values $0 < \beta < 1$. $\beta = 0$ is covered by Case 1, by symmetry.

Case 3 $\left[\sqrt{\gamma_1} \leqslant \alpha \leqslant 1\right]$: The possible values of β are supposed to be described by

$$\cos^{-1}\alpha + \cos^{-1}\beta \geqslant \cos^{-1}\sqrt{\gamma_1},$$

or what is the same, by

$$\beta \leqslant \alpha\sqrt{\gamma_1} + (1-\alpha^2)^{1/2}(1-\gamma_1)^{1/2},$$

with the single exception that $\beta = 0$ is not permitted if $\alpha = 1$. Pick f of unit length with $\|Af\|_2 = \alpha$, express it as the sum of a piece $c_1 Af + c_2 Bf$ in the plane of Af and Bf and a piece f^0 perpendicular to this plane:

$$f = c_1 Af + c_2 Bf + f^0,$$

and compute inner products:

$$1 = (f,f) = c_1 \alpha^2 + c_2 \beta^2 + (f,f^0),$$

$$\alpha^2 = (f, Af) = c_1 \alpha^2 + c_2 (Bf, Af),$$

$$\beta^2 = (f, Bf) = c_1 (Af, Bf) + c_2 \beta^2,$$

$$(f,f^0) = \|f^0\|_2^2.$$

The two middle identities can be solved for c_1 and c_2, since

$$\Delta = \det \begin{bmatrix} \alpha^2 & (Af, Bf) \\ (Af, Bf) & \beta^2 \end{bmatrix} = \alpha^2 \beta^2 - |(Af, Bf)|^2 > 0,$$

with the result that

$$c_1 = \Delta^{-1} \beta^2 [\alpha^2 - (Bf, Af)]$$

$$c_2 = \Delta^{-1} \alpha^2 [\beta^2 - (Af, Bf)],$$

and substituting back into the first identity, you find

$$1 = \Delta^{-1} \alpha^2 \beta^2 [\alpha^2 + \beta^2 - 2\operatorname{Re}(Af, Bf)] + \|f^0\|_2^2,$$

whence

$$\Delta = \alpha^2 \beta^2 - |(Af, Bf)|^2 \geqslant \alpha^2 \beta^2 [\alpha^2 + \beta^2 - 2\operatorname{Re}(Af, Bf)]$$

$$\geqslant \alpha^2 \beta^2 [\alpha^2 + \beta^2 - 2|(Af, Bf)|].$$

To put this into a more transparent form, let

$$|(Af, Bf)|/\alpha\beta = \gamma.$$

Then $0 \leqslant \gamma \leqslant 1$, and

$$\alpha^2 + \beta^2 - 2\alpha\beta\gamma \leqslant 1 - \gamma^2,$$

or what is the same, after a little manipulation,

$$\beta \leqslant \alpha\gamma + (1-\alpha)^{1/2} (1-\gamma^2)^{1/2}.$$

This upper bound is an increasing function of γ for $\gamma \leqslant \alpha$, so to finish the proof of

$$\beta \leqslant \alpha\sqrt{\gamma_1} + (1-\alpha^2)^{1/2} (1-\gamma_1)^{1/2},$$

it is enough to check that $\gamma \leqslant \sqrt{\gamma_1}\ (\leqslant \alpha)$. This is easy:

$$|(Af, Bf)| = |(Af, ABf)|$$
$$\leqslant \|Af\|_2 \|ABf\|_2$$
$$\leqslant \|Af\|_2 \sqrt{\gamma_1} \|Bf\|_2$$
$$= \alpha\beta\sqrt{\gamma_1}.$$

The final point is to check that the indicated range

$$\beta \leqslant \beta' = \alpha\sqrt{\gamma_1} + (1-\alpha^2)^{\frac{1}{2}}(1-\gamma_1)^{\frac{1}{2}}$$

is actually filled out. The upper bound $\beta = \beta'$ is achieved by the function

$$f = \frac{\alpha}{\sqrt{\gamma_1}} A e_1 + \left(\frac{1-\alpha^2}{1-\gamma_1}\right)^{\frac{1}{2}}(1-A)e_1,$$

the lemma fills in the intermediate values $0 < \beta < \beta'$, and by Case 1, $\beta = 0$ can also be achieved except for $\alpha = 1$. The proof is finished.

A fascinating aspect of the eigenfunctions of BA centers about a "folk-theorem" to the effect that *there are approximately*

$$4ab = 2a \times 2b = \text{duration} \times \text{bandwidth}$$

time- and band-limited signals. The statement is false as it stands ($f \equiv 0$ is the only such signal), *but do not be misled*: It is just a sloppy variant of a deep and important fact! A rough idea of what is intended can be obtained by expanding the Fourier transform $\hat{f}$ of a band-limited signal f into a Fourier series

$$\hat{f}(\gamma) = \sum \left[\frac{1}{2b}\int_{-b}^{b} \hat{f}(\gamma)\, e^{2\pi i \gamma n/2b}\, d\gamma\right] e^{-2\pi i \gamma n/2b}$$
$$= \sum (2b)^{-1} f(n/2b)\, e^{-2\pi i \gamma n/2b}$$

and then inverting:

$$f(t) = \int_{-b}^{b} \hat{f}(\gamma)\, e^{2\pi i \gamma t}\, d\gamma$$
$$= \sum \frac{1}{2b} f\left(\frac{n}{2b}\right) \int_{-b}^{b} e^{2\pi i \gamma (t - n/2b)}\, d\gamma$$
$$= \sum \frac{1}{2b} f\left(\frac{n}{2b}\right) \frac{\sin 2\pi b(t - n/2b)}{\pi(t - n/2b)}.$$

This is the so-called "sampling formula" for band-limited functions; it suggests that if f is small for $|t| > a$, then you should be able to neglect the tail $|n| > 2ab$ and have only about $4ab$ terms that really count. To put this on a more solid footing, permit a phase shift in the signal $[f \to f_\delta(t) = f(t+\delta)]$, assume for simplicity's sake that $2ab$ is integral, and use the Plancherel identity

$$\|f\|_2^2 = \|\hat{f}\|_2^2 = \int_{-b}^{b} |\hat{f}(\gamma)|^2 \, d\gamma = (2b)^{-1} \sum |f(n/2b)|^2$$

to estimate

$$\int_0^{1/2b} \left\| f_\delta - \sum_{|n| \leqslant 2ab} \frac{1}{2b} f_\delta\left(\frac{n}{2b}\right) \frac{\sin 2\pi b(t-n/2b)}{\pi(t-n/2b)} \right\|_2^2 d\delta$$

$$= \int_0^{1/2b} \frac{1}{2b} \sum_{|n| > 2ab} \left| f_\delta\left(\frac{n}{2b}\right) \right|^2 d\delta$$

$$= (2b)^{-1} \sum_{|n| > 2ab} \int_{n/2b}^{(n+1)/2b} |f(\delta)|^2 \, d\delta$$

$$\leqslant (2b)^{-1} \int_{|t| > a} |f(t)|^2 \, dt$$

$$= \frac{1-\alpha^2}{2b}.$$

The moral is that a "small" time lag in the amount δ, less than or equal to *the reciprocal bandwidth*, will bring the signal into a $(4ab+1)$-dimensional space, up to an error ("noise") of "small" total power less than or equal to *the reciprocal bandwidth* $\times (1-\alpha^2)$. This way of stating the "folk-theorem" is due to Shannon; see Shannon [1949] for discussion of this circle of ideas.

Landau and Pollack [1962] put the matter in a very much sharper form with the help of the eigenfunctions e_n of BA, as will now be stated, mostly without proofs. Define $\Delta[f_1, ..., f_n]$ to be the least upper bound of

$$\left\| f - \sum_{k=1}^{n} (f, f_k) f_k \right\|$$

over the class of band-limited signals f of unit length with $\|Af\| = \alpha$ for fixed $0 < \alpha < 1$, $f_k: k \leqslant n$ being any unit-perpendicular family of similarly band-limited signals. Then

(a) $\Delta[f_1, ..., f_n]$ is least for $f_1 = e_1, ..., f_n = e_n$, and this is so for every $n \geqslant 1$.

(b) $\Delta^2[e_1, ..., e_n] \leqslant 12(1-\alpha^2)$ for every $n > 4ab$.

(c) $\Delta^2[e_1, ..., e_{[4ab+1]+n}] \geqslant (0.916)^{-1}[1-\alpha^2 - 2\sqrt{2}\, e^{-\pi ab}]$ if $1-\alpha^2 < 0.916$, n is fixed and ab is sufficiently large. Thus the constant 12 in (b) cannot be reduced to 1 by adding a "few" extra eigenfunctions.

(d) $\Delta^2 [e_1, \ldots, e_n] \leqslant (1+\delta)(1-\alpha^2)$ for $n \geqslant 4ab+1+$ a constant multiple of $\log(4ab+1)$; the constant depends upon the number $\delta > 0$ only, and the latter can be made as small as you like.

The proof of (b) rests upon the simple estimate

$$\left\| f - \sum_{k=1}^{n} (f, e_k) e_k \right\|_2^2 = \sum_{k>n} |(f, e_k)|^2$$

$$\leqslant (1 - \gamma_{n+1})^{-1} \sum_{k>n} |(f, e_k)|^2 (1 - \gamma_k)$$

$$\leqslant (1 - \gamma_{n+1})^{-1} \sum_{k=1}^{\infty} |(f, e_k)|^2 (1 - \gamma_k)$$

$$= (1 - \alpha^2)/(1 - \gamma_{n+1})$$

and a study of the eigenvalues γ_n of BA. These decrease "sharply" from "near 1" to "near 0" as n passes beyond $4ab$; this accounts for the tightness of the bound. The number 12 comes from the estimate $\gamma_{n+1} < 0.916$ (valid for $n > 4ab$) applied to the last line $[(1-0.916)^{-1} < 12]$.

EXERCISE 2. Check that $\sum_{n=1}^{\infty} \gamma_n = 4ab$. This is consistent with the picture of $4ab$ eigenvalues "near 1" and the rest "near 0." *Hint:* Check that $\sum \gamma_n e_n(s) e_n(t)$ converges to

$$\frac{\sin 2\pi b(t-s)}{\pi(t-s)}$$

uniformly on the square $|s| \leqslant a$, $|t| \leqslant a$, and then look at Exercise 1.7.13. In case of difficulty with the first part, look up Mercer's theorem in Courant and Hilbert [1953].

A final item[1] refers to the (unit-perpendicular) "sampling functions"

$$f_n(t) = \frac{\sin 2\pi b(t - n/2b)}{(2b)^{1/2} \pi(t - n/2b)}$$

employed above. By the estimate

$$\int_0^{1/2b} \left\| f - \sum_{|n| \leqslant 2ab+1} (2b)^{-1/2} f_\delta(n/2b) (f_n)_{-\delta} \right\|_2^2 d\delta \leqslant (1 - \alpha^2)/2b$$

used for Shannon's theorem, you can guarantee the existence of a number $0 < \delta < 1/2b$ so as to make

$$\left\| f - \sum_{|n| \leqslant 2ab+1} (2b)^{-1/2} f_\delta \left(\frac{n}{2b} \right) (f_n)_{-\delta} \right\|_2^2 \leqslant 1 - \alpha^2.$$

---

[1] E. Arthurs, private communication.

The range of summation is now $|n| \leqslant 2ab+1$ in place of $|n| \leqslant 2ab$ because $2ab$ is no longer assumed to be integral. However, this estimate says that $2[2ab+1]+1 \leqslant 4ab+3$ of the shifted sampling functions $(f_n)_{-\delta}$ are even better approximants than the e_n's, in fact, by (c), the latter will only give you an error greater than or equal to $(0.916)^{-1}(1-\alpha^2-2\sqrt{2}e^{-\pi ab})$, which is more than $1.09(1-\alpha^2)$ for large ab. The practical problem here is that the optimal lag δ depends upon f in a very complicated nonlinear way, which is not easily determined.

2.10 SEVERAL-DIMENSIONAL FOURIER INTEGRALS

The Fourier integral of a function $f \in L^2(R^n)$ is defined by the recipe

$$\hat{f}(\gamma) = \int_{R^n} f(x) e^{-2\pi i \gamma \cdot x} d^n x.$$

The (formal) integral is shorthand for

$$\lim_{Q \uparrow R^n} \int_Q f(x) e^{-2\pi i \gamma \cdot x} d^n x.$$

Here, Q is a ball or a box expanding to R^n, $\gamma \cdot x$ is the customary inner product $\gamma_1 x_1 + \cdots + \gamma_n x_n$, $|x| = (x_1^2 + \cdots + x_n^2)^{1/2}$ as usual, and the limit is understood in the sense of $L^2(R^n)$. Because n-fold products

$$f_1(x_1) \cdots f_n(x_n)$$

of functions from $L^2(R^1)$ span $L^2(R^n)$, it is easy to see from the 1-dimensional Fourier inversion formula and Plancherel identity that

$$(\hat{f})^{\vee} = f$$

and that

$$\|\hat{f}\|_2 = \|f\|_2.$$

$\vee$ stands for the inverse transform

$$f^{\vee}(x) = \int_{R^n} \hat{f}(\gamma) e^{2\pi i \gamma \cdot x} d^n \gamma$$

with the interpretation of the integral as explained previously.

The case of radial functions is of special interest even for $n = 1$ or 2. For $n = 1$, a radial function is simply an *even* function, and you get the Fourier cosine transform [see Exercise 2.3.5]; for $n = 2$, you obtain a so-called "Bessel transform," as will now be explained.

Bring in the "Bessel function"

$$J_0(r) = \frac{1}{2\pi} \int_0^{2\pi} e^{ir\cos\theta}\, d\theta,$$

let $f \in L^2(R^2)$ be a radial function: $f(x) = f(|x|)$, and change to polar co-
ordinates $x = r(\cos\theta, \sin\theta)$. Then you can compute $\hat{f}$ as follows:

$$\begin{aligned}
\hat{f}(\gamma) &= \int_{R^2} f(|x|)\, e^{-2\pi i\gamma\cdot x}\, d^2 x \\
&= \int_0^\infty f(r)\, r\, dr \int_0^{2\pi} e^{-2\pi i|\gamma|r\cos\theta}\, d\theta \\
&= 2\pi \int_0^\infty f(r)\, J_0(2\pi|\gamma|r)\, r\, dr.
\end{aligned}$$

Clearly, $\hat{f}$ is likewise a radial function, and

$$f(x) = (\hat{f})^\vee = \int_{R^2} \hat{f}(\gamma)\, e^{2\pi i\gamma\cdot x}\, d\gamma = 2\pi \int_0^\infty \hat{f}(k)\, J_0(2\pi k|x|)\, k\, dk$$

with $k = |\gamma|$; in particular, $\wedge = \vee$. This is the Bessel transform cited above.
The corresponding Plancherel identity states that

$$\|f\|_2^2 = 2\pi \int_0^\infty |f(r)|^2 r\, dr = \|\hat{f}\|_2^2 = 2\pi \int_0^\infty |\hat{f}(k)|^2 k\, dk.$$

The same kind of thing can be done for dimensions $n \geqslant 3$. Now J_0 is
replaced by the "spherical Bessel function"

$$j_n(k) = \frac{\int_{|x|=1} e^{i\gamma\cdot x}\, d^{n-1}o}{\int_{|x|=1} d^{n-1}o} = \frac{\int_0^\pi e^{ik\cos\varphi} \sin^{n-2}\varphi\, d\varphi}{\int_0^\pi \sin^{n-2}\varphi\, d\varphi},$$

in which $d^{n-1}o$ stands for the element of surface area [or volume] on the
unit $(n-1)$-dimensional sphere $|x| = 1$ and the angle φ in the second integral
is the inclination of γ to x. To obtain the second formula for j_n from the first,
use the fact that the area of the $(n-2)$-dimensional sphere $\cos\varphi =$ constant
is proportional to $[(1-\cos^2\varphi)^{\frac{1}{2}}]^{n-2} = \sin^{n-2}\varphi$; the factor of proportionality is washed out by the integral on the bottom, so you do not need to
know it! The Bessel transform is

$$\hat{f}(k) = \int_0^\infty f(r)\, j_n(2\pi kr)\, r^{n-1}\, dr \times \text{the area of the unit sphere},$$

$\wedge = \vee$ as before, and the Plancherel identity states that

$$\int_0^\infty |f(r)|^2 r^{n-1}\, dr = \int_0^\infty |\hat{f}(k)|^2 k^{n-1}\, dk.$$

Bessel functions will come into play again in the study of Fourier integrals on noncommutative groups; see Section 4.15.

EXERCISE 1. Check that $j_2 = J_0$ and compute j_3. *Answer:* $j_3(k) = k^{-1} \sin k$.

EXERCISE 2. Check that the action of the n-dimensional Laplace operator $\Delta = \partial^2/\partial x_1^2 + \cdots + \partial^2/\partial x_n^2$ on radial functions $f = f((x_1^2 + \cdots + x_n^2)^{1/2})$ is

$$\Delta f = f'' + \frac{n-1}{r} f'.$$

EXERCISE 3. Use the integral formula to verify that j_n is a solution of Bessel's equation:

$$j_n'' + \frac{n-1}{r} j_n' = -j_n.$$

Hint: See Exercise 2.

EXERCISE 4. A permutation p of the numbers $1, \dots, n$ acts upon $x \in R^n$ according to the rule $x \to px = (x_{p1}, \dots, x_{pn})$. The signature of a permutation is ± 1 according as it involves an even or an odd number of transpositions $ij \to ji$; see Exercises 4.1.15 and 4.1.16. Check that for *antisymmetric* functions $[f(px) = \operatorname{sign} p \times f(x)]$,

$$\hat{f}(\gamma) = \int_{x_1 < \cdots < x_n} f(x) \det[\exp(-2\pi i \gamma_j x_k)] \, d^n x,$$

and

$$f(x) = \int_{\gamma_1 < \cdots < \gamma_n} \hat{f}(\gamma) \det[\exp(2\pi i \gamma_j x_k)] \, d^n \gamma.$$

Fourier integrals of this kind are closely allied to the one-dimensional sine transform [Exercise 2.3.5]. Find a similar (cosine-like) transform for *symmetric* functions $[f(px) = f(x)]$.

2.11 MISCELLANEOUS APPLICATIONS OF SEVERAL-DIMENSIONAL FOURIER INTEGRALS

The whole development of one-dimensional Fourier integrals for $C_\downarrow^\infty (R^1)$, $L^2(R^1)$, and $L^1(R^1)$ can now be carried over to n dimensions; it would be boring to do too much of this, so it may be left to the reader to think through as many such topics as seems profitable to him. Besides, it is always more instructive to see how a something is actually used. This section deals with a number of typical applications. Sommerfeld [1949] is recommended for additional topics.

1.   Spherical Waves in Two and Three Dimensions

The purpose of this subsection is to study the propagation of spherical waves in the plane and in three-dimensional space. This propagation is governed by the wave equation

$$\frac{\partial^2 u}{\partial t^2} = \Delta u = \frac{\partial^2 u}{\partial x_1^2} + \frac{\partial^2 u}{\partial x_2^2} \left[+ \frac{\partial^2 u}{\partial x_3^2} \right], \qquad t > 0, \quad x \in R^n,$$

and just as in one dimension, you must specify both u and $\partial u/\partial t$ at time $t = 0$. To simplify the discussion, let the initial data be $u(0+, \cdot) = 0$ and $\partial u(0+, \cdot)/\partial t = f \in C_{\downarrow}^{\infty}(R^n)$. Then

$$\hat{u}(t, \gamma) = \int u(t, x) e^{-2\pi i \gamma \cdot x} \, d^n x$$

is a solution of

$$\partial^2 \hat{u}/\partial t^2 = -4\pi^2 |\gamma|^2 \hat{u} .$$

As such, you see that

$$\hat{u}(t, \gamma) = \hat{f}(\gamma) \frac{\sin 2\pi |\gamma| t}{2\pi |\gamma|} ,$$

and inverting, you find

$$u(t, x) = \hat{u}^{\vee} = \int_{R^n} \hat{f}(\gamma) \frac{\sin 2\pi |\gamma| t}{2\pi |\gamma|} \, e^{2\pi i \gamma \cdot x} \, d^n \gamma .$$

The formula will now be put into a more transparent form for dimensions $n = 2$ and 3. For $n = 3$, let $d^2 o$ be the element of surface area $R^2 \sin \varphi \, d\varphi \, d\theta$ on the spherical surface $|x| = R$. Then

$$\int_{|x| = R} e^{2\pi i \gamma \cdot x} d^2 o = 4\pi R \frac{\sin 2\pi |\gamma| R}{2\pi |\gamma|} ,$$

as in Exercise 2.10.1, so

$$u(t, x) = \int_{R^3} \hat{f}(\gamma) e^{2\pi i \gamma \cdot x} d^3 \gamma \frac{1}{4\pi t} \int_{|y| = t} e^{2\pi i \gamma \cdot y} d^2 o$$

$$= (4\pi t)^{-1} \int_{|y| = t} d^2 o \int_{R^3} \hat{f}(\gamma) e^{2\pi i \gamma \cdot (x+y)} d^3 \gamma$$

$$= (4\pi t)^{-1} \int_{|y| = t} f(x+y) \, d^2 o ;$$

in particular, if you concentrate f in the vicinity of the origin keeping $\int f = 1$, you can make u approximate a sharp spherical wave front propagating outward from the origin at speed 1, as indicated in Fig. 1.

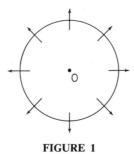

FIGURE 1

EXERCISE 1. Check by hand that the function u proposed above actually solves the problem of wave motion.

A nice formula for the two-dimensional case is easily deduced from the three-dimensional one by the "method of descent." Think of $\gamma \in R^2$ as a point of $R^2 \times 0 \subset R^3$. Then

$$4\pi R \frac{\sin 2\pi |\gamma| R}{2\pi |\gamma|}$$

is expressed as before as a surface integral in 3 dimensions:

$$\int_{|x| = R} e^{2\pi i \gamma \cdot x} d^2 o$$

$$= R^2 \int_0^{2\pi} d\theta \int_0^{\pi} \exp[2\pi i R \sin \varphi (\gamma_1 \cos \theta + \gamma_2 \sin \theta)] \sin \varphi \, d\varphi$$

$$= 2R \int_0^{2\pi} \int_0^R \exp[2\pi i r (\gamma_1 \cos \theta + \gamma_2 \sin \theta)] \frac{r \, dr \, d\theta}{(R^2 - r^2)^{\frac{1}{2}}},$$

as you can easily see by putting $x_1 = R \sin \varphi \cos \theta$ and $x_2 = R \sin \varphi \sin \theta$ in the first integral and then making the substitution

$$r = R \sin \varphi = (x_1^2 + x_2'^2)^{\frac{1}{2}}.$$

[*Be careful:* $d\varphi = \pm (R^2 - r^2)^{-\frac{1}{2}} dr$ according as $0 \leqslant \varphi < \pi/2$ or $\pi/2 < \varphi \leqslant \pi$.] But this is simply the two-dimensional integral

$$2R \int_{|x| \leqslant R} \frac{e^{2\pi i \gamma \cdot x}}{(R^2 - |x|^2)^{\frac{1}{2}}} d^2 x,$$

so

$$u(t,x) = \int_{R^2} \hat{f}(\gamma) e^{2\pi i \gamma \cdot x} \, d^2\gamma \, \frac{1}{2\pi} \int_{|y| \leqslant t} \frac{e^{2\pi i \gamma \cdot y}}{(t^2 - |y|^2)^{\frac{1}{2}}} \, d^2 y$$

$$= \frac{1}{2\pi} \int_{|y| \leqslant t} \frac{f(x+y)}{(t^2 - |y|^2)^{\frac{1}{2}}} \, d^2 y \,,$$

and if you concentrate f in the vicinity of the origin as before, u will approximate the (ideal) spherical wave:

$$u(t,x) = \begin{cases} (2\pi)^{-1}(t^2 - |x|^2)^{-\frac{1}{2}} & \text{if } |x| < t, \\ 0 & \text{if } |x| > t. \end{cases}$$

The wave front is still sharp and propagates outward at speed 1, but unlike the three-dimensional case, it has a trailing edge. This illustrates a very striking fact: If you sit at $x \in R^n$, then you do not "hear" the spherical wave until time $t = |x|$, and for $n = 3$, it passes outward and is never heard by you again, while for $n = 2$, you keep on hearing it forever, although the sound slowly fades away. Think how noisy a two-dimensional New York would be!

2*.   The Radon Transform

The "Radon transform" provides an elegant reduction of the problem of wave motion from three dimensions to one dimension, and that is the subject of this article; for other dimensions and additional information, see Ludwig [1966] and/or John [1955].

The Radon transform $f^\flat$ of a function $f \in C_\downarrow^\infty (R^3)$ is defined by the recipe

$$f^\flat(k,o) = -\frac{1}{4\pi^2} \frac{\partial^2}{\partial k^2} f^\#(k,o),$$

in which $f^\#$ is the integral over a moveable plane:

$$f^\#(k,o) = \int_{x \cdot o = k} f(x) \, d^2 x,$$

k being a real number, o being a direction from the unit sphere $|o| = 1$, and $d^2 x$ being the element of area on the plane $x \cdot o = k$. The original function f can be recovered from $f^\flat$ by the recipe

$$f(x) = \tfrac{1}{2} \int_{|o|=1} f^\flat(x \cdot o, o) \, d^2 o,$$

in which $d^2 o = \sin \varphi \, d\varphi \, d\theta$ is the element of surface area on $|o| = 1$. #, $\flat$, and the inversion formula can be extended to the whole of $L^2(R^3)$, but this is left aside.

PROOF OF THE INVERSION FORMULA. The function $\hat{f}(\gamma)$ is computed by integrating first over the plane $x \cdot o = k$, in which o is the direction of γ $[\gamma = |\gamma| o]$:

$$\hat{f}(\gamma) = \int_{R^3} f(x) e^{-2\pi i \gamma \cdot x} d^3 x$$

$$= \int_{-\infty}^{\infty} e^{-2\pi i |\gamma| k} dk \int_{x \cdot o = k} f(x) d^2 x$$

$$= \int_{-\infty}^{\infty} f^{\#}(k, o) e^{-2\pi i |\gamma| k} dk .$$

Now you perform the inversion in spherical polars $\gamma = (|\gamma|, o)$:

$$f(x) = \int_{R^3} \hat{f}(\gamma) e^{2\pi i \gamma \cdot x} d^3 \gamma$$

$$= \int_{|o|=1} d^2 o \int_{0}^{\infty} e^{2\pi i |\gamma| x \cdot o} d|\gamma| \, |\gamma|^2 \int_{-\infty}^{\infty} f^{\#}(k, o) e^{-2\pi i |\gamma| k} dk$$

$$= \frac{1}{2} \int_{|o|=1} d^2 o \int_{-\infty}^{\infty} e^{2\pi i y x \cdot o} dy \int_{-\infty}^{\infty} \left(-\frac{1}{4\pi^2} \right) \frac{\partial^2}{\partial k^2} f^{\#}(k, o) e^{-2\pi i y k} dk$$

$$= \frac{1}{2} \int_{|o|=1} f^{\flat}(x \cdot o, o) d^2 o .$$

The fact that o and $-o$ appear in the spherical integral with the same weight is used in line three to replace $\int_0^{\infty} d|\gamma|$ by $\frac{1}{2} \int_{-\infty}^{\infty} dy$.

EXERCISE 2. Check that $(\Delta f)^{\#} = \partial^2 f^{\#} / \partial k^2$ for nice functions. *Hint:* Use the formula for $\hat{f}$ in terms of $f^{\#}$.

EXERCISE 3. Check the Plancherel identity

$$8\pi^2 \|f\|_2^2 = \int_{|o|=1} d^2 o \int_{-\infty}^{\infty} dk \, |\partial f^{\#} / \partial k|^2 .$$

Hint: $\int e^{-2\pi i |\gamma| k} \partial f^{\#} / \partial k \, dk = 2\pi i |\gamma| \hat{f}(\gamma)$.

The solution of the three-dimensional wave equation

$$\frac{\partial^2 u}{\partial t^2} = \Delta u = \frac{\partial^2 u}{\partial x_1^2} + \frac{\partial^2 u}{\partial x_2^2} + \frac{\partial^2 u}{\partial x_3^2}$$

by means of the Radon transform is simplicity itself. By Exercise 2,

$$\frac{\partial^2 u^{\#}}{\partial t^2} = (\Delta u)^{\#} = \frac{\partial^2 u^{\#}}{\partial k^2}$$

for any fixed direction o, assuming u dies away fast enough, and this is just the problem of one-dimensional wave motion! Now $u^\sharp$ can be expressed in terms of the initial data $u(0+, \cdot)^\sharp = f^\sharp$ and $\partial u(0+, \cdot)^\sharp / \partial t = g^\sharp$ using d'Alembert's formula of Subsection 2.7.3:

$$u^\sharp(t,k) = \tfrac{1}{2}[f^\sharp(k+t,o) + f^\sharp(k-t,o)] + \tfrac{1}{2} \int_{k-t}^{k+t} g^\sharp(k', o)\, dk',$$

an application of $(-1/4\pi^2)\partial^2/\partial k^2$ converts the "sharps" to "flats", and the Radon inversion formula produces the three-dimensional analogue of the d'Alembert formula:

$$u(t,x) = \tfrac{1}{4} \int_{|o|=1} f^\flat(x \cdot o + t, o)\, d^2 o + \tfrac{1}{4} \int_{|o|=1} f^\flat(x \cdot o - t, o)\, d^2 o$$

$$+ \tfrac{1}{4} \int_{|o|=1} d^2 o \int_{x \cdot o - t}^{x \cdot o + t} g^\flat(k', o)\, dk',$$

expressing u as a superposition of plane waves traveling at speed 1 in the several directions o.

EXERCISE 4.   Use the three-dimensional d'Alembert formula to confirm the fact that in R^3 it is possible to have sharp spherical waves with no trailing edge; see Subsection 1.

3*. The Poisson Summation Formula

Given $-\infty < a < \infty$ and $b > 0$, let $Z \subset R^2$ be the lattice of points of the form

$$\omega = j(1,0) + k(a,b) \qquad \text{with integral } j \text{ and } k,$$

as in Subsection 1.10.3, and recall the dual lattice Z' of points

$$\omega' = j(1, -a/b) + k(0, 1/b) \qquad \text{with integral } j \text{ and } k.$$

The Poisson summation formula for Z states that

$$\sum_{\omega \in Z} f(\omega) = [\text{area}(T)]^{-1} \sum_{\omega' \in Z'} \hat{f}(\omega')$$

if, for example, $f \in C_\downarrow^\infty(R^2)$. Here, T is the (nonstandard) torus R^2/Z, and area$(T) = b$; for $a = 0$ and $b = 1$, it is just the standard lattice Z^2 of integral points of R^2 and $Z' = Z$. For $n = 1$, you recover the classical Poisson summation formula of Subsection 2.7.5.

PROOF.   Pick $f \in C_\downarrow^\infty(R^2)$ and look at

$$f^0(x) = \sum_Z f(x + \omega).$$

This function is periodic, with periods from Z, and belongs to $C^\infty(T)$. Therefore, you can expand it, as in Subsection 1.10.3, into a rapidly converging Fourier series

$$f^0(x) = \sum_{Z'} \left[[\text{area}(T)]^{-1} \int_T f^0(y) \exp(-2\pi i \omega' \cdot y) \, d^2 y \right] \exp(2\pi i \omega' \cdot x),$$

with coefficients

$$[\text{area}(T)]^{-1} \int_T f^0(x) \exp(-2\pi i \omega' \cdot x) \, d^2 x$$

$$= [\text{area}(T)]^{-1} \sum_Z \int_T f(x+\omega) \exp(-2\pi i \omega' \cdot x) \, d^2 x$$

$$= [\text{area}(T)]^{-1} \int_{R^2} f(x) \exp(-2\pi i \omega' \cdot x) \, d^2 x$$

$$= [\text{area}(T)]^{-1} \hat{f}(\omega').$$

To finish the proof, just evaluate at $x = 0$.

EXERCISE 5. Check the "theta-function" identity

$$(2\pi t)^{-1} \sum_Z \exp(-|\omega|^2/2t) = [\text{area}(T)]^{-1} \sum_{Z'} \exp(-2\pi^2 t |\omega'|^2).$$

EXERCISE 6. Use Exercise 5 to prove the formula of Gauss for the number $\#(n)$ of points of the standard lattice Z^2 at distance $\sqrt{n}$ from the origin:

$$t^{-\frac{1}{2}} \sum_{n=0}^\infty \#(n) \exp(-\pi n/t) = t^{\frac{1}{2}} \sum_{n=0}^\infty \#(n) \exp(-\pi n t).$$

4*. Siegel's Proof of Minkowski's Theorem

This proof of the celebrated theorem of Minkowski in the "geometry of numbers" is a beautiful application of the Poisson summation formula. Minkowski proved that a closed convex body $K \subset R^n$ which is symmetrically disposed about the origin contains a point $\omega \neq 0$ of the (nonstandard) lattice Z if

$$\text{volume}(K) \geq 2^n \times \text{volume}(R^n/Z).$$

EXERCISE 7. Give an example showing that the statement cannot be improved. *Hint:* $n = 2$ is enough.

PROOF.[1]   The plan is to check that if K contains no lattice point $\omega \neq 0$ then its volume is less than $2^n \times$ volume(R^n/Z). The indicator function f of K is real and even, so the same is true of $\hat{f}$. Keeping this fact in mind, apply the Poisson summation formula to $j = f \circ f$ on the lattice $2Z$. The dual lattice is $\frac{1}{2}Z'$, and the volume of the "fundamental cell" $R^n/2Z$ is $2^n \times$ volume(R^n/Z), so

$$\sum_Z j(2\omega) = [2^n \times \text{volume}(R^n/Z)]^{-1} \sum_{Z'} |\hat{f}(\omega'/2)|^2.$$

But

$$0 \neq j(2\omega) = \int_{R^n} f(2\omega - x) f(x) \, d^n x$$

means that both $2\omega - x$ and x belong to K for some $x \in R^n$. Because K is convex, this happens only if

$$\tfrac{1}{2}(2\omega - x) + \tfrac{1}{2}x = \omega$$

also belongs to K. This is ruled out for $\omega \neq 0$, so the left-hand summation reduces to a single term:

$$j(0) = \int_{R^n} f^2(x) \, dx = \text{volume}(K),$$

while on the right-hand side, you have

$$\frac{[\text{volume}(K)]^2}{2^n \times \text{volume}(R^n/Z)} + \sum_{\omega' \neq 0} \frac{|\hat{f}(\omega'/2)|^2}{2^n \times \text{volume}(R^n/Z)}.$$

Now the summation vanishes only if

$$f^0(x) = \sum_Z f(x + 2\omega)$$

is constant, and that is not the case: Namely, $f^0(0) = 1$, while $f^0 = 0$ at the points of Z closest to the origin. Therefore,

$$\text{volume}(K) > \frac{[\text{volume}(K)]^2}{2^n \times \text{volume}(R^n/Z)},$$

and the proof is finished. For additional information see Chandrasekharan [1968].

EXERCISE 8.   Actually, the proof is fake since f does not belong to $C_{\uparrow}^{\infty}(R^n)$. Fix it up. *Hint:* Approximate f from above by functions of class $C_{\downarrow}^{\infty}(R^n)$.

---

[1] Adapted from Rademacher [1956].

5*. Random Flights

Think of a particle moving in three-space according to the following rule: Start at $x = 0$, pick a direction e_1 from the unit sphere $|x| = 1$ according to the uniform distribution:

$$P(e_1 \in A) = (4\pi)^{-1} \times \text{area}(A),$$

and jump to the place $s_1 = e_1$ at time $n = 1$; pick an independent direction e_2 with the same distribution and jump to $s_2 = e_1 + e_2$ at time $n = 2$; and so on, arriving at $s_n = e_1 + \cdots + e_n$ at time $n \geqslant 3$. The problem is to estimate the distribution of s_n as $n \uparrow \infty$. The answer, discovered by Lord Rayleigh, is that for any (smooth) three-dimensional region Q,

$$\lim_{n \uparrow \infty} P(n^{-\frac{1}{2}} s_n \in Q) = (3/2\pi)^{\frac{3}{2}} \int_Q \exp(-(3/2)|x|^2) \, d^3 x.$$

There is a self-evident connection with the random walk of Subsection 1.10.2 and with the central limit problem of Subsection 2.7.7; for applications of Fourier integrals to similar problems of statistical physics, see Chandrasekhar [1943].

PROOF.   Because of the independence of the steps $e_k : k \leqslant n$, the mean value of

$$\exp(2\pi i \gamma \cdot s_n)$$

is the nth power of the mean value for 1 step:

$$\frac{1}{4\pi} \int_0^{2\pi} d\theta \int_0^\pi e^{2\pi i |\gamma| \cos\varphi} \sin \varphi \, d\varphi = \frac{\sin 2\pi |\gamma|}{2\pi |\gamma|};$$

in particular, the mean value of

$$\exp(2\pi i \gamma \cdot n^{-\frac{1}{2}}) s_n$$

is

$$\left[\frac{\sin 2\pi n^{-\frac{1}{2}} |\gamma|}{2\pi n^{-\frac{1}{2}} |\gamma|} \right]^n = \left[1 - \frac{4\pi^2}{6n} |\gamma|^2 [1 + o(1)] \right]^n,$$

in which $o(1)$ indicates an error that approaches 0 as $n \uparrow \infty$, and this whole expression is bounded (by 1) and tends to $\exp[-(2/3)\pi^2 |\gamma|^2]$. Therefore, if $f \in C_\uparrow^\infty(R^3)$, the mean value of

$$f(n^{-\frac{1}{2}} s_n) = \int \hat{f}(\gamma) \exp(2\pi i \gamma \cdot n^{-\frac{1}{2}} s_n) \, d^3 \gamma$$

tends to

$$\int \hat{f}(\gamma) \exp[-(2/3)\pi^2 |\gamma|^2] \, d^3\gamma = (3/2\pi)^{3/2} \int f(x) \exp[-(3/2)|x|^2] \, d^3 x$$

and the estimate for $P(n^{-1/2} \mathbf{s}_n \in Q)$ follows, as in Subsection 2.7.7, by approximating the indicator function of Q above and below by functions f from $\mathbf{C}_1^\infty (R^3)$.

EXERCISE 9.   Check that for the analogous two-dimensional problem of random flights,

$$\lim_{n\uparrow\infty} P(n^{-1/2} \mathbf{s}_n \in Q) = \pi^{-1} \int_Q \exp(-|x|^2) \, d^2 x.$$

Hint: The Bessel function J_0 is involved, and $J_0(r) = 1 - (r^2/4)\,[1 + o(1)]$ for small r.

Chapter 3 Fourier Integrals and Complex Function Theory

3.1 A SHORT COURSE IN FUNCTION THEORY

To delve more deeply into the subject of Fourier integrals, it is necessary to use the powerful tools of complex function theory, taking advantage of the lucky fact that the line sits in the complex plane. The interplay between complex function theory and Fourier integrals is a very rich subject; for additional information, see Paley and Wiener [1934], Zygmund [1959], and at a more elementary level, Rudin [1966]. Two fine elementary texts for the necessary function theory are Copson [1935] and Levinson and Redheffer [1970]. The present explanation is divided into a series of short articles.

1. Analytic Functions

A connected open piece of the plane R^2 is called a "domain" or "region." Take a complex-valued function $f = f(\gamma)$ of $\gamma = a + ib$ defined on such a domain D, and let it be of class $C^1(D)$, i.e., let $\partial f/\partial a$ and $\partial f/\partial b$ be continuous

144

in D. f is said to be "analytic" in D if the (complex) differential coefficient

$$f'(\alpha) = \lim_{\beta \to \alpha} \frac{f(\beta) - f(\alpha)}{\beta - \alpha}$$

exists *independently of the mode of approach of β to α, for every $\alpha \in D$.*

EXERCISE 1. Show that $f(\gamma) = \gamma^n$ is analytic in the whole plane for any integer $n \geqslant 0$ whereas $f(\gamma) = \gamma^* = a - ib$, is not analytic in *any* nonempty region.

EXERCISE 2. Show that $f(\gamma) = e^{2\pi i \gamma x}$ is analytic in the whole plane for any fixed x, real or complex.

EXERCISE 3. Show that $f(\gamma) = \log \gamma = \log R + i\theta$ is analytic in the cut plane $(\gamma = R \exp(i\theta): R > 0, |\theta| < \pi)$. The same is true of $\gamma^k = R^k e^{ik\theta}$ for any real exponent k, integral or not.

EXERCISE 4. Show that $\hat{f}(\gamma)$ is analytic in the whole complex plane if $f \in L^2(R^1)$ is compact, whereas it is analytic in the open strip $(\gamma = a + ib: |2\pi b| < B)$ if $|f(x)| \leqslant \text{constant} \times e^{-B|x|}$.

2. Cauchy's Theorem

A piecewise smooth, simple, closed, directed curve C in a region D is the image of the unit interval $0 \leqslant t \leqslant 1$ under a map

$$\Gamma: t \to \gamma(t) = a(t) + ib(t).$$

The proviso *piecewise smooth* signifies that Γ is of class $C^1[0, 1]$ apart from a finite number of corners;[1] *simple* means that the curve does not cross itself, i.e., $\gamma(t_1) \neq \gamma(t_2)$ for $0 \leqslant t_1 < t_2 < 1$; *closed* means that $\gamma(0) = \gamma(1)$; and *directed* means that you must respect the direction of travel dictated by making t increase from 0 to 1. Two such maps represent the *same* directed curve C if they sweep out the same geometrical figure and if the direction of travel is the same. C divides the plane into two regions D' and D''. C is *positively directed* with respect to D' if the points in D' always lie to your left as you travel along C in the indicated direction. Since C is continuous D' has no holes. The (positively directed) boundary of a nice region D with $n < \infty$ holes consists of $n+1$ such (positively) directed curves. For example, the boundary of the open ring $1 < |\gamma| < 2$ is the circle $C_2 = (\gamma: |\gamma| = 2)$,

---

[1] A corner is a point at which the left- and right-hand derivatives of Γ exist but are unequal.

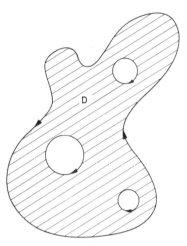

FIGURE 1

directed counterclockwise, plus the circle $C_1 = (\gamma: |\gamma| = 1)$, directed clock-wise. A second example with $n = 3$ is sketched in Fig. 1.

Warning: The boundary of a nice region is *always* assumed to be *positively-directed* if nothing is said to the contrary.

The line integral (about C) of a function $f \in C(D)$ is declared to be

$$\int_C f(\gamma)\, d\gamma = \int_0^1 f[\gamma(t)]\, \gamma'(t)\, dt,$$

and it is easy to see that this number depends only upon the directed curve C itself and not upon the particular map $\Gamma: [0,1] \to D$ employed. The point is that the line integral is well approximated by the Riemann sum

$$\sum_{k=0}^{n-1} f(\gamma_k)\, [\gamma_{k+1} - \gamma_k],$$

for any choice of $\gamma_n = \gamma_0, \ldots, \gamma_{n-1}$ on C, provided the labeling respects the direction of travel and the spacing is sufficiently fine; see Fig. 2 ($n = 7$). The direction *is* important; if you reverse it, the integral changes sign.

EXERCISE 5.   Check that $\left| \int_C f(\gamma)\, d\gamma \right| \leqslant \|f\|_\omega \times$ the length of C.

EXERCISE 6.   Check that if C is the counterclockwise directed circle $|\gamma| = R > 0$, then

$$\int_C \gamma^n\, d\gamma = \begin{cases} 0 & \text{if } n \neq -1 \quad \text{is integral,} \\ 2\pi i & \text{if } n = -1. \end{cases}$$

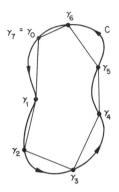

FIGURE 2

Repeat the calculations if C is the boundary of the square $D = (a + ib:$ $|a| < R,\ |b| < R)$. *Hint:* Take $\gamma(t) = R e^{2\pi i t}$ for the circle.

The source of all good things in complex function theory is

CAUCHY'S THEOREM: $\int_C f(\gamma)\, d\gamma = 0$ *if f is analytic in a domain D containing the piecewise smooth, simple, closed, directed curve C, provided the region D' enclosed by C lies wholly inside D.*

The proviso is essential: $f = \gamma^{-1}$ is analytic in the punctured plane $R^2 - 0$, but

$$\int_{|\gamma|=1} \gamma^{-1}\, d\gamma = 2\pi i \neq 0\, ;$$

see Exercise 6.

PROOF OF CAUCHY'S THEOREM. Represent C by a piecewise smooth map $\Gamma\colon [0,1] \to D$ and express the integral as

$$\int_C f(\gamma)\, d\gamma = \int_0^1 f[\gamma(t)]\, \gamma'(t)\, dt$$

$$= \int_0^1 [ifb' - (-f)a']\, dt\, .$$

An application of the divergence theorem

$$\iint_{D'} \operatorname{div} \mathfrak{f}\, da\, db = \iint_{D'} \left(\frac{\partial f_1}{\partial a} + \frac{\partial f_2}{\partial b} \right) da\, db = \int_0^1 (f_1 b' - f_2 a')\, dt$$

to the complex vector field $\mathfrak{f} = (f_1, f_2) = (if, -f)$ lets you rewrite the latter as a double integral

$$\iint_{D'} \left(i \frac{\partial f}{\partial a} - \frac{\partial f}{\partial b} \right) da \, db,$$

and to finish the proof, you only have to notice that the integrand vanishes by the analyticity of f:

$$\frac{\partial f}{\partial a} = \lim_{\varepsilon \to 0} \varepsilon^{-1} \left[f(\gamma + \varepsilon) - f(\gamma) \right]$$

$$= f'(\gamma) = \lim_{\varepsilon \to 0} (i\varepsilon)^{-1} \left[f(\gamma + i\varepsilon) - f(\gamma) \right]$$

$$= \frac{1}{i} \frac{\partial f}{\partial b}.$$

Amplification: The divergence theorem is customarily stated for *real* fields, but it also holds for complex fields. Why? The equation $\partial f/\partial a = i^{-1} \partial f/\partial b$ is equivalent to the pair of equations $\partial u/\partial a = \partial v/\partial b$ and $\partial u/\partial b = -\partial v/\partial a$ in which u and v denote the real and imaginary parts of f respectively. This pair is called the Cauchy–Riemann equations. With the help of the divergence theorem and Exercise 9 you can characterize the analytic functions f in $\mathbf{C}^1(D)$ as those for which the Cauchy–Riemann equations are satisfied.

EXERCISE 7.   Explain why the integrals of Exercise 6 are the same for circle and square.

3.   Cauchy's Formula

Take f, D, and C as in the statement of Cauchy's theorem and let us check that

$$f(\alpha) = \frac{1}{2\pi i} \int_C \frac{f(\beta)}{\beta - \alpha} \, d\beta$$

for any point α inside C. This is Cauchy's formula.

PROOF.   Draw a small circle K of radius $R > 0$ centered at α, as in Fig. 3, and integrate $(\beta - \alpha)^{-1} f(\beta)$ counterclockwise about C from the point marked by the black spot $\bullet$ and back again, then *in* along the indicated cut to K, clockwise around K, and finally *out* again along the cut to the starting point. By Cauchy's theorem, the integral over this composite curve is 0 since the integrand is analytic in the region so enclosed. Besides, the two

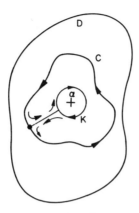

FIGURE 3

opposing integrals along the cut cancel by a change of sign, so

$$0 = \int \frac{f(\beta)}{\beta - \alpha} \, d\beta = \int_C \frac{f(\beta)}{\beta - \alpha} \, d\beta - \int_K \frac{f(\beta)}{\beta - \alpha} \, d\beta,$$

in which K is now directed counterclockwise. The integral around K is now expressed more concretely: K is represented by

$$\gamma(t) = \alpha + R e^{2\pi it} \colon 0 \leqslant t \leqslant 1,$$

and you get

$$\int_K \frac{f(\beta)}{\beta - \alpha} \, d\beta = 2\pi i \int_0^1 f(\alpha + R e^{2\pi it}) \, dt.$$

But this is independent of R since $\int_K = \int_C$, so you can evaluate it by making $R \downarrow 0$:

$$\int_K \frac{f(\beta)}{\beta - \alpha} \, d\beta = \lim_{R \downarrow 0} 2\pi i \int_0^1 f(\alpha + R e^{2\pi it}) \, dt = 2\pi i f(\alpha).$$

The proof is finished.

An important consequence of Cauchy's formula is that *if f is analytic in D, then it is actually infinitely differentiable, and its successive (complex) derivatives*

$$D^n f(\alpha) = \frac{n!}{2\pi i} \int_C \frac{f(\beta)}{(\beta - \alpha)^{n+1}} \, d\beta, \qquad n \geqslant 1$$

are also analytic in D.[2]

---

[2] It should be clear from the usage that D by itself stands for domain while Df stands for f', $D^2 f$ for f'', and so forth.

EXERCISE 8. Check this fact. *Hint for n = 1:* $\varepsilon^{-1}[(\beta - \alpha - \varepsilon)^{-1} - (\beta - \alpha)^{-1}]$ converges to $(\beta - \alpha)^2$ as $\varepsilon \to 0$, uniformly for β on C.

EXERCISE 9. Check that if f is of class $C(D)$ and if $\int_C f(\gamma)\, d\gamma = 0$ for every curve C of the class admitted into the statement of Cauchy's theorem, then f is analytic in D. This fact is called "Morera's theorem." *Hint:* The line integral $\int_\alpha^\beta f(\gamma)\, d\gamma$ is independent of the curve joining α to β and is analytic in β for fixed α.

EXERCISE 10. Check that for the curve C of Fig. 4

$$1 = \frac{1}{4\pi i} \int_C \frac{d\beta}{\beta - \alpha}.$$

What happened? What is the value of the integral for other locations of α?

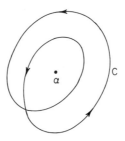

FIGURE 4

EXERCISE 11. Evaluate

$$\int_0^\infty (1 + x^2)^{-1}\, dx = \pi/2$$

by integrating $f = (1 + \gamma^2)^{-1}$ around the semicircle of Fig. 5 and checking that the integral along the semicircular arc tends to 0 as $R \uparrow \infty$.

EXERCISE 12. Evaluate the (improper) integral

$$\int_0^\infty x^{-1} \sin x\, dx = \pi/2$$

in a similar way by integrating $f = \gamma^{-1} e^{i\gamma}$ about the semiannulus of Fig. 6 and then making $\varepsilon \downarrow 0$ and $R \uparrow \infty$.

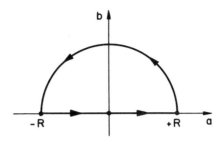

FIGURE 5

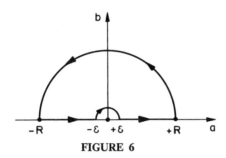

FIGURE 6

EXERCISE 13.   Check that if α belongs to the open upper half-plane $b > 0$, then

$$\frac{1}{2\pi i} \int_{-\infty}^{\infty} \frac{e^{2\pi i\gamma x}}{x - \alpha} \, dx = \begin{cases} e^{2\pi i\alpha\gamma} & \text{if } \gamma > 0, \\ 0 & \text{if } \gamma < 0. \end{cases}$$

The integral is improper: $\int_{-\infty}^{\infty} = \lim_{R \uparrow \infty} \int_{-R}^{R}$. *Hint:*   Use the semicircle of Fig. 5 for $\gamma > 0$. Do things in the lower half-plane for $\gamma < 0$.

4.   Power Series

A power series with center at α is an infinite sum of the form

$$p_\alpha(\beta) = \sum_{n=0}^{\infty} c_n (\beta - \alpha)^n$$

with complex coefficients $c_0, c_1, \dots$. The basic fact about such power series is that the sum converges (geometrically fast) *inside* the circle $|\beta - \alpha| = R$ and diverges *outside*, R being the so-called "radius of convergence":

$$0 \leqslant R = [\limsup_{n \uparrow \infty} (|c_n|)^{1/n}]^{-1} \leqslant \infty.$$

The first statement is vacuous if $R = 0$, the second if $R = \infty$. The proof is easy: If $|\beta - \alpha| < R$ then $|c_n| |\beta - \alpha|^n \leqslant k^n$ for some positive constant $k < 1$ and all sufficiently large n, which ensures geometrically fast convergence, while if $|\beta - \alpha| > R$ then $|c_n| |\beta - \alpha|^n \geqslant 1$ infinitely often, preventing convergence.

The function $p_\alpha(\beta)$ is analytic in the open disc $D' = (\beta: |\beta - \alpha| < R)$. A proof can be based on Morera's theorem [Exercise 9]:

$$\int_C p_\alpha(\beta)\, d\beta = \int_C \lim_{n \uparrow \infty} \sum_{k=0}^n c_k(\alpha - \beta)^k\, d\beta$$

$$= \lim_{n \uparrow \infty} \int_C \sum_{k=0}^n c_k(\alpha - \beta)^k\, d\beta = 0$$

for every nice curve C lying inside D'. The fact that p_α is of class $C(D')$ inside the disc and that the limit can be brought outside the integral is justified by the rapidity of the convergence of the sum. Check this for yourself. The important point is the converse: *If f is analytic in a domain D containing the open disc $D' = (\beta: |\beta - \alpha| < R)$ then f is expressible as a power series p_α in D'.*

PROOF.   Let C denote the boundary of D', directed counterclockwise, and use Cauchy's formula to express f at a point β inside C:

$$2\pi i f(\beta) = \int_C \frac{f(\gamma)}{\gamma - \beta}\, d\gamma = \int_C \frac{f(\gamma)}{\gamma - \alpha} \left(1 - \frac{\beta - \alpha}{\gamma - \alpha}\right)^{-1} d\gamma.$$

Because

$$\left| \frac{\beta - \alpha}{\gamma - \alpha} \right| = R^{-1} |\beta - \alpha| < 1$$

for γ on C, you can expand $[1 - (\beta - \alpha)/(\gamma - \alpha)]^{-1}$ into a geometric series

$$\sum_{n=0}^\infty \left(\frac{\beta - \alpha}{\gamma - \alpha}\right)^n.$$

The sum converges uniformly on C, so you can interchange sum and integral to express $f(\beta)$ as a power series:

$$f(\beta) = \frac{1}{2\pi i} \int_C \frac{f(\gamma)}{\gamma - \alpha} \left[\sum_{n=0}^\infty \left(\frac{\beta - \alpha}{\gamma - \alpha}\right)^n \right] d\gamma$$

$$= \sum_{n=0}^\infty \left[\frac{1}{2\pi i} \int_C \frac{f(\gamma)}{(\gamma - \alpha)^{n+1}}\, d\gamma \right] (\beta - \alpha)^n$$

$$= \sum_{n=0}^\infty c_n(\beta - \alpha)^n$$

$$= p_\alpha(\beta)$$

with radius of convergence R or more and coefficients

$$c_n = \frac{1}{2\pi i} \int_C \frac{f(\gamma)}{(\gamma - \alpha)^{n+1}} \, d\gamma,$$

The proof is finished.

By the formula of Subsection 3, $n! \, c_n = D^n f(\alpha)$. This identity relates the local to the global behavior of f, via the power series. This feature of the class of analytic functions is not shared by $C^\infty(D)$; compare Exercises 14–17.

EXERCISE 14. Check that if f and all its complex derivatives vanish at a single point of D, then $f \equiv 0$ in D. *Hint:* D is connected.

EXERCISE 15. Check that if f and g are analytic in D and if $f = g$ in an open subdisc, then $f = g$ throughout D.

EXERCISE 16. The Fourier transform of a time-limited signal f cannot vanish on any interval unless $f \equiv 0$; see Section 2.9 for a direct proof. *Hint:* See Exercises 4 and 14.

EXERCISE 17. Check that if $B > 0$ and if $|f(x)| \leqslant \exp(-B|x|)$ for $-\infty \leqslant a \leqslant x \leqslant b \leqslant \infty$, then the functions $x^n f: n \geqslant 0$ span $L^2[a,b]$. A simple example is provided by $f = e^{-x}$, $a = 0$, $b = \infty$; the functions of the associated unit-perpendicular family obtained by the Gram–Schmidt recipe of Section 1.3 are known as "Laguerre functions." *Hint:* Take $g \in L^2[a,b]$ perpendicular to $x^n f$ for every $n \geqslant 0$. Then

$$(fg^*)^\wedge(\gamma) = \int_a^b fg^* e^{-2\pi i \gamma x} \, dx$$

is analytic in the open strip $D: 2\pi |b| < B$ [see Exercise 4] and vanishes together with all its derivatives at $\gamma = 0$.

Besides analytic functions, it will occasionally be necessary to use functions with (isolated) poles. A function f, which is analytic in a punctured disc $0 < |\gamma - \alpha| < R$, has a "pole" of degree $n > 0$ at $\gamma = \alpha$ if $(\gamma - \alpha)^n f(\gamma)$ can be extended to $\gamma = \alpha$ so as to be analytic in the whole disk $|\gamma - \alpha| < R$ and the extended function does not vanish at α. This is the same as to say that in the punctured disk

$$f(\gamma) = \frac{c_{-n}}{(\gamma - \alpha)^n} + \cdots + \frac{c_{-1}}{\gamma - \alpha} + \text{a power series centered at } \alpha$$

with $c_{-n} \neq 0$. The number

$$c_{-1} = (2\pi i)^{-1} \int_{|\gamma - \alpha| = R'} f(\gamma) \, d\gamma, \qquad 0 < R' < R$$

is the "residue" of f at α.

5. Liouville's Theorem

A function f which is analytic on the whole complex plane is said to be "integral" or "entire." Liouville's theorem states that *an integral function of bounded modulus is constant.*

PROOF. Because f is integral, it can be expressed on the whole plane as a power series centered at $\gamma = 0$ with coefficients

$$c_n = \frac{1}{2\pi i} \int_{|\gamma| = R} \frac{f(\gamma)}{\gamma^{n+1}} \, d\gamma \qquad \text{for any} \quad R > 0.$$

But

$$|c_n| \leqslant (2\pi)^{-1} (\|f\|_\infty / R^{n+1}) \times 2\pi R = \|f\|_\infty R^{-n},$$

and since this tends to 0 as $R \uparrow \infty$ if $n \neq 0$, you find that all the coefficients vanish except for c_0, i.e., $f = c_0$. The proof is finished.

EXERCISE 18.   Check that if f is integral and if $|f(\gamma)| \leqslant \text{constant} \times |\gamma|^n$ far out, then f is a polynomial of degree less than or equal to n.

EXERCISE 19.   Prove the "fundamental theorem of algebra": a complex polynomial p of degree $n \geqslant 1$ has at least one (and therefore precisely n) complex roots. *Hint:* $f = 1/p$ is integral if p is root-free, and $|f| \leqslant 1$ far out.

EXERCISE 20.   Check the partial fraction expansion

$$\frac{(x - \alpha_1) \cdots (x - \alpha_n)}{(x - \beta_1) \cdots (x - \beta_m)} = \sum_{k=1}^{m} \frac{\prod_{j=1}^{n}(\beta_k - \alpha_j)}{\prod_{j \neq k}(\beta_k - \beta_j)} \frac{1}{x - \beta_k}$$

in case $n < m$ and all the α's and β's are different. *Hint:* The difference is an integral function of bounded modulus. Why? What then? A second proof can be made by calling the left-hand side $f(x)$ and evaluating

$$\int (\gamma - \gamma_0)^{-1} f(\gamma) \, d\gamma$$

around a suitable closed curve. Try it.

6.   The Maximum Principle

This refers to a function f which is analytic in a (bounded) region D and continuous on its closure $\bar{D}$. The statement is that *either f is constant or $|f| < \|f\|_\infty$ inside D; in particular, a nonconstant function f achieves its maximum modulus only on the boundary of D.*

PROOF. The problem is to check that if $|f(\alpha)| = \|f\|_\infty$ at a point α inside D, then f is constant. By Cauchy's formula,

$$f(\alpha) = \int_0^1 f(\alpha + R e^{2\pi i t}) \, dt$$

for small $R > 0$; in particular,

$$\|f\|_\infty = |f(\alpha)| \leqslant \int_0^1 |f(\alpha + R e^{2\pi i t})| \, dt.$$

But since $|f(\alpha + R e^{2\pi i t})|$ is itself less than or equal to $\|f\|_\infty$, this can happen only if

$$|f(\alpha + R e^{2\pi i t})| = \|f\|_\infty \qquad \text{for} \quad 0 \leqslant t \leqslant 1.$$

R can be made as small as you like, so $|f| = \|f\|_\infty$ on a whole disc about α, and this propagates throughout the (connected!) region D. But then f maps D into the circle of radius $\|f\|_\infty$, and it is easy to see that this cannot happen for nonconstant f: If f is not constant, you can find a point $\alpha \in D$ at which $f'(\alpha) \neq 0$, and then the modulus of [3]

$$f(\beta) = f(\alpha) + f'(\alpha)\,(\beta - \alpha)\,[1 + o(1)]$$

exceeds $|f(a)|$ for an infinite number of points β in every neighborhood of α.

7.   A Phragmén–Lindelöf Theorem

This is a variant of the maximum principle for use with functions f of exponential type in (unbounded) sectors D. The phrase "exponential type" means that

$$|f(\gamma)| \leqslant \text{constant} \times e^{T|\gamma|}$$

for some $T < \infty$. The Phragmén–Lindelöf theorem states that *if f is analytic and of exponential type in a sector D of (angular) opening less than π, if it is also continuous in the closed sector $\bar{D}$, and if $|f| \leqslant M$ on the boundary of D,*

---

[3] In the formula $o(1)$ tends to 0 as $\beta - \alpha \to 0$.

then $|f| \leqslant M$ *throughout D.* The condition $\beta - \alpha < \pi$ is essential: The function $f = e^{-iy}$ is bounded on the line $b = 0$, it is clearly of exponential type, but $f(ib) = e^{b}$ is unbounded as $b \uparrow \infty$.

PROOF. Take D symmetrically disposed about the positive half-line, as in Fig. 7. The half-opening ψ is less than $\pi/2$, so you can pick a number $B > 1$ and still have $B\psi < \pi/2$. Then for any $A > 0$, the modulus of the function $f_1 = f \exp(-A\gamma^{B})$ is bounded by

$$\text{constant} \times e^{TR} \exp(-AR^{B} \cos B\psi) \leqslant M$$

for $|\theta| \leqslant \psi$ and large R. At the same time, $|f_1| \leqslant M$ on the rays $\theta = \pm\psi$, so by the maximum principle, $|f_1| \leqslant M$ on the whole of $\bar{D}$. To finish the proof, recall that $A > 0$ was chosen at pleasure and infer from $f = \lim_{A \downarrow 0} f_1$ that $|f| \leqslant M$ in the whole sector as advertised.

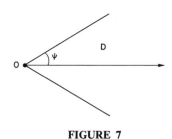

FIGURE 7

3.2 HARDY'S THEOREM

The first application of the machinery of complex function theory to Fourier integrals will be to prove the following beautiful theorem of G. H. Hardy: *If* α *and* β *are positive numbers, if*

$$|f(x)| \leqslant \text{constant} \times \exp(-\alpha x^{2})$$

on the line, and

$$|\hat{f}(\gamma)| \leqslant \text{constant} \times \exp(-\beta\gamma^{2}),$$

on the line, then either $f = 0$*, or* f *is a constant multiple of* $\exp(-\alpha x^{2})$*, or else there are infinitely many such functions* f*, according as* $\alpha\beta > \pi^{2}$*,* $\alpha\beta = \pi^{2}$*, or* $\alpha\beta < \pi^{2}$. This may be thought of as a companion to and an amplification of Heisenberg's inequality [see Section 2.8]. The moral is the same: A function and its Fourier transform cannot both decay too fast.

PROOF. The product $\alpha\beta$ is all that counts, as can be seen by a self-evident scaling, so you can take $\alpha = \beta$. The chief task is to check that f is a constant multiple of $\exp(-\pi x^2)$ if $\alpha = \beta = \pi$. Once this is known, the proof that $f = 0$ if $\alpha = \beta > \pi$ becomes self-evident. Also, the Hermite functions of Section 2.5 provide an infinite number of examples if $\alpha = \beta < \pi$. Therefore you may take $\alpha = \beta = \pi$, in which case $\hat{f}$ is an integral function of $\gamma = \alpha + ib$ [see Exercise 3.1.4] with

$$|\hat{f}(\gamma)| \leqslant \int |f(x)| e^{2\pi bx} dx$$

$$\leqslant \text{constant} \times \int \exp(-\pi x^2) e^{2\pi bx} dx$$

$$\leqslant \text{constant} \times \exp(\pi b^2).$$

Now suppose that f is even. Then $\hat{f}$ is also even, and you can expand it as a sum of even powers: $\hat{f} = \sum c_n \gamma^{2n}$. This implies that $h(\gamma) = \hat{f}(\sqrt{\gamma}) = \sum c_n \gamma^n$ is itself an integral function. The proof is finished by checking that $e^{\pi\gamma}h$ is constant by a clever application of the theorems of Phragmén–Lindelöf [Subsection 3.1.7] and Liouville [Subsection 3.1.5].

To begin with, by the global bound for $\hat{f}$,

$$|h(\gamma)| \leqslant M \exp[\pi(\text{Im}\sqrt{\gamma})^2] = M \exp[\pi R \sin^2(\theta/2)]$$

for $\gamma = R\exp(i\theta)$, so that h is of exponential type. At the same time, you have a much sharper bound for h on the half-line $\gamma = R > 0$:

$$|h(\gamma)| \leqslant M e^{-\pi R},$$

in which it is understood that the previous constant M has been increased if need be so that both bounds hold with the *same* M. But then

$$\left| \exp\left[\frac{i\pi\gamma \, e^{-i\delta/2}}{\sin(\delta/2)}\right] h(\gamma) \right| = \exp\left[\frac{-\pi R \sin(\theta - \delta/2)}{\sin(\delta/2)}\right] |h(Re^{i\theta})|$$

$$\leqslant M$$

for $\theta = 0$ and for $\theta = \delta$, provided only that $0 < \delta < \pi$. An application of Phragmén–Lindelöf to the intervening sector $0 < \theta < \delta$ lets you conclude that

$$|h(\gamma)| \leqslant M \exp\left[\frac{\pi R \sin(\theta - \delta/2)}{\sin(\delta/2)}\right]$$

in this sector, and making $\delta \uparrow \pi$, you find

$$|h(\gamma)| \leqslant M \exp(-\pi R \cos\theta) \qquad \text{for} \quad 0 \leqslant \theta \leqslant \pi.$$

A similar argument for the lower half-plane supplies the bound

$$|h(\gamma)| \leqslant M \exp(-\pi R \cos\theta) \qquad \text{for} \quad -\pi \leqslant \theta \leqslant 0;$$

in particular, $|e^{\pi\gamma}h|$ is bounded by M on the whole plane and is therefore constant, by Liouville's theorem. This finishes the proof for even functions f.

Now suppose that f is *odd*. Then $\hat{f}$ is also odd, $\hat{f}(0) = 0$, and the even proof can be applied to the integral function $\gamma^{-1}\hat{f}$. Thus, $\gamma^{-1}\hat{f}$ is a constant multiple of $\exp(-\pi\gamma^2)$, and for $\hat{f}$ itself to satisfy the stated bound, the constant multiplier can only be 0, i.e., $\hat{f} = f = 0$.

The final step is to split the general f into even and odd parts. The even and odd proofs can be applied to $\hat{f}_{\text{even}} = (f_{\text{even}})^{\wedge}$ and $\hat{f}_{\text{odd}} = (f_{\text{odd}})^{\wedge}$, separately, and you see that f is a constant multiple of $\exp(-\pi x^2)$. The proof of Hardy's theorem is finished.

3.3 THE PALEY–WIENER THEOREM

An integral function f is said to be of "exponential type" $T < \infty$ if

$$\limsup_{R\uparrow\infty} R^{-1} \log\left[\max_{0 \leqslant \theta < 2\pi} |f(Re^{i\theta})|\right] = T.$$

The number T cannot be negative unless $f = 0$. [Why?] An alternative statement is that f is of type less than or equal to T if for each $\varepsilon > 0$ you can find a constant M, independent of γ, such that

$$|f(\gamma)| \leqslant M e^{|\gamma|[T+\varepsilon]}$$

for *all* complex γ. f is said to be of "minimal type" if $T = 0$.

EXERCISE 1. Give examples of nonconstant integral functions of type 0, and of type 1.

The Paley–Wiener theorem describes the functions f of class $L^2(R^1)$ for which $\hat{f}$ is of exponential type. The precise statement is that $\hat{f}$ *can be extended off the line so as to be an integral function of exponential type less than or equal to $2\pi T$ iff $f = 0$ for $|x| > T$*. This provides an alternative description of the time-limited functions of Section 2.9.

PROOF. If $f = 0$ for $|x| > T$, then f is summable and

$$\hat{f}(\gamma) = \int_{-T}^{T} f(x) e^{-2\pi i\gamma x} dx$$

is an integral function of exponential type $2\pi T$:

$$|\hat{f}(\gamma)| \leqslant \|f\|_1 \exp(2\pi|\gamma| T).$$

The proof of the converse is not so cheap: $\hat{f}$ is an integral function of exponential type $2\pi T$, and it is to be proved that $f = 0$ for $|x| > T$. To do this, look first at the function

$$h(\gamma) = \int_{-\frac{1}{2}}^{\frac{1}{2}} \hat{f}(\gamma + y)\, dy .$$

The explicit evaluation

$$h(\gamma) = \int_{-\frac{1}{2}}^{\frac{1}{2}} \int_{-\infty}^{\infty} f(x)\, e^{-2\pi i (\gamma + y) x}\, dx\, dy$$

$$= \int_{-\infty}^{\infty} f(x)\, \frac{\sin \pi x}{\pi x}\, e^{-2\pi i \gamma x}\, dx$$

$$= \left[f(x)\, \frac{\sin \pi x}{\pi x} \right]^{\wedge}$$

is not needed until later. For the present, it is enough to know that h is an integral function of exponential type less than or equal to $2\pi T$ whose growth on R^1 is subject to the restraints

$$|h(\gamma)|^2 \leqslant \int_{-\frac{1}{2}}^{\frac{1}{2}} |\hat{f}(\gamma + y)|^2\, dy \leqslant \|\hat{f}\|_2^2 < \infty$$

and

$$\|h\|_2^2 \leqslant \int_{-\frac{1}{2}}^{\frac{1}{2}} \|\hat{f}\|_2^2\, dy = \|\hat{f}\|_2^2 < \infty .$$

Now pick $B > T$ and look at $e^{(2\pi i B \gamma)} h$. This function is of exponential type; it is bounded on R^1; it is also bounded on the positive imaginary half line, seeing as

$$|e^{2\pi i (ib) B} h(ib)| = e^{-2\pi Bb} |h(ib)|$$

tends to zero as $b \uparrow \infty$. A double application of the Phragmén–Lindelöf theorem now supplies you with the bound

$$|h(Re^{i\theta})| \leqslant \text{constant} \times \exp(2\pi BR \sin \theta)$$

on the (closed) upper half-plane $0 \leqslant \theta \leqslant \pi$. Cauchy's theorem for the semi-circle of Fig. 3.1.5 is now applied to $(1 - Ai\gamma)^{-1} e^{2\pi i x \gamma} h$ for fixed $A > 0$ and $x > B$, with the result that

$$\int_{-R}^{R} \frac{e^{2\pi i x \gamma}}{1 - Ai\gamma} h(\gamma)\, d\gamma = -i \int_{0}^{\pi} \frac{\exp(2\pi i x Re^{i\theta})}{1 - AiRe^{i\theta}} h(Re^{i\theta})\, Re^{i\theta}\, d\theta .$$

This allows you to bound the modulus of the left-hand integral by a constant multiple of

$$R \int_{0}^{\pi} \frac{e^{-2\pi x R \sin \theta}}{AR - 1} e^{2\pi BR \sin \theta}\, d\theta = \frac{R}{AR - 1} \int_{0}^{\pi} e^{-2\pi (x - B) R \sin \theta}\, d\theta ,$$

which approaches zero as $R \uparrow \infty$. However, this is the same as saying that the inverse Fourier transform of $(1 - Ai\gamma)^{-1} h$ vanishes for $x > B$, and since the latter tends to h in $L^2(R^1)$ as $A \downarrow 0$, the same is true for h:

$$0 = h^\vee(x) = f(x) \frac{\sin \pi x}{\pi x}$$

for almost every $x > B$. Because $B > T$ was chosen at pleasure, $f = 0$ for almost every $x > T$; similarly, $f = 0$ for almost every $x < -T$. The proof is finished.

EXERCISE 2. Check that if f is an integral function of exponential type $T = 0$ and if $\int (1+x^2)^{-1} |f(x)|^2 dx < \infty$, then f is constant. *Hint:* $(\gamma - i)^{-1} \times [f(\gamma) - f(i)]$ is of minimal type and of class $L^2(R^1)$.

EXERCISE 3. Check that if f is summable on R^1, if $\hat{f}$ is an integral function of exponential type less than or equal to $2\pi T$, and if $B > T$, then

$$|\hat{f}(\gamma)| \leqslant \text{constant} \times e^{2\pi BR \sin\theta}$$

on the closed upper half-plane. *Hint:* Apply Phragmén–Lindelöf to $e^{2\pi i B \gamma} \hat{f}(\gamma)$.

EXERCISE 4. Use Exercise 3 to check that the Paley–Wiener theorem is also valid for summable functions. *Hint:* Check that if f is summable and if $\hat{f}$ is of type less than or equal to $2\pi T$, then

$$\lim_{R \uparrow \infty} \int_{-R}^{R} (\gamma + i)^{-1} e^{2\pi i \gamma x} \hat{f}(\gamma) \, d\gamma = 0$$

and

$$[(\gamma + i)^{-1} \hat{f}]^\vee(x) = -2\pi i \int_{x}^{\infty} \exp[2\pi(x - y)] f(y) \, dy = 0$$

for $x > T$.

Warning: For the rest of this chapter, it will be convenient to use the following variant of the Fourier integral cited at the end of Section 2.2:

$$\hat{f}(\gamma) = \int_{-\infty}^{\infty} f(x) \, e^{i\gamma x} \, dx$$

$$f(x) = \hat{f}^\vee = (2\pi)^{-1} \int_{-\infty}^{\infty} \hat{f}(\gamma) \, e^{-i\gamma x} \, d\gamma .$$

This modification may cause you a little grief for a few pages, but you'll get used to it soon. The principal thing to keep in mind is that the Plancherel identity is modified by a factor $\sqrt{2\pi}$:

$$\|\hat{f}\|_2 = \sqrt{2\pi}\,\|f\|_2!$$

3.4 HARDY FUNCTIONS

The Paley–Wiener theorem deals with integral functions of exponential type less than ∞. The next topic is functions that are only analytic in the open upper half-plane

$$R^{2+} = (\omega = a + ib : b > 0).$$

Given a point $\omega \in R^{2+}$, the function $g(\gamma) = [2\pi i(\gamma - \omega)]^{-1}$ belongs to $L^2(R^1)$, and

$$2\pi g^{*\vee*}(x) = \int_{-\infty}^{\infty} \frac{e^{i\gamma x}}{2\pi i(\gamma - \omega)}\,d\gamma = \begin{cases} e^{i\omega x} & \text{for } x > 0, \\ 0 & \text{for } x < 0, \end{cases}$$

by Exercise 3.1.13. Therefore, if $f \in L^2(R^1)$, then

$$h(\omega) = \frac{1}{2\pi i}\int_{-\infty}^{\infty} \frac{\hat{f}(\gamma)}{\gamma - \omega}\,d\gamma = (\hat{f}, g^*) = 2\pi(f, g^{*\vee}) = \int_0^{\infty} f(x)\,e^{i\omega x}\,dx,$$

or what is the same

$$h_b(a) \equiv h(a+ib) = \int_0^{\infty} f(x)\,e^{iax}\,e^{-bx}\,dx.$$

Bring in the "half-transform"

$$f^+(a) = \int_0^{\infty} f(x)\,e^{iax}\,dx.$$

Then the following facts are self-evident:

(a) h is analytic on R^{2+}.

(b) $\|h_b\|_2^2 = 2\pi\int_0^{\infty} e^{-2bx}\,|f(x)|^2\,dx$

$$\leqslant 2\pi\int_0^{\infty} |f(x)|^2\,dx = \|f^+\|_2^2 < \infty.$$

(c) $\|h_b - f^+\|_2^2 = 2\pi\int_0^{\infty} (1 - e^{-bx})^2\,|f(x)|^2\,dx \downarrow 0$ as $b \downarrow 0$.

A function h which is analytic on R^{2+} and satisfies

$$\sup_{b>0} \|h_b\|_2^2 = \sup_{b>0}\int |h(a+ib)|^2\,da < \infty$$

is called a "Hardy function" of class H^{2+}. By items (a) and (b) above,

$$h(\omega) = \frac{1}{2\pi i} \int \frac{\hat{f}(\gamma)}{\gamma - \omega} \, d\gamma = \int_0^\infty f(x) \, e^{i\omega x} \, dx$$

belongs to H^{2+} for any $f \in L^2(R^1)$. By the second integral, h is unchanged if $f = \hat{f}^\vee$ is modified on the left-hand half-line $x < 0$, so you can take $f = 0$ there if you like. The content of the following theorem of Paley and Wiener [1934] is that every $h \in H^{2+}$ arises in this way.

THEOREM 1. *A function h is of class H^{2+} iff*

$$h(\omega) = \frac{1}{2\pi i} \int \frac{\hat{f}(\gamma)}{\gamma - \omega} \, d\gamma = \int_0^\infty f(x) \, e^{i\omega x} \, dx$$

on R^{2+} for some $f \in L^2(R^1)$ vanishing on the left-hand half-line.

PROOF. The point at issue is to verify that if h is analytic on R^{2+} and if $\sup_{b>0} \|h_b\|_2 < \infty$, then

$$h(\omega) = \int_0^\infty f(x) \, e^{i\omega x} \, dx$$

for some function f with $\int_0^\infty |f|^2 < \infty$. The first step is to check that if $\omega \in R^{2+}$ and if $b > 0$, then

$$h_b(\omega) = h(\omega + ib) = \frac{1}{2\pi i} \int \frac{h_b(\gamma)}{\gamma - \omega} \, d\gamma.$$

This looks like a self-evident application of Cauchy's formula, but you do not have a sharp enough bound for h to do it in the naive way. To overcome this difficulty, let

$$k(\omega) = e^{iA\omega} A^{-1} \int_0^A h_b(\omega + y) \, dy$$

for fixed $A > 0$. Then k is analytic on R^{2+}, and

$$|k(Re^{i\theta})| \leq \frac{e^{-AR\sin\theta}}{A} \left(\int_0^A dy \right)^{1/2} \left(\int_0^A |h_b(\omega + y)|^2 \, dy \right)^{1/2}$$

$$\leq \frac{e^{-AR\sin\theta}}{\sqrt{A}} \|h_{b+R\sin\theta}\|_2$$

$$\leq \text{constant} \times \frac{e^{-AR\sin\theta}}{\sqrt{A}},$$

for $0 \leqslant \theta \leqslant \pi$, so by Cauchy's formula applied to the semicircle of Fig. 3.1.5,

$$2\pi i k(\omega) = \int_{-R}^{R} \frac{k(\gamma)}{\gamma - \omega} d\gamma + \int_{0}^{\pi} \frac{k(Re^{i\theta}) R i e^{i\theta}}{Re^{i\theta} - \omega} d\theta$$

for $R > |\omega|$. The modulus of the second integral does not exceed a constant multiple of

$$\int_{0}^{\pi} e^{-AR \sin\theta} d\theta \leqslant \text{constant} \times R^{-1},$$

so for $R \uparrow \infty$, you find the Cauchy formula

$$2\pi i k(\omega) = \int_{-\infty}^{\infty} \frac{k(\gamma) \, d\gamma}{\gamma - \omega},$$

and as $A \downarrow 0$, this goes over into the stated Cauchy formula for h_b because k approximates h_b locally uniformly and the tails of the integral are small:

$$\int_{|\gamma| > n} \frac{|k(\gamma)|}{|\gamma - \omega|} d\gamma \leqslant \|k\|_2 \left(\int_{|\gamma| > n} |\gamma - \omega|^{-2} \, d\gamma \right)^{\frac{1}{2}}$$

$$\leqslant \|h_b\|_2 \times \text{a constant multiple of } n^{-1}.$$

The rest of the proof is easy. The Cauchy formula for h_b implies that for $\omega = a + ic \in R^{2+}$,

$$h_b(\omega) = h_b(a + ic) = h_{b+c}(a) = \int_{0}^{\infty} h_b{}^{\vee}(x) e^{i\omega x} \, dx$$

$$= \int_{0}^{\infty} h_b{}^{\vee}(x) e^{iax} e^{-cx} \, dx,$$

so that

$$h_{b+c}^{\vee} = e^{-cx} h_b{}^{\vee}$$

for any $c > 0$ and $b > 0$. Therefore, the function

$$e^{bx} h_b{}^{\vee} = f$$

is independent of $b > 0$, and

$$2\pi \int_{0}^{\infty} |f|^2 = \lim_{b \downarrow 0} 2\pi \int_{0}^{\infty} e^{-2bx} |f|^2$$

$$= \lim_{b \downarrow 0} 2\pi \|h_b{}^{\vee}\|_2^2$$

$$= \lim_{b \downarrow 0} \|h_b\|_2^2 < \infty.$$

To finish the proof, you have only to make $b \downarrow 0$ in the formula

$$h(\omega + ib) = h_b(\omega) = \int_0^\infty f(x)\, e^{i\omega x}\, e^{-bx}\, dx,$$

keeping $\omega = a + ic$ fixed.

H^{2+} can now be identified with $\mathsf{L}^2[0, \infty)^\wedge$. The point is that for $f = 0$ on the left half-line,

$$h(\omega) = \frac{1}{2\pi i} \int \frac{\hat{f}(\gamma)}{\gamma - \omega}\, d\gamma = \int_0^\infty f(x)\, e^{i\omega x}\, dx$$

can be thought of as an analytic extension to R^{2+} of $\lim_{b \downarrow 0} h_b = \hat{f} = f^+$, so that you can permit yourself the luxury of confusing h and $\hat{f}$ and speak also of $\hat{f}$ as a "Hardy function"; in short, you can write

$$\mathsf{H}^{2+} = (\hat{f} \in L^2(R^1): \check{f} = f = 0 \quad \text{for} \quad x < 0).$$

The map

$$\hat{f} \to f^+ = \lim_{b \downarrow 0} \frac{1}{2\pi i} \int \frac{\hat{f}(\gamma)}{\gamma - \omega}\, d\gamma = \int_0^\infty f(x)\, e^{iax}\, dx,$$

for the general $\hat{f} \in L^2(R^1)$ now appears as projection of $\hat{f}$ onto H^{2+}. The allied map

$$\hat{f} \to f^- = -\lim_{b \uparrow 0} \frac{1}{2\pi i} \int \frac{\hat{f}(\gamma)}{\gamma - \omega}\, d\gamma = \int_{-\infty}^0 f(x)\, e^{iax}\, dx$$

is the projection onto the class H^{2-} of Hardy functions on the *lower* half-plane:

$$\mathsf{H}^{2-} = (\hat{f} \in L^2(R^1): f = \check{f} = 0 \quad \text{for} \quad x > 0).$$

In particular, the perpendicular splitting $L^2(R^1) = L^2(-\infty, 0] \oplus L^2[0, \infty)$ is mapped by $\wedge$ into the dual perpendicular splitting $L^2(R^1) = \mathsf{H}^{2-} \oplus \mathsf{H}^{2+}$.

Amplification: The observation just above yields the so-called Poisson formula for H^{2+} as a bonus: If $f \in L^2(R^1)$ vanishes on the left half-line, then

$$\hat{f}(\omega) = \frac{1}{2\pi i} \int \frac{\hat{f}(\gamma)}{\gamma - \omega}\, d\gamma$$

$$= \frac{1}{2\pi i} \int \frac{\hat{f}(\gamma)}{\gamma - \omega}\, d\gamma - \frac{1}{2\pi i} \int \frac{\hat{f}(\gamma)}{\gamma - \omega^*}\, d\gamma$$

$$= \frac{b}{\pi} \int \frac{\hat{f}(\gamma)}{(\gamma - a)^2 + b^2}\, d\gamma$$

for every $\omega = a + ib \in R^{2+}$.

EXERCISE 1. Check that a rational function $\hat{f} \in L^2(R^1)$ is Hardy iff all its poles lie in the open lower half-plane. *Hint:* Use Cauchy's theorem to prove that $f = \hat{f}^{\vee} = 0$ for $x < 0$ if the poles lie in the lower half-plane.

EXERCISE 2. Check that the functions

$$e_n = \frac{1}{\sqrt{\pi}(i\gamma - 1)} \left(\frac{i\gamma + 1}{i\gamma - 1} \right)^n, \qquad n = 0, \pm 1, \pm 2, \ldots$$

form a unit-perpendicular family. *Hint:* Use Cauchy's formula.

EXERCISE 3. Check that e_n: $n \geq 0$ spans H^{2+}, while e_n: $n < 0$ spans H^{2-}. *Hint:* Use Exercise 1 to check that e_n belongs to H^{2+} for $n \geq 0$ and to H^{2-} for $n < 0$. Then prove that the whole family spans $L^2(R^1)$. The fact that $(i\gamma - 1)^{-1}(i\gamma + 1)$ maps R^1 onto the unit circle is helpful.

EXERCISE 4. Compute the inverse transforms of e_0, e_1, and e_2. *Answer:* $e_0^{\vee} = e^{-x}$, $e_1^{\vee} = e^{-x}(2x - 1)$, $e_2^{\vee} = e^{-x}(2x^2 - 4x + 1)$ up to factors of $\pi^{-\frac{1}{2}}$. These are the first three "Laguerre functions."

EXERCISE 5. Pick a nonnegative summable weight function Δ on R^1. Check that the span A of the family $e^{i\gamma x}$: $x \geq 0$ in $L^2(\Delta) = L^2(R^1, \Delta(\gamma) \, d\gamma)$ is the same as the span B of e_n: $n \geq 0$. *Hint:* B is the same as the span of $(i\gamma - 1)^{-n}$: $n \geq 1$. Approximate the integral

$$\frac{1}{(1 - i\gamma)^n} = \frac{1}{(n-1)!} \int_0^{\infty} x^{n-1} e^{-x} e^{i\gamma x} \, dx$$

by Riemann sums to verify that $B \subset A$. Now pick $h \in A$ perpendicular to B. Check that

$$\int \frac{h(\gamma)}{\gamma - \omega} \Delta(\gamma) \, d\gamma$$

is analytic on R^{2+} and vanishes there:

$$\int \frac{h(\gamma)}{\gamma - \omega} \Delta(\gamma) \, d\gamma = \sum_{n=0}^{\infty} (\omega - i)^n \int \frac{h(\gamma)}{(\gamma - i)^{n+1}} \Delta(\gamma) \, d\gamma = 0$$

for $|\omega - i| < 1$. Deduce that h is perpendicular to A and therefore $h = 0$. *Warning:* $h\Delta$ is summable but need not be of class $L^2(R^1)$.

A very striking and important fact about Hardy functions is embodied in the following theorem.

THEOREM 2.   *For any Hardy function* $h \not\equiv 0$

$$\int_{-\infty}^{+\infty} \frac{\log|h(\gamma)|}{1+\gamma^2}\, d\gamma > -\infty .$$

Amplification: $\log|h| \leqslant |h|$, so $(1+\gamma^2)^{-1}\log|h|$ is bounded above by a summable function, and $\int(1+\gamma^2)^{-1}\log|h|$ can only diverge to $-\infty$. The significance of this theorem is explained in Section 3.5; see also Theorem 3.

PROOF OF THEOREM 2.   Pick a rational Hardy function h with no roots on the line and let h^0 be the rational function obtained by swinging all the roots of $h=0$ that fall inside R^{2+} to their conjugate positions in the open lower half plane. Then

$$\log|h^0(\gamma)| = \sum \log|\gamma-\alpha| - \sum \log|\gamma-\beta| + \text{a constant},$$

the first sum running over the roots α of h^0 and the second over its poles β, all of which lie in R^{2-}. But for any $\omega \in R^{2-}$,

$$\frac{1}{\pi}\int \frac{\log|\gamma-\omega|}{1+\gamma^2}\, d\gamma = \log|i-\omega|$$

[see Exercise 6], so

$$\frac{1}{\pi}\int \frac{\log|h(\gamma)|}{1+\gamma^2}\, d\gamma = \frac{1}{\pi}\int \frac{\log|h^0(\gamma)|}{1+\gamma^2}\, d\gamma$$

$$= \log|h^0(i)|$$

$$\geqslant \log|h(i)|,$$

in view of $\pi^{-1}\int(1+\gamma^2)^{-1}=1$ [see Exercise 3.1.11]. Now the same holds for *any* Hardy function h. To see this, use the functions $e_n\colon n\geqslant 0$ of Exercise 3 to approximate h by rational Hardy functions h_n with no roots on the line, in such a way that $\lim_{n\uparrow\infty}\|h_n-h\|_2=0$ and also $\lim_{n\uparrow\infty}h_n=h$ at almost every point of the line. Then

$$\lim_{n\uparrow\infty}h_n(i) = \lim_{n\uparrow\infty}\frac{1}{2\pi i}\int\frac{h_n(\gamma)}{\gamma-i}\, d\gamma = \frac{1}{2\pi i}\int\frac{h(\gamma)}{\gamma-i}\, d\gamma = h(i).$$

But also, if you define

$$\log^+ x = \begin{cases} \log x & \text{for } 1\leqslant x\leqslant\infty, \\ 0 & \text{for } x<1, \end{cases}$$

$$\log^- x = \begin{cases} -\log x & \text{for } 0\leqslant x\leqslant 1, \\ 0 & \text{for } x>1, \end{cases}$$

you will have

$$\log|h(i)| = \lim_{n\uparrow\infty}\log|h_n(i)| \leqslant \lim_{n\uparrow\infty}\frac{1}{\pi}\int\frac{\log|h_n|}{1+\gamma^2}$$

$$= \lim_{n\uparrow\infty}\left[\frac{1}{\pi}\int\frac{\log^+|h_n|}{1+\gamma^2} - \frac{1}{\pi}\int\frac{\log^-|h_n|}{1+\gamma^2}\right]$$

$$\leqslant \frac{1}{\pi}\int\frac{\log^+|h|}{1+\gamma^2} + \limsup_{n\uparrow\infty}\frac{1}{\pi}\int\frac{|h-h_n|}{1+\gamma^2} - \liminf_{n\uparrow\infty}\frac{1}{\pi}\int\frac{\log^-|h_n|}{1+\gamma^2}$$

$$\leqslant \frac{1}{\pi}\int\frac{\log^+|h|}{1+\gamma^2} - \frac{1}{\pi}\int\frac{\log^-|h|}{1+\gamma^2}$$

$$\leqslant \frac{1}{\pi}\int\frac{\log|h|}{1+\gamma^2},$$

as stated. The estimate $|\log^+ x - \log^+ y| \leqslant |x-y|$ is used to get the second integral in line three and Fatou's lemma, applied to the third integral in that line, brings you to line four. You see now that $(1+\gamma^2)^{-1}\log|h|$ is summable if $h(i) \neq 0$. In general, if $h \not\equiv 0$, then the degree n of its root at $\gamma = i$ is less than ∞, and the above estimate may be applied to the Hardy function

$$k = \left(\frac{\gamma+i}{\gamma-i}\right)^n h$$

with the result that

$$\frac{1}{\pi}\int\frac{\log|h|}{1+\gamma^2} = \frac{1}{\pi}\int\frac{\log|k|}{1+\gamma^2} \geqslant \log|k(i)| > -\infty.$$

The proof is finished.

EXERCISE 6

$$\frac{1}{\pi}\int\frac{\log|\gamma-\omega|}{1+\gamma^2}\,d\gamma = \log|i-\omega|$$

for $\omega \in R^{2^-}$. *Hint:*   Use Cauchy's formula to evaluate

$$\frac{1}{2\pi i}\int\frac{\log(\gamma-\omega)}{\gamma+i}\frac{d\gamma}{\gamma-i}$$

by integrating over the semicircle of Fig. 3.1.5; see Exercise 3.1.3 for the meaning of the logarithm. Then take real parts.

There is a beautiful converse to Theorem 2.

THEOREM 3.   *Any nonnegative function $G \in L^2(R^1)$ with $\int (1+\gamma^2)^{-1} \times \log G > -\infty$ is the modulus of some Hardy function h, that is to say, $G = |h|$ a.e. on the line.*

Before explaining the proof, it is convenient to prepare

EXERCISE 7.   Prove "Jensen's inequality": For any nonnegative weight function Δ on an interval Q with total mass $\int_Q \Delta(\gamma) \, d\gamma = 1$,

$$\int_Q [\log k(\gamma)] \Delta(\gamma) \, d\gamma \leqslant \log\left(\int_Q k(\gamma) \Delta(\gamma) \, d\gamma\right)$$

for nonnegative functions k.   *Hint:*   Begin from the familiar inequality between the arithmetical and geometrical means of the positive numbers $x_1, \ldots, x_n$:

$$(x_1 \cdots x_n)^{1/n} \leqslant n^{-1}(x_1 + \cdots + x_n).$$

Deduce that Jensen's inequality holds in the special case that Q is divided into n nonoverlapping intervals I on each of which k is constant and $\int_I \Delta = 1/n$.

PROOF OF THEOREM 3.   Define an analytic function of $\omega \in R^{2+}$ by the formula

$$h(\omega) = \exp\left[\frac{1}{\pi i}\int_{-\infty}^{\infty} \frac{\gamma\omega+1}{\gamma-\omega} \log G(\gamma) \frac{d\gamma}{1+\gamma^2}\right],$$

using the condition $\int (1+\gamma^2)^{-1}|\log G| < \infty$ to ensure that the integral converges nicely. Then for fixed $\omega = a + ib$,

$$|h(\omega)|^2 = \exp\left[\frac{b}{\pi}\int_{-\infty}^{\infty} \frac{\log G^2(\gamma)}{(\gamma-a)^2+b^2} \, d\gamma\right] = \exp\left[\int (\log G^2(\gamma)) \Delta(\gamma) \, d\gamma\right]$$

with a self-evident notation. Because $\Delta \geqslant 0$ and $\int \Delta = 1$ [see Exercise 3.1.11], you may apply Jensen's inequality [Exercise 7]:

$$|h(\omega)|^2 \leqslant \exp\left[\log\left(\int G^2 \Delta\right)\right] = \int G^2 \Delta = \frac{b}{\pi}\int \frac{G^2(\gamma)}{(\gamma-a)^2+b^2} \, d\gamma,$$

and then you can integrate over a to obtain

$$\|h_b\|_2^2 = \int \frac{b}{\pi}\int \frac{G^2(\gamma)}{(\gamma-a)^2+b^2} \, d\gamma \, da = \int G^2(\gamma) \frac{b}{\pi}\int \frac{da}{(\gamma-a)^2+b^2} \, d\gamma$$

$$= \int G^2(\gamma) \, d\gamma = \|G\|_2^2 < \infty,$$

independently of $b > 0$. This proves that $h \in H^{2+}$. In particular, $\lim_{b\downarrow 0} h_b = h_{0+}$ exists in $L^2(R^1)$, and so to finish the proof, you have only to check that

$|h_{0+}| = G$, a.e.; in fact, since

$$\int |h_{0+}|^2 = \lim_{b \downarrow 0} \int |h_b|^2 \leqslant \int G^2,$$

it is even enough to check that $|h_{0+}| \geqslant G$, a.e. To do this, pick a nonnegative compact function k from $C(R^1)$ and make $b \downarrow 0$ in such a way that $\lim_{b \downarrow 0} h_b = h_{0+}$, a.e., in the ordinary numerical sense. Then

$$\liminf_{b \downarrow 0} \int k \log^- |h_b| \geqslant \int k \log^- |h_{0+}|$$

by Fatou's lemma, and

$$\lim_{b \downarrow 0} \int k \log^+ |h_b| = \int k \log^+ |h_{0+}|,$$

much as in the proof of Theorem 2. Therefore

$$\limsup_{b \downarrow 0} \int k \log |h_b| \leqslant \int k \log |h_{0+}|.$$

At the same time,

$$\int k \log |h_b| = \int k(a) \frac{b}{\pi} \int \frac{\log G(\gamma)}{(\gamma - a)^2 + b^2} \, d\gamma \, da$$

$$= \int \log G(\gamma) \frac{b}{\pi} \int \frac{k(a)}{(\gamma - a)^2 + b^2} \, da \, d\gamma$$

$$= \int \log G(\gamma) \frac{1}{\pi} \int \frac{k(\gamma + ab)}{1 + a^2} \, da \, d\gamma$$

tends to $\int k \log G$ as $b \downarrow 0$, by dominated convergence, seeing as the inside integral is bounded by a constant multiple of $(1 + \gamma^2)^{-1}$ and $(1 + \gamma^2)^{-1} \log G$ is summable. But then

$$\int k \log |h_{0+}| \geqslant \limsup_{b \downarrow 0} \int k \log |h_b| = \int k \log G,$$

and $|h_{0+}| \geqslant G$, as advertised. The proof is finished.

The subject of Hardy functions is not as commonly known as it should be. A few of its deeper aspects will be developed in Sections 3.5–3.9. The advanced student will find additional information in Beurling [1949], Carleman [1944], Duren [1970], Helson [1964], Hoffman [1962], and Paley and Wiener [1934].

3.5* HARDY FUNCTIONS AND FILTERS

Hardy functions have important applications to electrical engineering. Think of a "filter" K acting upon an input signal f, as in Fig. 1, with $k \in L^2(R^1)$. The response of K to a (complex) sinusoid $f(t) = e^{-i\gamma t}$ is

$$Kf(t) = \int_{-\infty}^{\infty} k(t-s) e^{-i\gamma s} ds = \hat{k}(\gamma) e^{-i\gamma t} = \hat{k}(\gamma) f(t),$$

so $G(\gamma) = |\hat{k}(\gamma)|$ is the "gain" [alias the amplification] of K and the argument of $\hat{k}(\gamma)$ is the "phase shift."

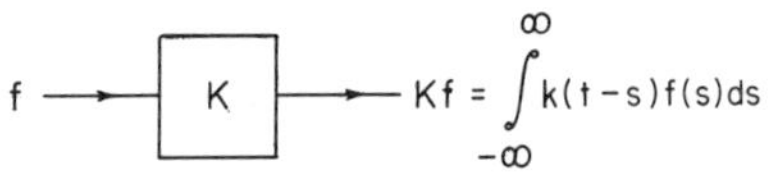

$$f \longrightarrow \boxed{K} \longrightarrow Kf = \int_{-\infty}^{\infty} k(t-s)f(s)ds$$

FIGURE 1

An actual filter K must have $k(t) = 0$ for $t < 0$; otherwise, K would respond to the *future* of the signal. This means that $\hat{k}$ has to be a Hardy function; especially, the gain $G = |\hat{k}|$ must satisfy

$$\int_{-\infty}^{\infty} \frac{\log G(\gamma)}{1+\gamma^2} d\gamma > -\infty,$$

by Theorem 3.4.2. The converse is also true: by Theorem 3.4.3, if $G \geqslant 0$, if $\int G^2 < \infty$, and if $\int (1+\gamma^2)^{-1} \log G > -\infty$, then G is the gain of a Hardy filter.

Under the additional condition that k is real $[\hat{k}^* = \hat{k}(-\cdot)$ on the real line], it is easy to make an actual circuit to do the filtering, or approximately so, as will now be explained.

To begin with, $\hat{k}$ can be expanded in a Fourier series relative to the unit-perpendicular family

$$e_n = \frac{1}{\sqrt{\pi}(i\gamma - 1)} \left(\frac{i\gamma + 1}{i\gamma - 1} \right)^n$$

of Exercise 3.4.2. The condition $\hat{k}^* = \hat{k}(-\cdot)$ means that the coefficients are real, and since $e_n \in H^{2-}$ for $n < 0$ while $e_n \in H^{2+}$ for $n \geqslant 0$, the terms with negative index are absent:

$$\hat{k} = \sum_{n=0}^{\infty} k_n e_n.$$

The circuit for K is depicted in terms of $\hat{k}$ in Fig. 2.

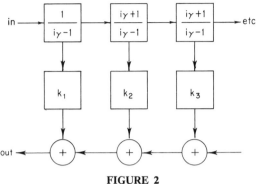

FIGURE 2

The circled plus $\oplus$ signifies addition of signals, while the boxes $\boxed{k_n}$ indicate multiplication by the (known) coefficients (amplification/attenuation). The box $\boxed{(i\gamma-1)^{-1}}$ can be mimicked by the simple circuit of Fig. 3

FIGURE 3

with unit resistance and capacitance and input voltage e. The current[1] j in this circuit satisfies

$$\frac{dj}{dt}+j = \frac{de}{dt},$$

or what is the same,

$$(-i\gamma+1)\,\hat{j} = -i\gamma\hat{e},$$

while the voltage $e_1(t)=\int_{-\infty}^{t} j$ across the capacitance satisfies $-i\gamma\hat{e}_1 = \hat{j}$. Therefore,

$$-\hat{e}_1 = (i\gamma-1)^{-1}\hat{e},$$

which implies that the box $\boxed{(i\gamma-1)^{-1}}$ can be simulated by measuring $-e_1$ (Fig. 4). At the same time, the voltage $e_2 = j$ across the resistance satisfies

$$\hat{e}_2 = \frac{i\gamma}{i\gamma-1}\,\hat{e},$$

---

[1] Electrical engineers: Beware of the notation.

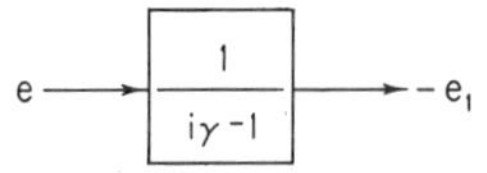

FIGURE 4

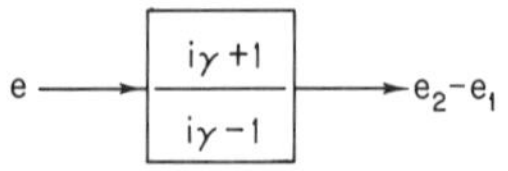

FIGURE 5

so the action of the box $\boxed{(i\gamma-1)^{-1}(i\gamma+1)}$ is to subtract e_1 from e_2 (Fig. 5). The circuit of Fig. 2 can now be put together out of such pieces, provided the expansion of $\hat{k}$ breaks off at some index $n < \infty$; for additional information about such networks, see Lee [1960, pp. 481–501].

The rest of this section is devoted to deeper properties of *rational* Hardy filters; excepting Subsection 5, everything carries over to the general Hardy filter, though the proofs become technically more complicated.

1. Inner and Outer Filters

To any rational Hardy function $\hat{k}$, you can associate the function k^0, introduced in the proof of Theorem 3.4.2, which is obtained by swinging all the roots of $\hat{k} = 0$ that fall in the open upper half-plane to their conjugate places in the lower half-plane. k^0 is still Hardy, it is root free on R^{2+}, it has the same gain as $\hat{k}$ $[|k^0| = |\hat{k}|$ on the axis], and the ratio $j = \hat{k}/k^0$ is of modulus < 1 on R^{2+} [unless $\hat{k} = k^0$] and of modulus $= 1$ on the axis. k^0 is the "outer" part of $\hat{k}$; the "inner" part j is a "phase-shift" [gain $\equiv 1$].

EXERCISE 1.   Show that k^0 is completely specified by its gain $|k^0| = |\hat{k}|$ on the axis. *Hint:* $k^0/|\hat{k}|$ is a rational function.

EXERCISE 2.   A nonnegative rational function G with $\int G^2 < \infty$ is always the gain of a rational Hardy filter $[\int(1+\gamma^2)^{-1} \log G > -\infty$ is automatic]. Give an explicit recipe for $\hat{k}$ in terms of G. *Hint:* $\hat{k}/G$ is a rational function.

EXERCISE 3.   Show that $\hat{k}$ is an outer function $[\hat{k} = k^0]$ iff

$$\log|\hat{k}(i)| = \frac{1}{\pi} \int_{-\infty}^{\infty} \frac{\log|\hat{k}(\gamma)|}{1+\gamma^2} \, d\gamma .$$

Hint: See Exercise 3.4.6.

2. Outer Filters are Amenable

The response of K to a unit "spike" arriving at time T is $k_T(t) = k(t-T)$. The purpose of this article is to prove that *the family* $k_T: T \geqslant 0$ *spans* $L^2[0, \infty)$ *iff* $\hat{k}$ *is outer.* This may be intrepreted as saying that an outer filter is "amenable": if you don't like the output, you can alter it at pleasure by feeding in a suitable train of "spikes." Any other kind of filter may balk. The Fourier transform of k_T is $e^{i\gamma T} \hat{k}$, so an alternative statement is that *the family* $e^{i\gamma T} \hat{k}: T \geqslant 0$ *spans* $H^{2+} = L^2[0, \infty)^{\wedge}$ *iff* $\hat{k}$ *is outer.*

PROOF.   Pick $h \in H^{2+}$ perpendicular to $e^{i\gamma t} \hat{k}$ for every $t \geqslant 0$. Then

$$\int h(\gamma) \, \hat{k}^*(\gamma) \, e^{-i\gamma t} d\gamma = 0$$

for $t \geqslant 0$, so $h\hat{k}^* \in H^{2-}$, and

$$\frac{1}{2\pi i} \int \frac{h(\gamma) \, \hat{k}^*(\gamma)}{\gamma - \omega} \, d\gamma = 0$$

for any $\omega \in R^{2+}$. Now suppose, for simplicity's sake, that the poles $\omega_1^*, \ldots, \omega_n^* \in R^{2-}$ of $\hat{k}$ are simple. Then, by Exercise 3.1.20,

$$\hat{k} = \sum_{i=1}^{n} c_i(\gamma - \omega_i^*)^{-1},$$

and by Cauchy's formula,

$$0 = \frac{1}{2\pi i} \int \frac{h(\gamma) \, \hat{k}^*(\gamma)}{\gamma - \omega} \, d\gamma$$

$$= \sum_{i=1}^{n} c_i^*(\omega - \omega_i)^{-1} \frac{1}{2\pi i} \int h(\gamma) \left(\frac{1}{\gamma - \omega} - \frac{1}{\gamma - \omega_i} \right) d\gamma$$

$$= \sum_{i=1}^{n} c_i^*(\omega - \omega_i)^{-1} [h(\omega) - h(\omega_i)]$$

for $\omega \in R^{2+}$. This identity may be expressed as $hk^{\#} = g$ with

$$k^{\#}(\omega) = \sum_{i=1}^{n} c_i^*(\omega - \omega_i)^{-1}$$

and

$$g(\omega) = \sum_{i=1}^{n} c_i^* h(\omega_i) (\omega - \omega_i)^{-1},$$

and you may now draw several important conclusions. To begin with, $h = g/k^{\#}$ is a rational function, so by Exercise 3.4.1, its poles lie in R^{2-}. Because g is pole-free in R^{2-}, this forces $k^{\#}$ to vanish at all the lower half-plane poles of h. However, this is not possible if $\hat{k}$ is outer: Namely, the roots of $k^{\#}$ are the conjugates of the roots of $\hat{k}$ and therefore lie in the *closed* upper half-plane. The only escape from this dilemma is to admit that h has no poles at all. Then it is a polynomial of class $L^2(R^1)$. As such, $h = 0$. The case of multiple poles is much the same and is left to the reader. This finishes the proof that $e^{i\gamma}\hat{k}: t \geqslant 0$ spans H^{2+} if $\hat{k}$ is outer. As to the converse, if $\hat{k}$ is not outer, it has a root $\omega \in R^{2+}$, and $(\gamma - \omega^*)^{-1} \in H^{2+}$ is perpendicular to $e^{i\gamma t}\hat{k}$ for every $t \geqslant 0$:

$$\frac{1}{2\pi i} \int \frac{e^{i\gamma t}\hat{k}(\gamma)}{\gamma - \omega}\, d\gamma = e^{i\omega t}\hat{k}(\omega) = 0.$$

EXERCISE 4.   Check that $e^{i\gamma t}\hat{k}$ belongs to $j H^{2+}$ for any $t \geqslant 0$, j being the inner factor of $\hat{k}$, and that the span of the family $e^{i\gamma t}\hat{k}: t \geqslant 0$ is exactly $j H^{2+}$.

3.   Outer Filters are Nonsingular

The fact alluded to is that *if K is an outer filter and if the input f is of class* $L^2(R^1)$, *then Kf vanishes for* $t \leqslant 0$ *only if f does so; only outer filters have this property.* To put it another way, an outer filter is a "perfect detector"; any other kind of filter knocks out some nontrivial signal.

Amplification: $\hat{k}$ has no poles on the line, so $\|\hat{k}\|_\infty < \infty$, and $\|\hat{k}\hat{f}\|_2 < \|\hat{k}\|_\infty \sqrt{2\pi}\|f\|_2 < \infty$. Therefore, $Kf = [\hat{k}\hat{f}]^\vee$ makes sense.

PROOF. $\hat{k}\hat{f} = h \in H^{2+}$ if Kf vanishes for $t \leqslant 0$. Now suppose that $\hat{k}$ is an outer function; as such, its roots $\alpha_1, \ldots, \alpha_m$ and its poles $\beta_1, \ldots, \beta_n$ $(n > m)$ lie in the closed/open lower half-plane, respectively. The point at issue is whether

$$\hat{f}(\gamma) = \frac{h(\gamma)}{\hat{k}(\gamma)} = \text{constant} \times h(\gamma) \frac{\prod_{i=1}^{n}(\gamma - \beta_i)}{\prod_{i=1}^{m}(\gamma - \alpha_i)}$$

belongs to H^{2+}. But this is easy to see. To begin with $\hat{f}$ is analytic on R^{2+}. Besides, as you can see by inspection,

$$\hat{f}_\varepsilon(\gamma) = \text{constant} \times h(\gamma) \frac{\prod_{i=1}^{n}(\gamma - \beta_i)}{\prod_{i=1}^{m}(\gamma + i\varepsilon - \alpha_i)(1 - i\varepsilon\gamma)^{n-m}}$$

belongs to H^{2+} for any $\varepsilon > 0$, and $|\hat{f}_\varepsilon| \leqslant |\hat{f}|$ on the line. Therefore, by Fatou's lemma,

$$\int |\hat{f}(a+ib)|^2\, da \leqslant \liminf_{\varepsilon \downarrow 0} \int |\hat{f}_\varepsilon(a+ib)|^2\, da$$

$$\leqslant \liminf_{\varepsilon \downarrow 0} \int |\hat{f}_\varepsilon(a)|^2\, da \leqslant \|\hat{f}\|^2$$

for any $b > 0$. This finishes one half of the proof. As to the other half, if $\hat{k}$ is not outer, then it has a root $\omega \in R^{2+}$, and $\hat{f} = (\gamma - \omega)^{-1}$ provides the necessary counterexample: Namely, $\hat{k}\hat{f}$ belongs to H^{2+}, but $\hat{f}$ does not.

4.   Outer Filters Have the Biggest Power

The fact alluded to is that *among all filters K with the same gain* $|\hat{k}|$, *the outer filter* k^0 *makes* $\int_0^T |k|^2$ *as big as possible, and it does so for every* $T > 0$, *simultaneously:*

$$\int_0^T |(k^0)^{\vee}|^2 \, dt \geqslant \int_0^T |k|^2 \, dt$$

for every $T > 0$.

This beautiful fact is due to Robinson [1962]; it does not seem to have found its way into the purely mathematical literature.

PROOF.   The simplest way of proving this is to look at the projection

$$\int_{-\infty}^0 [e^{-i\gamma T}\hat{k}]^{\vee} e^{i\gamma t} \, dt = \int_0^T e^{i\gamma(t-T)} k(t) \, dt$$

of $e^{-i\gamma T}\hat{k}$ upon H^{2-}. By the Plancherel identity applied to the second integral, the length of this projection is

$$\left(2\pi \int_0^T |k(t)|^2 \, dt \right)^{\frac{1}{2}}.$$

The projection of $e^{-i\gamma T}\hat{k} = j(e^{-i\gamma T}k^0)$ upon the bigger space jH^{2-} is naturally longer. But the latter is of the *same* length as the projection of $e^{-i\gamma T}k^0$ upon H^{2-} since the gain of j is 1, i.e., $\int_0^T |k(t)|^2 \, dt$ is increased if $\hat{k}$ is replaced by k^0. The proof is finished. If you don't understand it do Exercise 6 and then try again.

EXERCISE 5.   Prove the converse: If, for fixed gain, $\int_0^T |k(t)|^2 \, dt$ is as big as possible for every $T > 0$, simultaneously, then $\hat{k}$ is outer.

EXERCISE 6.   Show that $Q jf = jPf$ if Pf [Qf] is the projection of $f \in L^2(R^1)$ onto H^{2-} [jH^{2-}]. *Hint:* $(Q jf, jg) = (jf, jg) = (f, g) = (Pf, g)$ for every $g \in H^{2-}$.

5.   Outer Filters Have the Smallest Delay

The "phase lag" (alias the "delay") $\varphi(\gamma)$ of the filter K is defined by $\hat{k}(\gamma) = |\hat{k}(\gamma)| e^{i\varphi(\gamma)}$, with the understanding that, in passing a real root of

$\hat{k} = 0$, φ jumps as if the root lay just a hair below the axis. Otherwise it is continuous. This specifies φ completely, up to an additive integral multiple of 2π, independent of γ. The name "phase lag" is justified by the fact that for $\gamma > 0$

$$Ke^{-i\gamma t} = \hat{k}(\gamma)\, e^{-i\gamma t} = |\hat{k}(\gamma)| \times e^{-i[\gamma t - \varphi(\gamma)]}.$$

Bring in the roots $\omega_1, \ldots, \omega_n$ of $\hat{k} = 0$ in the open upper half-plane. You have

$$\frac{\hat{k}(\gamma)}{k^0(\gamma)} = e^{i[\varphi(\gamma) - \varphi^0(\gamma)]} = \prod_{l=1}^{n} \frac{\gamma - \omega_l}{\gamma - \omega_l^*} = \exp\left[2i \sum_{l=1}^{n} \theta_l(\gamma)\right]$$

with a self-evident notation, $0 < \theta_l < \pi$ being the angle depicted in Fig. 6.

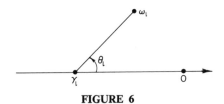

FIGURE 6

Consequently,

$$\varphi(\gamma) - \varphi(-\infty) = \varphi^0(\gamma) - \varphi^0(-\infty) + 2 \sum_{l=1}^{n} [\theta_l(\gamma) - \theta_l(-\infty)]$$

$$> \varphi^0(\gamma) - \varphi^0(-\infty)$$

for every γ unless $\hat{k}$ is outer. It is in this sense that, for fixed gain, the outer filter has the smallest delay. See Robinson [1962] for additional information on this subject.

3.6* WIENER–HOPF FACTORIZATION: MILNE'S EQUATION

A typical application of the Fourier transform in the complex domain is to Milne's equation:

$$f(x) = \int_0^\infty k(x-y)\, f(y)\, dy, \qquad x > 0,$$

in which the kernel k is defined by the integral

$$k(x) = \frac{1}{2} \int_{|x|}^\infty e^{-y}\, \frac{dy}{y} = \frac{1}{2} \int_0^{\pi/2} e^{-|x|\sec\varphi} \tan\varphi\, d\varphi.$$

The method of solution employed below is due to Hopf and Wiener [1931] and bears their names. The problem arises from a model of (equilibrium) radiation in stars, as will now be explained. The purely mathematical reader can skip directly to Subsection 2.

1.   Radiative Equilibrium

Think of the star as so big that its curvature can be neglected and introduce, as coordinates, the depth $0 \leqslant x < \infty$ into the stellar interior and the inclination $0 \leqslant \varphi \leqslant \pi$ to the downward direction, as in Fig. 1.

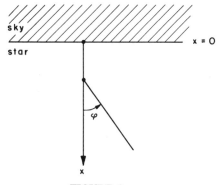

FIGURE 1

The distribution of radiation as regards depth and inclination is described by a "radiation density" $e = e(x, \varphi)$ so that the amount of radiation in a slice $a \leqslant x \leqslant b$ traveling at inclinations $\alpha \leqslant \varphi \leqslant \beta$ is

$$\int_a^b dx \int_\alpha^\beta e(x, \varphi) \sin \varphi \, d\varphi.$$

The equilibrium is produced and maintained by streaming (at speed 1, say) and by scattering, and a detailed balancing of these two mechanisms leads to the following law:

$$\cos \varphi \, \frac{\partial e}{\partial x}(x, \varphi) + e(x, \varphi) = \frac{1}{2} \int_0^\pi e(x, \psi) \sin \psi \, d\psi;$$

see Hopf [1934] for the actual derivation. Milne's problem is to compute the angular distribution of radiation at the stellar surface:

$$\frac{e(0, \varphi)}{\frac{1}{2} \int_0^\pi e(0, \psi) \sin \psi \, d\psi},$$

under the condition that no radiation is coming in from the sky [$e(0, \varphi) = 0$

for $0 \leqslant \varphi \leqslant \pi/2$]; this is the so-called "law of darkening." To do this, you bring in the radiation intensity

$$f(x) = \tfrac{1}{2} \int_0^\pi e(x, \psi) \sin \psi \, d\psi$$

and solve

$$\frac{\partial e}{\partial x} + \sec \varphi \, e = e^{-x \sec \varphi} \frac{\partial (e^{x \sec \varphi} e)}{\partial x} = \sec \varphi f$$

for $\varphi < \pi/2$ and for $\varphi > \pi/2$, separately. This is an easy integration, aside from a possible contribution from $x = \infty$, and as the latter turns out to be absent, you find that

$$e(x, \varphi) = \sec \varphi \int_0^x e^{(y - x) \sec \varphi} f(y) \, dy \qquad \text{for} \quad \varphi < \pi/2,$$

$$= -\sec \varphi \int_x^\infty e^{(y - x) \sec \varphi} f(y) \, dy \qquad \text{for} \quad \varphi > \pi/2.$$

The angular distribution of the outward radiation of the stellar surface is now computed from f by means of the second of these two formulas, and f itself is seen to be a solution of Milne's equation:

$$f(x) = \tfrac{1}{2} \int_0^{\pi/2} e(x, \psi) \sin \psi \, d\psi + \tfrac{1}{2} \int_{\pi/2}^\pi e(x, \psi) \sin \psi \, d\psi$$

$$= \tfrac{1}{2} \int_0^{\pi/2} \left[\sec \psi \int_0^x e^{(y - x) \sec \psi} f(y) \, dy \right] \sin \psi \, d\psi$$

$$+ \tfrac{1}{2} \int_{\pi/2}^\pi \left[-\sec \psi \int_x^\infty e^{(y - x) \sec \psi} f(y) \, dy \right] \sin \psi \, d\psi$$

$$= \tfrac{1}{2} \int_0^\infty f(y) \int_0^{\pi/2} e^{-|x - y| \sec \psi} \tan \psi \, d\psi \, dy$$

$$= \int_0^\infty k(x - y) f(y) \, dy.$$

The Fourier transform in the half-plane comes into play in solving for f.

2. A Simpler Problem

To understand the plan of attack, look first at the simpler problem

$$f(x) = \tfrac{1}{2} \int_0^\infty e^{-|x - y|} f(y) \, dy, \qquad x \geqslant 0.$$

As it happens, it is easy to solve this problem directly.

EXERCISE 1. Check that the only solutions are constant multiples of $1+x$. *Hint:* Differentiate both sides of the integral equation twice with respect to x.

Now to illustrate the method of Hopf and Wiener, put $f(x) = 0$ for $x < 0$ and set $k(x) = \frac{1}{2}e^{-|x|}$. The problem is to find a solution of $f = k \circ f$ for $x \geqslant 0$. In view of Exercise 1, it is pointless to search for solutions which decay at ∞. Notice, however, that the integral

$$\int_0^\infty e^{-|x-y|} f(y)\, dy = e^{-x} \int_0^x e^y f(y)\, dy + e^x \int_x^\infty e^{-y} f(y)\, dy$$

converges even if f grows exponentially fast provided that

$$|f(x)| \leqslant \text{constant} \times e^{(1-\delta)x}$$

for some choice of $0 < \delta < 1$. It is therefore reasonable to search for solutions within the class of f which meet this constraint. This strategem is adopted without further ado. Now, put

$$g(x) = \begin{cases} (k \circ f)(x) = \displaystyle\int_0^\infty k(x-y) f(y)\, dy & \text{for} \quad x < 0, \\[2ex] 0 & \text{for} \quad x \geqslant 0, \end{cases}$$

so that

$$k \circ f = f + g$$

on the whole line. It is tempting to take the Fourier transform of both sides and in so doing to conclude that

$$[1 - \hat{k}(\gamma)] \hat{f}(\gamma) = -\hat{g}(\gamma).$$

However, this may not make sense on the line: You have no assurance that the integrals involved will converge there. The trick is to shift γ to the upper half-plane and to integrate along appropriately chosen horizontal lines, as will now be explained. To begin with,

$$\hat{k}(\gamma) = \frac{1}{2} \int_{-\infty}^\infty e^{-|x|} e^{i\gamma x}\, dx = \frac{1}{2}\left[\frac{1}{1+i\gamma} + \frac{1}{1-i\gamma} \right] = \frac{1}{1+\gamma^2}$$

is analytic in the strip $|b| < 1$. Besides

$$\hat{f}(\gamma) = \int_0^\infty f(x)\, e^{i\gamma x}\, dx$$

is analytic in the half-plane $b > 1 - \delta$, in view of the bound on f, while

$$\hat{g}(\gamma) = \int_{-\infty}^0 g(x)\, e^{i\gamma x}\, dx$$

is analytic in the half-plane $b < 1$ in view of the bound

$$|g(x)| = \left| \tfrac{1}{2} e^x \int_0^\infty e^{-y} f(y)\, dy \right| \leqslant \text{constant} \times e^x$$

for $x < 0$. Now take $\gamma = a + ib$ with $1 - \delta < b < 1$ and compute the Fourier transform

$$(k \circ f)^\wedge(\gamma) = \int_{-\infty}^\infty \left[\int_0^\infty k(x-y)\, e^{by}\, e^{-by} f(y)\, dy \right] e^{iax}\, e^{-bx}\, dx$$

$$= \int_0^\infty \left[\int_{-\infty}^\infty k(x-y)\, e^{-b(x-y)}\, e^{ia(x-y)}\, dx \right] f(y)\, e^{iay}\, e^{-by}\, dy$$

$$= \hat{k}(\gamma)\, \hat{f}(\gamma),$$

at that place. You should convince yourself that the necessary interchange of integrals is fully justified by Fubini's theorem and infer that

$$(1 - \hat{k})\hat{f} = -\hat{g}$$

is indeed valid *in the strip* $1 - \delta < b < 1$. Now factor $1 - \hat{k}(\gamma) = \gamma^2 (1 + \gamma^2)^{-1}$ into

$$\frac{\gamma^2}{\gamma + i} \times \frac{1}{\gamma - i}$$

and observe that

$$\frac{\gamma^2}{\gamma + i}\hat{f} = -(\gamma - i)\,\hat{g}$$

defines an analytic function on the whole complex plane since the left-hand side is analytic *above* $b = 1 - \delta$ and the right-hand side is analytic *below* $b = 1$, while the identity holds in the common strip $1 - \delta < b < 1$. The modulus of this function is bounded below $b = 1 - \delta/3$ by a constant multiple of

$$(|\gamma| + 1) \int_{-\infty}^0 e^{-bx} e^x\, dx = \frac{|\gamma| + 1}{1 - b} \leqslant 3\delta^{-1}(|\gamma| + 1),$$

as you see by looking at the right-hand side, while above $b = 1 - 2\delta/3$, it is bounded by a constant multiple of

$$|\gamma| \int_0^\infty e^{-bx} e^{(1-\delta)x}\, dx = \frac{|\gamma|}{b - (1 - \delta)} \leqslant 3\delta^{-1}|\gamma|,$$

as you see by looking at the left-hand side. Therefore, it is a polynomial of degree less than or equal to 1 by Exercise 3.1.18, and you see that the degree is actually 0 by looking on the imaginary axis. But this means that $\gamma^2(\gamma+i)^{-1}\hat{f}$ is constant in the strip $1-\delta<b<1$, and you may as well put the constant equal to i, so as to make

$$\hat{f}(\gamma) = (i\gamma-1)/\gamma^2.$$

Finally to recover f from

$$\hat{f}(a+ib) = \int_{-\infty}^{\infty} [f(x)\,e^{-bx}]\,e^{iax}\,dx,$$

apply the Fourier inversion formula to obtain

$$f(x)\,e^{-bx} = \lim_{R\uparrow\infty} \frac{1}{2\pi} \int_{-R}^{R} \frac{i(a+ib)-1}{(a+ib)^2}\,e^{-iax}\,da.$$

The integral exists in the L^2 sense, and can be evaluated by integrating around a semicircle in the lower [upper] half-plane for $x>0$ $[x<0]$ and applying Cauchy's formula. For $x>0$, you find

$$f(x)\,e^{-bx} = (1+x)\,e^{-bx},$$

and you conclude that constant multiples of $1+x$ are the only solutions of the stated problem that grow no faster than $e^{-(1-\delta)x}$.

3.   Solution of Milne's Problem

The situation for Milne's problem is quite similar: you put $f=0$ for $x<0$, and you have $k\circ f=f$ for $x>0$ and $k\circ f=g$ for $x<0$, with an extra unknown function g vanishing for $x>0$. Assume $f\geqslant 0$ as this is the only case of physical interest. Then as

$$k(x) = \frac{1}{2}\int_{|x|}^{\infty} e^{-y}\,\frac{dy}{y} \leqslant \frac{1}{2}e^{-|x|}$$

for $|x|\geqslant 1$, you have the bounds

$$0\leqslant g(x) = \int_{0}^{\infty} k(x-y)\,f(y)\,dy \leqslant \begin{cases} \frac{1}{2}e^{x}\displaystyle\int_{0}^{\infty} e^{-y}f(y)\,dy \\ \qquad\qquad\qquad\qquad \text{for}\quad x<-1, \\ \displaystyle\int_{0}^{\infty} k(-x-y)\,f(y)\,dy = f(-x) \\ \qquad\qquad\qquad\qquad \text{for}\quad -1\leqslant x<0. \end{cases}$$

Now if you assume that $f(x)\leqslant e^{(1-\delta)x}$ for some $0<\delta<1$, you will be in exactly the same situation as before, only with a more complicated function $\hat{k}$:

$$\hat{k}(\gamma) = \frac{1}{2} \int_{-\infty}^{\infty} \left[\int_{|x|}^{\infty} \frac{e^{-y}}{y} \, dy \right] e^{i\gamma x} \, dx$$

$$= \int_0^1 \frac{dx}{1+\gamma^2 x^2} = \gamma^{-1} \tan^{-1} \gamma.$$

EXERCISE 2.   Check the formula for $\hat{k}$. *Hint:*

$$\hat{k}(\gamma) = \int_{-\infty}^{\infty} e^{i\gamma x} \, dx \, \frac{1}{2} \int_1^{\infty} e^{-y|x|} \frac{dy}{y};$$

now use Exercise 2.6.6, but remember that the transform has been modified by a factor of 2π.

EXERCISE 3.   Show that $\hat{k}$ is analytic in the strip $|b| < 1$. *Hint:*   Use $\hat{k}(\gamma) = \int_0^1 (1+\gamma^2 x^2)^{-1} \, dx$.

EXERCISE 4.   Show that $\hat{k} = 1$ has a root of degree 2 at $\gamma = 0$ and no other roots in the strip $|b| < 1$. *Hint:* $\hat{k}(\gamma) = \int_0^1 (1+\gamma^{*2} x^2) |1+\gamma^2 x^2|^{-2} \, dx = 1$ only if $ab = 0$.

EXERCISE 5.   Verify the inequalities:

$$|\hat{k}(a+ib)| \leqslant (a^2-b^2)^{-\frac12} \tan^{-1}(a^2-b^2)^{\frac12} \leqslant (\pi/2)\,(a^2-b^2)^{-\frac12} \quad \text{for} \quad a^2 > b^2.$$

The rest of the computation depends upon factoring the auxiliary function

$$k^0 = \gamma^{-2}(1+\gamma^2)\,(1-\hat{k})$$

as a quotient k^+/k^-, in which k^+ [k^-] is analytic, root-free, and bounded on any closed half-plane in $b > -1$ [$b < +1$].

EXERCISE 6.   Check that the real part of k^0 is positive in the strip $|b| < 1$. *Hint:*

$$k^0(\gamma) = \frac{1+\gamma^2}{\gamma^2} \left[1 - \int_0^1 \frac{dx}{1+\gamma^2 x^2} \right] = \int_0^1 (1+\gamma^{*2} x^2)\,(1+\gamma^2) \, \frac{x^2 \, dx}{|1+\gamma^2 x^2|^2};$$

now compute the real part of $(1+\gamma^{*2} x^2)(1+\gamma^2)$ in the strip.

By Exercises 6 and 3.1.3, $\log k^0$ is analytic in the strip, and it is easy to justify the Cauchy formula:

$$\log k^0(\omega) = \frac{1}{2\pi i} \int_+ \frac{\log k^0(\gamma)}{\gamma - \omega} \, d\gamma - \frac{1}{2\pi i} \int_- \frac{\log k^0(\gamma)}{\gamma - \omega} \, d\gamma$$

$$= K^+(\omega) - K^-(\omega),$$

in which $\int_+$ $[\int_-]$ stands for integration (in the positive sense) along any horizontal line $b = -B$ $[b = B]$ just above $b = -1$ [below $b = +1$], as in Fig. 2.

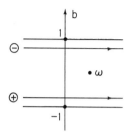

FIGURE 2

This is done by integrating $(\gamma - \omega)^{-1} \log k^0(\gamma)$ around the box marked off by $|a| = A$ and $|b| = B$ and checking that the contribution of the vertical sides tends to 0 as $A \uparrow \infty$; see Exercise 5, which permits you to bound $|\log k^0|$ by a constant multiple of $|a|^{-1}$ far out. The function K^+ $[K^-]$ so defined is analytic and bounded on any closed half-plane inside $b > -1$ $[b < +1]$, and

$$k^0 = \exp(\log k^0) = \frac{\exp(K^+)}{\exp(K^-)} \equiv \frac{k^+}{k^-}$$

is the desired factorization of k^0.

The rest is easy. The identity $\hat{f}(1 - \hat{k}) = -\hat{g}$, valid in the little strip $1 - \delta < b < 1$, can now be expressed as

$$\frac{\gamma^2}{\gamma + i} \hat{f} k^+ = -(\gamma - i)\, \hat{g} k^- .$$

As before, this defines an integral function which is easily seen to be constant [equal to i, say], so

$$\hat{f} = (i\gamma - 1)/\gamma^2 k^+$$

for $b > 1 - \delta$, and the "law of darkening" is found from the formula

$$e(0, \varphi) = -\sec \varphi \int_0^\infty e^{y\,\sec\varphi} f(y)\, dy$$

$$= -\sec \varphi \hat{f}(-i \sec \varphi)$$

$$= -\sec \varphi \times \frac{\sec \varphi - 1}{(-\sec^2 \varphi)\, k^+(-i \sec \varphi)}$$

$$= (1 - \cos \varphi) \times \exp\left[\frac{-1}{2\pi i} \int_+ \frac{\log[\gamma^{-2}(1 - \gamma^2)(1 - \hat{k})]}{\gamma + i \sec \varphi}\, d\gamma \right]$$

for $\varphi > \pi/2$. Unfortunately, this is not an elementary function, but it *is* tabulated; see Case *et al.* [1953, p. 147, Fig. 41a] for a picture and additional information in the context of neutron scattering. For additional information on problems of stellar radiation, see Hopf [1934].

EXERCISE 7. $\log k^0$ is of class $L^2(R^1)$ on the line $b = 0$. Use this to check that

$$K^-(\omega) = \frac{1}{2\pi i} \int_- \frac{\log k^0(\gamma)}{\gamma - \omega} \, d\gamma = \int_0^\infty (\log k^0)^\vee(x) \, e^{i\omega x} \, dx$$

in the strip $0 < b < 1$. $(\log k^0)^\vee$ is the customary inverse transform on the line:

$$(\log k^0)^\vee(x) = (2\pi)^{-1} \int_{-\infty}^\infty \log k^0(\gamma) \, e^{-i\gamma x} \, d\gamma .$$

3.7★ SPITZER'S IDENTITY

The ideas behind Hardy functions can be used in the context of summable functions to prove a beautiful identity of Spitzer [1956] with deep applications to probability theory. Actually, the Fourier transform is superfluous, and the only thing you need is the very simplest facts about the projection

$$f \to f^+ = f \times \text{ the indicator function of } [0, \infty).$$

The express Spitzer's identity in a slick way, bring in the "exponential"

$$Ef = \sum_{n=0}^\infty \frac{f^n}{n!}$$

and the "logarithm"

$$-L(1-f) = \sum_{n=1}^\infty \frac{f^n}{n},$$

in which f^n stands for the n-fold product $f \circ \cdots \circ f$ of the summable function f with itself, and the logarithm is defined for $\|f\|_1 < 1$, only. The leading term in the sum for Ef is "1," which you may think of as a (formal) unit: $f \circ 1 = 1 \circ f = f$. If you do not like to do that, you may prefer to deal with the bona fide summable function

$$\text{"}Ef - 1\text{"} = \sum_{n=1}^\infty \frac{f^n}{n!}.$$

To verify identities involving Ef and $L(1-f)$, you *must* resort to the defining sums.

EXERCISE 1.   Check the identities $[Ef-1]^\wedge = e^{\hat{f}} - 1$ and
$$[-L(1-f)]^\wedge = -\log(1-\hat{f}).$$

Hint: $\wedge$ is a homomorphism mapping $f_1 \circ f_2$ into the ordinary numerical product $\hat{f_1}\hat{f_2}$.

EXERCISE 2.   Check the further identities
$$Ef^+ - 1 = 0 \qquad\qquad \text{for}\quad x < 0,$$
$$E(f^+ - f) - 1 = 0 \qquad\qquad \text{for}\quad x \geqslant 0,$$
$$Ef^+ \circ E(-f) = E(f^+ - f),$$
$$E[L(1-f)] = 1 - f \qquad\qquad \text{for}\quad \|f\|_1 < 1,$$
$$E[-L(1-f)] = \sum_{n=0}^{\infty} f^n = (1-f)^{-1} \qquad \text{for}\quad \|f\|_1 < 1.$$

Spitzer's identity states that
$$1 + f^+ + (f \circ f^+)^+ + (f \circ (f \circ f^+)^+)^+ + \cdots = E(-[L(1-f)]^+),$$
provided $\|f\|_1 < 1$. By Exercise 1 this is the same as
$$[f^+ + (f \circ f^+)^+ + (f \circ (f \circ f^+)^+)^+ + \cdots]^\wedge = [E(-[L(1-f)]^+) - 1]^\wedge$$
$$= \exp\left[\sum_{n=1}^{\infty} n^{-1} \int_0^{\infty} f^n(x)\, e^{iyx}\, dx\right] - 1.$$

The probabilistic content of the identity will now be explained.
   Think of a "random walk" on the line: The position of the particle at time $n \geqslant 1$ is a sum $s_n = e_1 + \cdots + e_n$ of statistically independent steps $e_k : k \leqslant n$ with common distribution
$$P(a \leqslant e_1 \leqslant b) = \int_a^b f(x)\, dx,$$
as in Subsection 2.7.7; naturally, f is nonnegative and $\int f(x)\, dx = 1$. To understand the left-hand side of Spitzer's identity, notice that
$$\int_a^b f^+\, dx = \int_{\substack{x \geqslant 0 \\ a \leqslant x \leqslant b}} f(x)\, dx$$
$$= P(s_1 \geqslant 0, \quad \text{and}\quad a \leqslant s_1 \leqslant b),$$
$$\int_a^b (f \circ f^+)^+\, dx = \int_{\substack{x_1 \geqslant 0 \\ x_1 + x_2 \geqslant 0 \\ a \leqslant x_1 + x_2 \leqslant b}} \int f(x_1) f(x_2)\, d^2x$$
$$= P(s_1 \geqslant 0,\ s_2 \geqslant 0, \quad \text{and}\quad a \leqslant s_2 \leqslant b),$$

and, at the nth stage,

$$\int_a^b (f \circ \cdots (f \circ f^+)^+)^+ \, dx = \int_{\substack{x_1 \geqslant 0 \\ x_1 + x_2 \geqslant 0 \\ \vdots \\ x_1 + \cdots + x_n \geqslant 0 \\ a \leqslant x_1 + \cdots + x_n \leqslant b}} \int f(x_1) \cdots f(x_n) \, d^n x$$

$$= P(\mathbf{s}_1 \geqslant 0, \ \mathbf{s}_2 \geqslant 0, ..., \mathbf{s}_n \geqslant 0, \ \text{and} \ a \leqslant \mathbf{s}_n \leqslant b).$$

Spitzer's identity can now be applied to εf for $|\varepsilon| < 1$ with the result that

$$\sum_{n=1}^\infty \varepsilon^n P(\mathbf{s}_1 \geqslant 0, ..., \mathbf{s}_n \geqslant 0, \ a \leqslant \mathbf{s}_n \leqslant b) = \int_a^b [E(-[L(1-\varepsilon f)]^+) - 1] \, dx$$

$$= \int_a^b \left[E\left(\sum_{n=1}^\infty \frac{\varepsilon^n}{n} (f^n)^+ \right) - 1 \right] dx.$$

The point is that f^n is relatively easy to compute, but usually, the left-hand side is *not!*

To illustrate the use of Spitzer's identity, let us compute the distribution of the "entrance time"

$$T = \min(n > 0 : \mathbf{s}_n < 0).$$

To do this, let

$$p_n = P(T > n) = P(\mathbf{s}_1 \geqslant 0, ..., \mathbf{s}_n \geqslant 0).$$

Then

$$\sum_{n=1}^\infty p_n \varepsilon^n = \int_{-\infty}^\infty \left[E\left(\sum_{n=1}^\infty \frac{\varepsilon^n}{n} (f^n)^+ \right) - 1 \right] dx$$

$$= \exp\left[\sum_{n=1}^\infty \frac{\varepsilon^n}{n} \int_0^\infty f^n(x) \, dx \right] - 1$$

$$= \exp\left[\sum_{n=1}^\infty \frac{\varepsilon^n}{n} P(\mathbf{s}_n \geqslant 0) \right] - 1.$$

Now suppose that f is *even*. Then $P(\mathbf{s}_n \geqslant 0) = \frac{1}{2}$ for every $n \geqslant 1$, and you have

$$\sum_{n=1}^\infty p_n \varepsilon^n = \exp\left[\sum_{n=1}^\infty \frac{\varepsilon^n}{2n} \right] - 1$$

$$= \exp\left[-\frac{1}{2} \log(1-\varepsilon) \right] - 1$$

$$= \frac{1}{\sqrt{1-\varepsilon}} - 1$$

$$= \sum_{n=1}^\infty (-1)^n \binom{-\frac{1}{2}}{n} \varepsilon^n,$$

i.e.,

$$p_n = P(T > n) = (-1)^n \begin{pmatrix} -\frac{1}{2} \\ n \end{pmatrix} = 4^{-n} \begin{pmatrix} 2n \\ n \end{pmatrix},$$

independently of f! This remarkable fact is due to Sparre-Anderson. The advanced student will find additional material related to this circle of ideas in Feller [1966, Vol. 2, Chapters 12 and 18], Kingman [1962], Rota [1969], Spitzer [1964, Chapter 4], and Wendel [1958].

PROOF OF SPITZER'S IDENTITY. Pick a summable function f and look at the self-evident identity

$$Ef^+ - 1 = E(f^+ - f) - 1 + (Ef^+) \circ (1 - E[-f]).$$

By Exercise 2,

$$Ef^+ - 1 = 0 \qquad \text{for} \quad x < 0,$$

while

$$E(f^+ - f) - 1 = 0 \qquad \text{for} \quad x \geqslant 0,$$

and so

$$Ef^+ - 1 = [Ef^+ - 1]^+ = [(1 - E[-f]) \circ (Ef^+)]^+,$$

by an application of the projection $+$ to both sides. Now replace f by $-L(1-f)$ for $\|f\|_1 < 1$. Then $1 - E(-f)$ is replaced by f, and you find

$$E(-[L(1-f)]^+) = 1 + [f \circ E(-[L(1-f)]^+)]^+,$$

which you may think of as a functional identity for $E(-[L(1-f)]^+)$. The sum

$$1 + f^+ + (f \circ f^+)^+ + (f \circ (f \circ f^+)^+)^+ + \cdots$$

satisfies the *same* identity, so the difference k between the sum and $E(-[L(1-f)]^+)$ is a solution of $k = (f \circ k)^+$. Therefore, $\|k\|_1 \leqslant \|f\|_1 \|k\|_1$. But, since $\|f\|_1 < 1$, this can happen only if $k = 0$. The proof is finished.

3.8* HARDY FUNCTIONS IN THE DISK AND SZEGÖ'S THEOREM

Hardy functions are also important in the context of Fourier series. Think of the circle S^1 as the boundary of the unit disk $D: |\gamma| < 1$, parametrize it by the angle $0 \leqslant \theta < 2\pi$, and expand $f \in L^2(S^1)$ into a Fourier series

$$f = \sum_{n=-\infty}^{\infty} \hat{f}(n) e^{in\theta}$$

with coefficients

$$\hat{f}(n) = (2\pi)^{-1} \int_0^{2\pi} f(\theta) \, e^{in\theta} \, d\theta \, .$$

The sum can be split into 2 pieces:

$$f^+ = \sum_{n \geqslant 0} \hat{f}(n) \, e^{in\theta} \, ,$$

$$f^- = \sum_{n < 0} \hat{f}(n) \, e^{in\theta} \, .$$

The first sum has a self-evident analytic extension into D:

$$h(\gamma) = \sum_{n \geqslant 0} \hat{f}(n) \, \gamma^n \, ,$$

and with the notation $h_R(\theta) = h(Re^{i\theta})$ for $R < 1$, you have

$$\|h_R\|_2^2 = \int_0^{2\pi} |h_R(\theta)|^2 \, d\theta = 2\pi \sum_{n \geqslant 0} |\hat{f}(n)|^2 \, R^{2n}$$

and

$$\sup_{R < 1} \|h_R\|_2^2 = 2\pi \sum_{n \geqslant 0} |\hat{f}(n)|^2 = \|f^+\|_2^2 \leqslant \|f\|_2^2 \, .$$

EXERCISE 1.   Check that $\lim_{R \uparrow 1} \|h_R - f^+\|_2 = 0$.

A "Hardy function" h of class H^{2+} is a function that is analytic in D and satisfies

$$\sup_{R < 1} \|h_R\|_2^2 = \sup_{R < 1} \int_0^{2\pi} |h(Re^{i\theta})|^2 \, d\theta < \infty \, .$$

The recipe

$$h = \sum_{n \geqslant 0} \hat{f}(n) \, \gamma^n$$

associates a Hardy function with each $f \in L^2(S^1)$.

EXERCISE 2.   Check that every Hardy function arises in this way. *Hint:*   Expand h into a power series with center at 0 and compute $\|h_R\|_2$.

The map $f^+ \to h$ from half-series to Hardy functions now appears as an isomorphism onto H^{2+} if you take

$$\|h\|_2 = \sup_{R < 1} \|h_R\|_2 = \|f^+\|_2 \, .$$

As in Section 3.4, it is convenient to allow yourself the luxury of identifying

h and f^+ and to speak of f^+ as a "Hardy function" also. The map

$$f \to f^+ = \sum_{n \geq 0} \hat{f}(n)\, e^{in\theta}$$

then appears as the projection of $L^2(S^1)$ onto H^{2+}.

EXERCISE 3.   Check the evaluation

$$\frac{1}{2\pi}\int_0^{2\pi} \log|e^{i\theta}-\omega|\, d\theta = \begin{cases} \log|\omega| & \text{if } |\omega| > 1, \\ 0 & \text{if } |\omega| < 1. \end{cases}$$

Hint: Use Cauchy's formula to evaluate $(2\pi i)^{-1}\int_{|\gamma|=1} \gamma^{-1}\log(\gamma-\omega)\,d\gamma$ for $|\omega| > 1$; see Exercise 3.1.3 for help with the logarithm. The second formula is an easy consequence of the first.

EXERCISE 4.   Prove that for any nontrivial $f \in H^{2+}$, $\log|f|$ is summable and

$$\frac{1}{2\pi}\int_0^{2\pi} \log|f(\theta)|\, d\theta \geq \log|\hat{f}(0)| = \log\left|\frac{1}{2\pi}\int_0^{2\pi} f(\theta)\,d\theta\right|.$$

Hint: Suppose first that

$$f = \hat{f}(0) + \hat{f}(1)\, e^{i\theta} + \cdots + \hat{f}(n)\, e^{in\theta}$$

is a trigonometric polynomial with $\hat{f}(0) \neq 0$. The associated function $h = \sum \hat{f}(k)\gamma^k$ is also a polynomial and $h(0) = \hat{f}(0) \neq 0$. Deduce that

$$\frac{1}{2\pi}\int_0^{2\pi} \log|f(\theta)|\, d\theta = \frac{1}{2\pi}\int_0^{2\pi} \log|h(e^{i\theta})|\, d\theta \geq \log|h(0)|,$$

using the evaluation provided in Exercise 3. The rest of the proof can be modeled upon that of Theorem 3.4.2.

The rest of this section is devoted to the proof of the following remarkable fact.

SZEGÖ'S THEOREM.   *For any nonnegative summable weight function* Δ *on the circle*

$$\inf \frac{1}{2\pi}\int_0^{2\pi} |p(\theta)|^2 \Delta(\theta)\, d\theta = \exp\left[\frac{1}{2\pi}\int_0^{2\pi} \log\Delta(\theta)\, d\theta\right],$$

in which the infemum is taken over all trigonometric polynomials p *with leading coefficient* 1:

$$p = 1 + c_1 e^{i\theta} + \cdots + c_n e^{in\theta}.$$

Amplification: $\log^+ \Delta \leqslant \Delta$, so $\int \log \Delta$ can only diverge to $-\infty$;

$$\exp\left[\frac{1}{2\pi} \int^{\cdot} \log \Delta\right]$$

is interpreted as 0 in this case, and Szegö's theorem states that 1 belongs to the span of

$$e^{in\theta} : n \geqslant 1$$

in $L^2(\Delta) = L^2(S^1, \Delta(\theta)\, d\theta)$. Also, however, in the case of divergence,

$\quad\quad e^{-i\theta}$ belongs to the span of $e^{in\theta} : \quad n \geqslant 0$,

$\quad\quad e^{-2i\theta}$ belongs to the span of $e^{in\theta} : \quad n \geqslant -1$,

and so on; in short, *the family*

$$e^{in\theta} : \quad n \geqslant 0$$

spans the whole of $L^2(\Delta)$ *iff* $\int \log \Delta = -\infty$!

EXERCISE 5.   The span of $e^{in\theta} : n \geqslant 0$ is merely H^{2+} if $\Delta = 1$, but you *do* get the whole space if you take $\Delta = 1$ on an arc of length less than 2π and 0 otherwise. Give a proof of this fact using Exercise 4, only. *Hint:* $f * \Delta$ is a Hardy function if f is perpendicular to $e^{in\theta} : n \geqslant 0$ in $L^2(\Delta)$.

EXERCISE 6.   Use the amplification to Szegö's theorem to prove that if Δ is a nonnegative summable function on the line, then the family $e^{i\gamma t}$: $t \geqslant 0$ spans $L^2(\Delta) = L^2(R^1, \Delta(\gamma)\, d\gamma)$ iff $\int (1+\gamma^2)^{-1} \log \Delta = -\infty$. *Hint:* Put $\omega = (i\gamma - 1)^{-1}(i\gamma + 1)$. By Exercise 3.4.5, it is enough to study the span of $e_n = \pi^{-\frac{1}{2}}(i\gamma - 1)^{-1}\omega^n : n \geqslant 0$. The rest is plain sailing if you notice that $\gamma \to \omega = e^{i\theta}$ maps the line onto the circle.

PROOF OF SZEGÖ'S THEOREM.[1]   To begin with, let $\Delta \in C^\infty(S^1)$ be positive, and let

$$f = \exp[(\log \Delta)^+ - \tfrac{1}{2}(\log \Delta)^\wedge(0)]$$

$$= \exp\left[\sum_{n \geqslant 0} (\log \Delta)^\wedge(n)\, e^{in\theta} - \tfrac{1}{2}(\log \Delta)^\wedge(0)\right].$$

Both f and $1/f$ belong to H^{2+}, and you have

$$\hat{f}(0) = \exp[\tfrac{1}{2}(\log \Delta)^\wedge(0)] = \exp\left[\frac{1}{4\pi} \int_0^{2\pi} \log \Delta\right],$$

$$(1/f)^\wedge(0) = \exp[-\tfrac{1}{2}(\log \Delta)^\wedge(0)] = \exp\left[-\frac{1}{4\pi} \int_0^{2\pi} \log \Delta\right],$$

---

[1] Adapted from Szegö [1920].

and

$$|f|^2 = ff^* = \exp\left[\sum_{n \geq 0} (\log \Delta)^\wedge(n)\, e^{in\theta} + \sum_{n \leq 0} (\log \Delta)^\wedge(n)\, e^{in\theta} - (\log \Delta)^\wedge(0)\right]$$

$$= \exp[\log \Delta]$$

$$= \Delta.$$

Because

$$\frac{1}{2\pi} \int pf = \hat{f}(0),$$

$pf - \hat{f}(0)$ is perpendicular to constants, so

$$\frac{1}{2\pi} \int |p|^2 \Delta = \frac{1}{2\pi} \int |pf|^2$$

$$= \frac{1}{2\pi} \int |pf - \hat{f}(0) + \hat{f}(0)|^2$$

$$= \frac{1}{2\pi} \int |pf - \hat{f}(0)|^2 + |\hat{f}(0)|^2$$

$$= \frac{1}{2\pi} \int |p - (\hat{f}(0)/f)|^2 \Delta + \exp\left[\frac{1}{2\pi} \int_0^{2\pi} \log \Delta\right],$$

and if you now recall that $1/f$ is a Hardy function, you can make the first integral in the last line as small as you like by adjusting p:

$$\frac{1}{2\pi} \int \left|p - \frac{\hat{f}(0)}{f}\right|^2 \Delta \leq \|\Delta\|_\infty \sum_{n \geq 1} \left|\hat{p}(n) - \hat{f}(0)\left(\frac{1}{f}\right)^\wedge(n)\right|^2.$$

Szegö's formula is now proved for this special case. The proof for the general weight Δ follows by approximation. Given positive numbers α and β, pick a weight function $\square \geq \alpha$ of class $C^\infty(S^1)$ so as to make $\|\Delta + \alpha - \square\|_1 < \beta$. Then

$$\|\log(\Delta + \alpha) - \log \square\|_1 = \left\|\int_\square^{\Delta + \alpha} \frac{dx}{x}\right\|_1$$

$$\leq \frac{1}{\alpha} \|\Delta + \alpha - \square\|_1 < \frac{\beta}{\alpha}.$$

Hence,

$$\int \log \square - \frac{\beta}{\alpha} \leq \int \log(\Delta + \alpha) \leq \int \log \square + \frac{\beta}{\alpha},$$

and

$$\frac{1}{2\pi}\int |p|^2\Delta = \frac{1}{2\pi}\int |p|^2(\Delta+\alpha-\square) - \frac{\alpha}{2\pi}\int |p|^2 + \frac{1}{2\pi}\int |p|^2\square$$

$$\geqslant -\|p\|_\infty^2 \frac{\beta}{2\pi} - \alpha\|p\|_\infty^2 + \exp\left[\frac{1}{2\pi}\int \log\square\right]$$

$$\geqslant -\left[\frac{\beta}{2\pi}+\alpha\right]\|p\|_\infty^2 + \exp\left[\frac{1}{2\pi}\int \log(\Delta+\alpha) - \frac{\beta}{2\pi\alpha}\right].$$

The latter goes over into

$$\frac{1}{2\pi}\int |p|^2\Delta \geqslant \exp\left[\frac{1}{2\pi}\int \log\Delta\right]$$

as $\beta\downarrow 0$ and $\alpha\downarrow 0$, in that order, which is half the battle. But also, the reciprocal of

$$f = \exp\left[(\log\square)^+ - \tfrac{1}{2}(\log\square)^\wedge(0)\right]$$

belongs to $H^{2+} \cap C^\infty(S^1)$ and $|f|^2 = \square$, so

$$\inf \frac{1}{2\pi}\int |p|^2\Delta \leqslant \frac{1}{2\pi}\int \left|\frac{\hat{f}(0)}{f}\right|^2\Delta$$

$$\leqslant \frac{|\hat{f}(0)|^2}{2\pi}\int \frac{\Delta+\alpha}{|f|^2}$$

$$= \frac{|\hat{f}(0)|^2}{2\pi}\int \frac{\Delta+\alpha-\square}{\square} + |\hat{f}(0)|^2$$

$$\leqslant \frac{|\hat{f}(0)|^2}{2\pi}\frac{\beta}{\alpha} + \exp\left[\frac{1}{2\pi}\int \log\square\right]$$

$$\leqslant \frac{|\hat{f}(0)|^2}{2\pi}\frac{\beta}{\alpha} + \exp\left[\frac{1}{2\pi}\int \log(\Delta+\alpha) + \frac{\beta}{2\pi\alpha}\right],$$

and as $\beta\downarrow 0$ and $\alpha\downarrow 0$ in the same order, this goes over into

$$\inf \frac{1}{2\pi}\int |p|^2\Delta \leqslant \exp\left[\frac{1}{2\pi}\int \log\Delta\right].$$

The proof of Szegö's theorem is finished.

EXERCISE 7. Use Jensen's inequality [Exercise 3.4.7] to make a new proof that

$$\inf \frac{1}{2\pi} \int_0^{2\pi} |p|^2 \Delta \geq \exp\left[\frac{1}{2\pi} \int_0^{2\pi} \log \Delta\right]$$

for any trigonometric polynomial p with leading coefficient 1. *Hint:* See Exercise 4.

The following supplementary exercises deal with "outer" functions of class H^{2+} The adjective "outer" means that

$$\frac{1}{2\pi} \int_0^{2\pi} \log|f(\theta)| \, d\theta = \log|\hat{f}(0)| > -\infty;$$

compare Section 3.5 and Exercise 4.

EXERCISE 8. Any nonnegative function G with $\int G^2 < \infty$ and $\int \log G > -\infty$ is the "gain" of an outer function f, i.e., $G = |f|$ a.e., on the circle for some such f. *Hint:* Check that

$$h(\gamma) = \exp\left[\frac{1}{2\pi} \int_0^{2\pi} \frac{e^{i\theta}+\gamma}{e^{i\theta}-\gamma} \log G(\theta) \, d\theta\right]$$

for $|\gamma| < 1$, belongs to H^{2+} with the help of Jensen's inequality [Exercise 3.4.7]. Check next that h is an outer function in the sense that

$$f = \lim_{R \uparrow 1} h(Re^{i\theta})$$

is one. The proof of Theorem 3.4.3 may be used as a model.

EXERCISE 9. Any function $h \in H^{2+}$ may be factored into the product of an "inner" function j and an outer function h^0. The adjective "inner" means that j is analytic and of modulus $|j| \leq 1$ in the open disc with $|j| = 1$ a.e. on the boundary; compare Subsection 3.5.1. *Hint:* The outer factor h^0 is specified by the gain $G = \lim_{R \uparrow 1} |h(Re^{i\theta})|$ as in Exercise 8. The presumptive inner factor $j = h/h^0$ is analytic in the disc and of modulus 1 a.e. on the boundary. The only moot point is whether $|j|$ is less than or equal to 1 inside, or what is the same, whether $|h| \leq |h^0|$ inside. Do this first for polynomial h, and then approximate the general case to finish. You will need the integral of Exercise 10.

EXERCISE 10. Check the evaluation

$$\log|Re^{i\theta} - \omega| = \frac{1}{2\pi} \int_0^{2\pi} \frac{1-R^2}{1-2R\cos(\varphi-\theta)+R^2} \log|e^{i\varphi} - \omega| \, d\varphi$$

for $R < 1 < |\omega|$; compare Exercise 3.4.6. *Hint for a direct evaluation:* Expand $\log(Re^{i\theta} - \omega)$ as a power series $\sum_{n=0}^{\infty} c_n(\omega)(Re^{i\theta})^n$, take real parts to obtain $\log|Re^{i\theta} - \omega|$ as a Fourier series $\sum_{n=-\infty}^{\infty} k_n(\omega) R^{|n|} e^{in\theta}$, compute k_n and stick the resulting integral back into the sum.

EXERCISE 11. Check that h is outer iff its inner factor j is constant. *Hint:* h is outer iff $|j(0)| = 1$; now use the maximum principle of Subsection 3.1.6.

EXERCISE 12. Check that the family $e^{in\theta}f(\theta): n \geqslant 0$ spans H^{2+} iff f is outer; compare Subsection 3.5.2. *Hint:* $\hat{f}(0) = c \neq 0$ is necessary if the family is to span, and you must also have

$$
\begin{aligned}
0 &= \inf \| 1 - (p + c^{-1})f \|_2^2 \\
&= -2\pi + \inf \| (p + c^{-1})f \|_2^2 \\
&= -2\pi + |c|^{-2} \inf \| (p+1)f \|_2^2,
\end{aligned}
$$

in which the infemum is taken over the class of trigonometric polynomials p with $\hat{p}(0) = 0$. Now apply Szegö's theorem to finish one half of the proof. The other half is done by running the argument backward.

EXERCISE 13. Check that a nontrivial function $f \in H^{2+}$ if outer iff $\sum_{k \leqslant n} |\hat{f}(k)|^2$ is maximal for every $0 \leqslant n < \infty$, over the class of Hardy functions with fixed gain $G = |f|$; compare Subsection 3.5.4.

The advanced student will find additional information about H^{2+} in Akhiezer [1956], de Branges and Rovnyak [1966], Duren [1970], Helson [1964], and Hoffman [1962].

3.9*   POLYNOMIAL APPROXIMATION:
THE SZÁSZ–MÜNTZ THEOREM

The Weierstrass approximation theorem of Subsection 1.7.3 implies that the family of powers $x^n: n \geqslant 0$ spans $L^2[0, 1]$. The purpose of this section is to use Hardy functions in the half-plane to prove the following remarkable refinement.

SZÁSZ–MÜNTZ THEOREM. *Pick a series of integers $1 \leqslant n_1 \leqslant n_2 < \cdots$ tending to ∞. Then the family $x^{n_k}: k \geqslant 1$ spans $L^2[0, 1]$ iff*

$$
\sum_{k=1}^{\infty} n_k^{-1} = \infty.
$$

PROOF.   Pick $f \in L^2[0,1]$ perpendicular to the family $x^{n_k}: k \geqslant 1$ and look at the function

$$h(\gamma) = \int_0^1 f(x) \, x^{(-i\gamma - 1/2)} \, dx$$

$$= \int_0^\infty f(e^{-y}) \, e^{i\gamma y} \, e^{-y/2} \, dy \qquad [x = e^{-y}]$$

in the open upper half-plane. Because

$$\int_0^\infty |f(e^{-y})|^2 \, e^{-y} \, dy = \int_0^1 |f(x)|^2 \, dx = \|f\|_2^2 < \infty,$$

this is a Hardy function, and it has a root at each of the points

$$\omega = \omega_k = i(n_k + \tfrac{1}{2}).$$

Besides,

$$|h_{1/2}(\gamma)| = |h(\gamma + i\tfrac{1}{2})| \leqslant \int_0^\infty |f(y)| \, e^{-y} \, dy \leqslant \|f\|_2$$

on R^{2+}. Therefore, by the maximum principle of Subsection 3.1.6,

$$h_{1/2}(\gamma) \prod_{k \leqslant m} \frac{\gamma + i n_k}{\gamma - i n_k}$$

lies under the same bound for $b \geqslant 0$, and so

$$|h_{1/2}(ib)| = \|f\|_2 \prod_{k \leqslant m} \left| \frac{ib - i n_k}{ib + i n_k} \right|$$

$$= \|f\|_2 \prod_{k \leqslant m} [1 - 2b(n_k + b)^{-1}].$$

An application of the inequality $1 - x \leqslant e^{-x}$ shows that this product diverges to 0 as $m \uparrow \infty$ if $\sum n_k^{-1} = \infty$, and you find that $h = f = 0$. This is half of the proof. The other half is similar: If $\sum n_k^{-1} < \infty$, then the product

$$j(\gamma) = \prod_{k=1}^\infty \frac{\gamma - \omega_k}{\gamma - \omega_k{}^*}$$

is analytic and of modulus < 1 for $b > 0$, so that $h = (1 - i\gamma)^{-1} j \in H^{2+}$, and

$$h(\gamma) = \int_0^\infty f(e^{-y}) \, e^{i\gamma y} \, e^{-y/2} \, dy$$

for some nontrivial function $f(e^{-y}) e^{-y/2}$ of class $L^2[0, \infty)$. But then

$$\int_0^1 |f(x)|^2 \, dx < \infty,$$

and

$$\int_0^1 f(x)\, x^n\, dx = \int_0^1 f(x)\, x^{(-i\omega - \frac{1}{2})}\, dx \qquad [\omega = i(n+\tfrac{1}{2})]$$

$$= \int_0^\infty f(e^{-y})\, e^{i\omega y}\, e^{-y/2}\, dy$$

$$= h(\omega)$$

vanishes at the points $\omega_k = i(n_k + \frac{1}{2})$ for every $k \geqslant 1$. This is the same as to say that f is perpendicular to the family $x^{n_k} \colon k \geqslant 1$. Consequently, the latter does *not* span $L^2[0,1]$. The proof is finished.

The interested reader will find an elementary (but more complicated) proof in Akhiezer [1956]; additional information can be found in Feller [1968b] and Schwarz [1959.]

EXERCISE 1. Deduce from the Szász–Müntz theorem that the powers $x^{n_k} \colon k \geqslant 1$, augmented by $x^0 = 1$, span $C[0,1]$ iff $\sum n_k^{-1} = \infty$; "span" is understood to mean that

$$\inf \left\| f - c_0 - \sum c_k x^{n_k} \right\|_\infty = 0$$

for every $f \in C[0,1]$. *Hint:* $\|f\|_\infty \leqslant \|f'\|_2$ for smooth functions vanishing at $x = 0$.

3.10*   THE PRIME NUMBER THEOREM

The prime number theorem states that *the number of primes* $2 \leqslant p \leqslant n$ *is approximately* $n/\log n$; more precisely, if $\#(n)$ is the number of such primes, then the prime number theorem says that

$$\lim_{n \uparrow \infty} \frac{\#(n)}{n/\log n} = 1.$$

The fact was conjectured from extensive numerical evidence by Gauss, but not proved until the end of the 19th century, and then only by means of complex function theory! An "elementary" (arithmetic) but very complicated proof was discovered by Erdös and Selberg in 1949; for an updated proof along such lines see Levinson [1969]. The proof presented below is adapted from Ikehara [1931]; properly speaking, it is a little to one side of the developments of this chapter, but the fact that it combines function theory and Fourier transforms in a striking way may serve as an excuse for its inclusion. A simple way of making the prime number theorem plausible will be found in Courant and Robbins [1941]; for improvements upon the prime number theorem, see Ingham [1932].

Step 1:   Verify that the prime number theorem is equivalent to

$$\lim_{x \uparrow \infty} Q(x)/x = 1,$$

in which[1]

$$Q(x) = \sum_{n \leqslant x} \begin{bmatrix} \log p & \text{if} \quad n \geqslant 2 \text{ is a power of a prime } p, \\ 0 & \text{otherwise} \end{bmatrix}.$$

PROOF.   For a fixed prime p, the number of integers $n \leqslant x$ of the form p^i is the largest integer j such that $j \log p \leqslant \log x$, so you have

$$Q(x) = \sum_{p \leqslant x} j \log p \leqslant \sum_{p \leqslant x} \frac{\log x}{\log p} \log p \leqslant \#(x) \log x.$$

A bound from the other side is just as easy: For $0 < \delta < 1$ and $y = x^{1-\delta}$, you have

$$\#(x) = \#(y) + \sum_{y < p \leqslant x} 1 \leqslant y + \frac{Q(x)}{\log y} \leqslant x^{1-\delta} + \frac{1}{(1-\delta)} \frac{Q(x)}{\log x}.$$

The rest of the proof is self-evident.

Step 2:   Connect Q with the Riemann "zeta function." The latter is defined by the sum

$$Z(\gamma) = \sum_{n=1}^{\infty} n^{-\gamma}$$

in the half-plane $\gamma = a + ib$, $a > 1$. Z is analytic there since $|n^{-\gamma}| = n^{-a}$ is the general term of a convergent sum, and you can differentiate Z termwise to obtain the formula

$$-Z'(\gamma) = \sum_{n=1}^{\infty} \log n \, n^{-\gamma},$$

or what is better for the present purpose,

$$-Z'(\gamma) = \sum_{n=1}^{\infty} n^{-\gamma} \log n$$

$$= \sum_{n=1}^{\infty} n^{-\gamma} \sum_{p} \sum_{k=1}^{\infty} \begin{bmatrix} \log p & \text{if} \quad p^k \quad \text{divides} \quad n, \\ 0 & \text{otherwise} \end{bmatrix}$$

$$= \sum_{p} \log p \sum_{k=1}^{\infty} \sum_{n \text{ a multiple of } p^k} n^{-\gamma}$$

$$= \sum_{p} \log p \sum_{k=1}^{\infty} \sum_{n=1}^{\infty} (p^k n)^{-\gamma}$$

$$= \sum_{p} \log p \sum_{k=1}^{\infty} p^{-k\gamma} \times Z(\gamma).$$

---

[1] You can give the quantity inside the big brackets in the defining sum for Q a name, say $\Lambda(n)$ if you like. Then $Q(x) = \sum_{n \leqslant x} \Lambda(n)$, which is perhaps a little easier to digest.

This formula contains the fact that $Z(\gamma) \neq 0$ for $a > 1$, [see Exercise 3.1.14], so you can divide through by it. The result may be expressed in terms of Q:

$$\frac{-Z'(\gamma)}{Z(\gamma)} = \sum_p \log p \sum_{k=1}^{\infty} p^{-k\gamma}$$

$$= \sum_{n=2}^{\infty} n^{-\gamma} \begin{bmatrix} \log p & \text{if } n \text{ is a power of a prime } p, \\ 0 & \text{otherwise} \end{bmatrix}$$

$$= \sum_{n=2}^{\infty} n^{-\gamma} [Q(n) - Q(n-1)]$$

$$= \sum_{n=1}^{\infty} Q(n) [n^{-\gamma} - (n+1)^{-\gamma}]$$

$$= \sum_{n=1}^{\infty} Q(n) \gamma \int_n^{n+1} x^{-\gamma-1} dx$$

$$= \gamma \int_1^{\infty} Q(x) x^{-\gamma-1} dx.$$

Step 3: The strategy of the proof can now be explained. Bring in the function

$$f_\delta(b) = -\frac{Z'(\gamma)}{\gamma Z(\gamma)} - \frac{1}{\gamma - 1},$$

evaluated at $\gamma = 1 + \delta + ib$, for $0 < \delta < 1$. The formula of Step 2 can be expressed as

$$f_\delta(b) = \int_0^{\infty} [e^{-y} Q(e^y) - 1] e^{-\delta y} e^{-iby} dy$$

after the substitution $x \to e^y$ in the integral $\int Q x^{-\gamma-1}$. Now multiply both sides by $e^{ibx} k(b)$ with $k \in C_\downarrow^{\infty}(R^1)$ and integrate over $-\infty < b < \infty$. Using the bound $Q(x) \leqslant x \log x$, you easily justify

$$\int e^{ibx} k(b) f_\delta(b) \, db = \int_0^{\infty} \hat{k}(x-y) [e^{-y} Q(e^y) - 1] e^{-\delta y} dy.$$

The content of Step 4 is that $|f_\delta(b)|$ is bounded by a constant multiple of $1 + b^4$ and that it tends to a function f_{0+} as $\delta \downarrow 0$. Granting this for now, you find

$$\int e^{ibx} k(b) f_{0+}(b) \, db = \int_0^{\infty} \hat{k}(x-y) [e^{-y} Q(e^y) - 1] dy,$$

and since kf_{0+} is summable, the Riemann–Lebesgue lemma tells you that

$$\lim_{x\uparrow\infty}\int_0^\infty \hat{k}(x-y)\, e^{-y} Q(e^y)\, dy = \lim_{x\uparrow\infty}\int_0^\infty \hat{k}(x-y)\, dy = \int_{-\infty}^\infty \hat{k}(-y)\, dy.$$

Now pick $\hat{k}$ as in Fig. 1 with adjustable $0 < \alpha < \beta < \infty$. Then

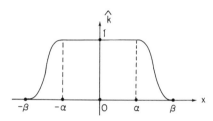

FIGURE 1

$$2\alpha < \int_{-\infty}^\infty \hat{k}(-y)\, dy$$

$$= \lim_{x\uparrow\infty}\int_0^\infty \hat{k}(x-y)\, e^{-y} Q(e^y)\, dy$$

$$\leqslant \liminf_{x\uparrow\infty}\int_{x-\beta}^{x+\beta} e^{-y} Q(e^y)\, dy$$

$$\leqslant \liminf_{x\uparrow\infty} e^{-x+\beta} Q(e^{x+\beta})\, 2\beta$$

$$= \liminf_{x\uparrow\infty}\, [Q(x)/x]\, e^{2\beta}\, 2\beta,$$

while

$$2\beta > \int_{-\infty}^\infty \hat{k}(-y)\, dy$$

$$= \lim_{x\uparrow\infty}\int_0^\infty \hat{k}(x-y)\, e^{-y} Q(e^y)\, dy$$

$$\geqslant \limsup_{x\uparrow\infty}\int_{x-\alpha}^{x+\alpha} e^{-y} Q(e^y)\, dy$$

$$\geqslant \limsup_{x\uparrow\infty} e^{-x-\alpha} Q(e^{x-\alpha})\, 2\alpha$$

$$\geqslant \limsup_{x\uparrow\infty}\, [Q(x)/x]\, e^{-2\alpha}\, 2\alpha,$$

so that

$$\frac{\alpha}{\beta}\, e^{-2\beta} \leqslant \liminf_{x\uparrow\infty}\frac{Q(x)}{x} \leqslant \limsup_{x\uparrow\infty}\frac{Q(x)}{x} \leqslant \frac{\beta}{\alpha}\, e^{2\alpha}.$$

The proof is finished by making $\beta \downarrow \alpha$ and $\alpha \downarrow 0$, in that order.

Step 4: Obtain the necessary information about the function f_δ, as will now be done in a series of short paragraphs.

1.
$$Z(\gamma) = \gamma \int_1^\infty \frac{[x]-x+\tfrac{1}{2}}{x^{\gamma+1}} \, dx + \frac{1}{\gamma-1} + \frac{1}{2},$$

in which $[x]$ is the largest integer less than or equal to x; in particular, Z is analytic in the half-plane $a > 0$, apart from a simple pole at $\gamma = 1$.

PROOF.   Just compute
$$\gamma \int_n^{n+1} \frac{n-x+\tfrac{1}{2}}{x^{\gamma+1}} \, dx = \left(n+\frac{1}{2}\right)\left[\frac{1}{n^\gamma} - \frac{1}{(n+1)^\gamma}\right] - \frac{\gamma}{\gamma-1}\left[\frac{1}{n^{\gamma-1}} - \frac{1}{(n+1)^{\gamma-1}}\right]$$

and add over $n \geqslant 1$:

$$\begin{aligned}
\gamma \int_1^\infty \frac{[x]-x+\tfrac{1}{2}}{x^{\gamma+1}} \, dx &= \sum_{n=1}^\infty \left(n+\frac{1}{2}\right)n^{-\gamma} - \sum_{n=2}^\infty \left(n-\frac{1}{2}\right)n^{-\gamma} - \frac{\gamma}{\gamma-1} \\
&= \sum_{n=2}^\infty n^{-\gamma} + \frac{3}{2} - \frac{\gamma}{\gamma-1} \\
&= Z(\gamma) - 1 + \frac{3}{2} - \frac{\gamma-1}{\gamma-1} - \frac{1}{\gamma-1} \\
&= Z(\gamma) - \frac{1}{\gamma-1} - \frac{1}{2}.
\end{aligned}$$

The rest is plain sailing.

2. $Z(\gamma)$ can be expressed by Euler's product $\prod_p (1-p^{-\gamma})^{-1}$ for $a > 1$.

PROOF.   The idea is to "sieve" out of the sum for Z the multiples of each prime $p = 2, 3, 5, \ldots$ in the following way:

$$(1-2^{-\gamma}) \sum_{n=1}^\infty n^{-\gamma} = \sum_{\substack{n \text{ indivisible} \\ \text{by } 2}} n^{-\gamma}$$

$$(1-2^{-\gamma})(1-3^{-\gamma}) \sum_{n=1}^\infty n^{-\gamma} = \sum_{\substack{n \text{ indivisible} \\ \text{by } 2 \text{ and } 3}} n^{-\gamma}$$

$$(1-2^{-\gamma}) \cdots (1-p^{-\gamma}) \sum_{n=1}^\infty n^{-\gamma} = \sum_{\substack{n \text{ indivisible} \\ \text{by } p \text{ or any} \\ \text{smaller prime}}} n^{-\gamma}.$$

As $p \uparrow \infty$, the right-hand side reduces to the single summand 1. The proof is finished.

3. $Z(a)^3 \, |Z(a+ib)|^4 \, |Z(a+2ib)| \geqslant 1$ for $a > 1$.

PROOF. This follows easily from the product formula of **2** and the elementary identity

$$3 + 4 \cos \theta + \cos 2\theta = 2(1 + \cos \theta)^2 .$$

By the product formula, the modulus of

$$Z(\gamma) = \exp \left[\sum_p - \log(1 - p^{-\gamma}) \right] = \exp \left[\sum_p \sum_{n=1}^{\infty} n^{-1} p^{-n\gamma} \right]$$

is

$$|Z(\gamma)| = \exp \left[\sum_p \sum_{n=1}^{\infty} n^{-1} p^{-na} \cos(nb \log p) \right],$$

so

$$Z(a)^3 \, |Z(a+ib)|^4 \, |Z(a+2ib)|$$

$$= \exp \left[\sum_p \sum_{n=1}^{\infty} n^{-1} p^{-na} [3 + 4 \cos(nb \log p) + \cos(2nb \log p)] \right],$$

which is greater than or equal to 1 since the exponent is nonnegative.

4. $Z \neq 0$ for $a = 1$ and $b \neq 0$.

PROOF. Suppose $\gamma = 1 + ib$ $(b \neq 0)$ is a root of Z. Then

$$|Z(1 + \delta + ib)| \leqslant \text{constant} \times \delta$$

for $0 < \delta < \frac{1}{2}$, say, since Z is analytic at γ. But now by **1** and **3**,

$$1 \leqslant Z(1+\delta)^3 \, |Z(1+\delta+ib)|^4 \, |Z(1+\delta+2ib)| \leqslant \text{constant} \times \delta^{-3} \times \delta^4$$

and that leads to a contradiction for small δ.

5. f_{0+} exists.

PROOF. $Z_1 = Z - (\gamma - 1)^{-1}$ is analytic in the half-plane $a > 0$ by **1**, and $(\gamma - 1) Z \neq 0$ for $a = 1$ by **1** and **4**, so

$$f_{\delta}(b) = \frac{-Z'(\gamma)}{\gamma Z(\gamma)} - \frac{1}{\gamma - 1} = \frac{(\gamma - 1) Z_1'(\gamma) + \gamma Z_1(\gamma) + 1}{\gamma (\gamma - 1) Z(\gamma)}, \qquad \gamma = 1 + \delta + ib,$$

is analytic in an open region including the closed half-plane $a \geqslant 1$. The existence of f_{0+} is now automatic.

6. $|f_\delta(b)|$ is bounded by a constant multiple of $1+b^4$ for $0 < \delta < 1$.

PROOF. $|f_\delta(b)|$ is bounded by a constant for $0 < \delta < 1$ and $|b| \leqslant 1$ by the proof of **5**. Now look at the rest of the strip: $0 < \delta < 1$, $|b| > 1$, and let $c_1, c_2, \ldots$ be positive constants. By **1**, $|Z(a+ib)| \leqslant c_1|b|$ in this region while $Z(a) \leqslant c_2(a-1)^{-1}$, so by **3**,

$$|Z(a+ib)| \geqslant Z(a)^{-\frac34}|Z(a+2ib)|^{-\frac14} \geqslant c_3(x-1)^{\frac34}|b|^{-\frac14}$$

for $1 < x \leqslant a < 2$. Besides, $|Z'(a+ib)| \leqslant c_4|b|$ by a second application of **1**. This is combined with the bound just proved to obtain

$$|Z(a+ib)| \geqslant |Z(x+ib)| - \int_a^x |Z'(y+ib)|\, dy$$

$$\geqslant c_3(x-1)^{\frac34}|b|^{-\frac14} - c_4(x-1)|b|$$

for $1 < a \leqslant x < 2$. To sum up,

$$|Z(a+ib)| \geqslant c_3(x-1)^{\frac34}|b|^{-\frac14}[1 - c_5(x-1)^{\frac14}|b|^{\frac54}]$$

for *any* $1 < a < 2$ and $1 < x < 2$. Now pick $x = 1 + c_6|b|^{-5}$ with $c_6 < 1$ so that $1 < x < 2$ for $|b| \geqslant 1$. Then

$$|Z(a+ib)| \geqslant c_3 c_6^{3/4}|b|^{-15/4}|b|^{-1/4}[1 - c_5 c_6^{1/4}] \geqslant c_7|b|^{-4}$$

for small c_6, and so

$$|f_\delta(b)| \leqslant \left|\frac{Z'(\gamma)}{\gamma Z(\gamma)}\right| + \frac{1}{|\gamma - 1|} \leqslant \frac{c_4|b|}{|b|\,c_7|b|^{-4}} + 1 \leqslant c_8(1+b^4),$$

as advertised.

Chapter 4 *Fourier Series and Integrals on Groups*

4.1 GROUPS

Up to now, you have studied Fourier series and integrals for assorted classes of functions, mostly on the circle and the line. Both of these spaces are commutative groups (under addition). The purpose of this chapter is, first, to take a new look at the old series and integrals from the point of view of groups, and second, to develop similar ways of expressing functions on other important (commutative and noncommutative) groups. The knowledge of groups that you will need is prepared in the present section and at appropriate places later. Birkhoff and MacLane [1965] is recommended for additional information; for a fascinating elementary account of finite groups and symmetry, see Weyl [1952].

1. Groups as Such

A *group* G is a class of objects g which is equipped with a *multiplication*, that is, a map of $(g_1, g_2) \in G \times G$ into G expressed as a (formal) product "$g_1 g_2$," in such a way that

(a) the multiplication is associative: $g_1(g_2 g_3) = (g_1 g_2) g_3$,
(b) there is an identity 1: $1g = g1 = g$.
(c) every $g \in G$ has an inverse g^{-1}: $g^{-1}g = gg^{-1} = 1$.

G is *commutative* if $g_2 g_1$ is always the same as $g_1 g_2$; in this circumstance, it is customary to use addition $[g_1 + g_2]$ in place of multiplication $[g_1 g_2]$ and to denote the identity by 0 instead of by 1.

EXAMPLES

(1) S^1:   the circle $[0, 1)$ under addition modulo 1.
(2) R^1:   the line under addition.
(3) R^{1+}: the positive real numbers under multiplication.
(4) Z^1:   the integers under addition.
(5) Z_m^+: the integers $0, 1, ..., m-1$ under addition modulo a positive integer m.
(6) $Z_p^\times$: the positive integers $1, 2, ..., p-1$ under multiplication modulo a prime p.
(7) the permutation of $n \geqslant 2$ letters under composition of permutations.
(8) the symmetries of the square: counterclockwise rotation by $0°$, $90°$, $180°$, $270°$, reflection in either diagonal, horizontal and vertical reflection, under composition.
(9) $SO(3)$: the special orthogonal group of 3×3 real orthogonal matrices [transpose $=$ inverse] with determinant $+1$, under the conventional multiplication of matrices.
(10) $M(2)$: the rigid motions of the plane, i.e., translations and rotations, *alias* the Euclidean congruences.

EXERCISE 1.   Example 6 is an actual group only if p is prime. *Hint:* If p is prime, then $0 < n < p$ and p are relatively prime, so $in + jp = 1$ for some integral i and j.

EXERCISE 2.   Example 7 is noncommutative if $n \geqslant 3$.

EXERCISE 3.   Examples 8–10 are noncommutative groups.

2. Subgroups and Homomorphisms

Let G be a group. Then $H \subset G$ is a *subgroup* if it is a group in its own right. G/H stands for the family of *cosets* $gH = (gh: h \in H)$ as g runs over G. A *homomorphism* j of G is a map of G into a second group, which preserves the multiplication $[j(g_1 g_2) = j(g_1)j(g_2)]$; j is an *isomorphism* if it is also $1:1$ and onto.

EXERCISE 4. Check that two cosets are either identical or disjoint, i.e., the nonidentical cosets cover G simply (no overlapping).

EXERCISE 5. Check that G/H is a group under the multiplication $(g_1 H)(g_2 H) = (g_1 g_2) H$ iff $g^{-1} H g \subset H$ for every $g \in G$, and that in that case the natural map j: $g \to gH$ of G into G/H is a homomorphism. *Hint:* Check first that the proposed rule for multiplication makes sense, i.e., if $g_1 H = g_3 H$ and $g_2 H = g_4 H$, then $g_1 g_2 H = g_3 g_4 H$.

EXERCISE 6. Every homomorphism j of G arises as in Exercise 5: Namely, if j is a homomorphism of G into a second group G', then $j(G)$ is a subgroup of G', $H = j^{-1}(1') = (g \in G: j(g) = 1')$ is a subgroup of G, and $j(G)$ is isomorphic to G/H. Here, $1'$ stands for the identity of G'. Check all these claims.

EXERCISE 7. The symmetries of the square [Example 8] which leave the upper left-hand corner fixed form a subgroup H. Show that G/H can be identified with the corners of the square, but is *not* a group. *Hint:* Identify $g \in G$ with the place to which it sends the upper left-hand corner.

EXERCISE 8. Check that the groups of Examples 2 and 3 are isomorphic.

EXERCISE 9. Check that any finite group is isomorphic to a subgroup of the permutations of n letters for sufficiently large n. *Hint:* Pick $k \in G$. Then you can identify the map $g \to kg$ with a permutation of the "letters" $g \in G$.

EXERCISE 10. Check that R^1/Z^1 is isomorphic to S^1.

3. Characters

A *character* of a group G is a homomorphism e of G into the (multiplicative) group of complex numbers of modulus 1:

(a) $|e(g)| = 1$, $g \in G$,

(b) $e(g_1 g_2) = e(g_1) e(g_2)$;

in the case of continuous groups, such as Examples 1, 2, and 3, it is customary to insist that e be a continuous function on G.

EXERCISE 11. Check that $e(1) = 1$.

EXERCISE 12.   Check that $e(g^{-1}) = e(g)^{-1} = e(g)^*$.

$\hat{G}$ stands for the class of characters of the group G.

EXERCISE 13.   Check that $\hat{G}$ is a commutative group under the multiplication $(e_1 e_2)(g) = e_1(g) e_2(g)$. What is the identity of $\hat{G}$? $\hat{G}$ is the so-called *dual group* of G.

EXERCISE 14.   Compute the dual group of $Z_m{}^+$. *Hint:* $Z_m{}^+$ may be identified with the integers $0 \leqslant n < m$. Check that $(Z_m{}^+)^\wedge = Z_m{}^+$.

EXERCISE 15.   The "signature" of a permutation is $(-1)^\#$ in which $\#$ is the number of transposition $ij \to ji$ figuring in the permutation. Check that this signature is a character of the group of permutations of $n = 3$ letters. *Hint:* Every permutation is a product of transpositions but can be so expressed in many ways. The point is that the *parity* of the number of transpositions involved is an attribute of the permutation itself.

EXERCISE 16.   The only character of the group of permutations of $n \geqslant 2$ letters besides the signature is $e \equiv 1$. The moral is that characters are probably not much use for *noncommutative* groups; see Subsection 4.8.4 for a proof that $e \equiv 1$ is the *only* character of the group $SO(3)$ of Example 9. *Hint:* $e(ij) = \pm 1$ for any transposition ij. Why? Besides, $(ik)(ij)(ik) = kj$, and so $e(ij)$ is either always $+1$ or always -1 for every ij.

4.2   FOURIER SERIES ON THE CIRCLE

The purpose of this section is to "explain" the exponentials

$$e_n(x) = e^{2\pi i n x}$$

figuring in the standard Fourier series

$$f = \sum \hat{f}(n)\, e_n$$

from the point of view of the circle group $S^1 = R^1/Z^1$.

1.   Characters

$e_n : n \in Z^1$ *is a complete list of the characters of the circle group.*

PROOF. e_n is a character for every integer n: Namely, $|e_n| = 1$ and $e_n(x+y) = e_n(x) e_n(y)$. Now let e be any character and think of it as a character of R^1 of period 1 by putting $e(x+n) = e(x)$ for $0 \leqslant x < 1$ and integral n. Because $|e| = 1$,

$$e(x) = e^{i\varphi(x)}$$

with a real phase φ, and since e is multiplicative $[e(x+y) = e(x)e(y)]$, φ is additive $[\varphi(x+y) = \varphi(x)+\varphi(y)$, modulo $2\pi]$. But then $\varphi(jx) = j\varphi(x)$, modulo 2π, for any integral j, so $\varphi(x) = x\varphi(1)$, modulo 2π, for rational $x = k/j$, and

$$e(x) = e^{ix\varphi(1)},$$

first for rational $x = k/j$, and then for every real x, by continuity. To finish the proof, you infer from

$$1 = e(0) = e(1) = e^{i\varphi(1)}$$

that $\varphi(1)$ can only be an integral multiple of 2π, i.e., $e = e_n$ for some $n \in Z^1$.

EXERCISE 1.   The map $e_n \to n \in Z^1$ is an isomorphism between the dual group $(S^1)^\wedge$ and Z^1. Check that $(S^1)^{\wedge\wedge}$ [the dual group of $(S^1)^\wedge$] is isomorphic to S^1 [$(Z^1)^\wedge$ up to isomorphism]. This is a simple instance of the so-called "Pontrjagin duality"; see also Exercises 4.1.10 and 4.1.14, and Section 4.5, especially the comment on the Poisson summation formula.

2.   Invariant Subspaces

A second group-theoretical way of getting at the exponentials e_n is via translation invariant subspaces of $L^2(S^1)$. A closed subspace M is translation invariant if it is closed under translations, i.e., if

$$f_y(x) = f(x+y)$$

belongs to M for every $f \in$ M and every $y \in S^1$. A simple example is provided by the class M_n of complex multiples of e_n.

Warning: In this book invariant subspaces are always closed subspaces.

EXERCISE 2.   Check that $e_n \circ f = e_n \hat{f}(n)$ belongs to M for any $f \in$ M. Hint: $e_n \circ f$ may be approximated in $L^2(S^1)$ by the Riemann sum

$$\sum_{k=0}^{m-1} f\left(x - \frac{k}{m}\right) \int_{k/m}^{(k+1)/m} e_n(y)\, dy.$$

EXERCISE 3.   The "spectrum" of M is the class of integers n such that $\hat{f}(n) \neq 0$ for some $f \in$ M. Check that M is the (perpendicular) sum $\oplus M_n$, n running over the spectrum of M.

The content of Exercise 3 is that M_n: $n \in Z^1$ *is a complete list of minimal invariant subspaces of* $L^2(S^1)$. The self-evident perpendicular splitting

$$L^2(S^1) = \bigoplus_{|n| < \infty} M_n$$

is a special case; you may regard it as a new statement of the Plancherel identity.

EXERCISE 4.   Check that the family f_y: $0 \leqslant y < 1$ spans $L^2(S^1)$ iff $\hat{f}(n)$ never vanishes.

EXERCISE 5.   Check that as n runs over Z^1, $M^n = (f \in L^2(S^1): \hat{f}(n) = 0)$ runs through the maximal invariant subspaces of $L^2(S^1)$. The adjective "maximal" means that there are no other invariant subspaces between M^n and $L^2(S^1)$.

3.   Eigenfunctions

Additional interest attaches to the exponentials e_n as eigenfunctions of the differential operator $Kf = f''$. What is not self-evident beforehand is that the minimal invariant subspaces should be eigenspaces of such a nice differential operator! A link is provided by

EXERCISE 6.   Check that any linear operator K acting on $C^\infty(S^1)$ and commuting with translations $[x \to x+y]$ and reflection $[x \to -x]$ acts like multiplication by a constant on $M_n \oplus M_{-n}$, i.e., $M_n \oplus M_{-n}$ is an eigenspace of K. Check that if $K = c_0(x) + c_1(x)D + \cdots + c_n(x)D^n$ is a differential operator with coefficients from $C^\infty(S^1)$ which commutes with translations and reflections, then it must be a polynomial in D^2.

4.   Homomorphisms

Think of summable functions on the circle as an algebra under the customary product

$$f_1 \circ f_2(x) = \int_0^1 f_1(x-y) f_2(y) \, dy.$$

A homomorphism of this algebra is a mapping $j \not\equiv 0$ into the complex numbers which respects

(a) complex multiplication: $j(\text{constant} \times f) = \text{constant} \times j(f)$,
(b) addition: $j(f_1 + f_2) = j(f_1) + j(f_2)$, and
(c) multiplication: $j(f_1 \circ f_2) = j(f_1) j(f_2)$,
   subject to the technical condition
(d) $|j(f)| \leqslant \text{constant} \times \|f\|_1$, with a constant independent of f.

A simple example is provided by the nth Fourier coefficient:

$$j_n(f) = \hat{f}(n) = \int_0^1 f e_n{}^*.$$

EXERCISE 7. Check that if j is a homomorphism then there is an integer n such that $j(e_m) = 1$ or 0 according as $m = n$ or not. *Hint:* $e_n \circ f = e_n \hat{f}(n)$; now apply j to both sides.

Any summable function f can be well-approximated by sums of exponentials, so by (d) and Exercise 7, either $j(f) = \hat{f}(n)$ for some n, or $j(f) \equiv 0$, which is not allowed; in short, $j_n : n \in Z^1$ *is a complete list of the homomorphisms of* $L^1(S^1)$.

EXERCISE 8. Give a second proof that $j_n(f) = \hat{f}(n)$ is a complete list of homomorphisms using Exercise 1.5.6 to represent $j(f)$ as $\int_0^1 f(x) e^*(x) dx$ with a bounded function e. *Hint:* Think of e and f as periodic functions on R^1. Then $j(f_1 \circ f_2) = j(f_1) j(f_2)$ implies $e(x+y) = e(x) e(y)$ a.e. on the plane, and therefore

$$e(x) \int_a^b e(y)\, dy = \int_a^b e(x+y)\, dy = \int_{a+x}^{b+x} e(y)\, dy$$

for almost all x and any a and b. Conclude that $e \in C^1(S^1)$ is a solution of $e'(x) = e'(0) e(x)$.

5. Summary

To sum up, the exponentials $e_n : n \in Z^1$ play four different roles: (a) They are the characters of the circle group; (b) they span the minimal invariant subspaces; (c) they are eigenfunctions of $Kf = f''$; (d) they can be identified with the homomorphisms of the algebra of summable functions.

This is the simplest statement of the four principal themes of the present chapter. You will see that, with appropriate modifications, they recur for (samples of) a wide class of important groups, and provide you with a powerful and flexible arsenal of Fourier methods, specially adapted to group-allied problems.

EXERCISE 9. Redo Subsections 1–4 for the standard $(n \geqslant 2)$-dimensional torus $T^n = R^n / Z^n$ of Subsection 1.10.1.

4.3 FOURIER INTEGRALS ON THE LINE

A group-theoretical interpretation is also available for Fourier integrals on the line, but with technical complications.

1. Characters

The exponentials

$$e_\gamma(x) = e^{2\pi i \gamma x}$$

are the characters of R^1, and the map $e_\gamma \to \gamma \in R^1$ is an isomorphism between the dual group $(R^1)^\wedge$ and R^1 itself, that is to say, R^1 is self-dual. The proof is made as before; the only difference is that no restriction is placed upon the phrase $\varphi(1) = 2\pi\gamma$.

2. Invariant Subspaces

The business of invariant subspaces is complicated by the fact that non-zero minimal closed subspaces do not exist. But there *is* a perfectly satisfactory analogue of Exercise 4.2.3.

EXERCISE 1.   Any invariant subspace M of $L^2(R^1)$ can be expressed as

$$\mathsf{M} = L^2(Q)^\wedge = (f \in L^2(R^1): \hat{f} = 0 \text{ off } Q)$$

for some measurable set $Q \subset R^1$. *Hint:*   By Exercise 1.3.13, M is separable. Pick $f_n: n \geqslant 1$ dense in M and let $Q = \bigcup_{n=1}^\infty (\gamma: \hat{f}_n(\gamma) \neq 0)$. Check that $\mathsf{M}^\wedge \subset L^2(Q)$. The fact that $\mathsf{M}^\wedge = L^2(Q)$ is now verified by picking k from the annihilator of $\mathsf{M}^\wedge$ in $L^2(Q)$ and concluding that $k = 0$ from

$$0 = \int e^{2\pi i \gamma y} \hat{f}(\gamma) k^*(\gamma) \, d\gamma$$

for every $y \in R^1$ and $f \in \mathsf{M}$.

The content of Exercise 1 can be expressed in a formal but suggestive way. The class M_γ of complex multiples of e_γ is closed under translations, and while it is *not* a subspace of $L^2(R^1)$, it is only "a little way out": namely,

$$\int_\alpha^\beta e_\gamma(x) \, d\gamma = \frac{e_\beta(x) - e_\alpha(x)}{2\pi i x}$$

belongs to $L^2(R^1)$ for any small interval $[\alpha, \beta]$ so you may think of $\mathsf{M}_\gamma \, d\gamma$ as a "thin slice" of $L^2(R^1)$. The content of Exercise 1 is that any closed invariant subspace M is the (perpendicular) sum (or better, the integral) of the "slices" in its "spectrum":

$$\mathsf{M} = \bigoplus_Q \mathsf{M}_\gamma \, d\gamma = \int_Q \mathsf{M}_\gamma \, d\gamma.$$

A special case of Exercise 1 is the fact that the translates of $f \in L^2(R^1)$ span $L^2(R^1)$ iff $(\gamma: \hat{f}(\gamma) = 0)$ is of measure 0; see Exercise 2.5.11 for an application to Hermite functions, and Exercise 4.2.4 for the analogue for the circle. The present statement is known as "Wiener's Tauberian theorem."

EXERCISE 2.[1] The translates of a summable function f span $\mathsf{L}^1(R^1)$ iff $\hat{f}(\gamma) \neq 0$ for *any* $\gamma \in R^1$. Check this for rapidly decreasing f. The general fact is known as "Wiener's Tauberian theorem for $\mathsf{L}^1(R^1)$." *Hint:* Any compact function of class $\mathbf{C}^\infty(R^1)$ can be expressed as $\hat{f}\hat{k}$ for some compact $\hat{k} \in \mathbf{C}^\infty(R^1)$. Why? Now look at $(\hat{f}\hat{k})^\vee = f \circ k$. The rest should be plain sailing.

3. Eigenfunctions

As before, it is very satisfactory to notice that the exponentials

$$e^{2\pi i \gamma x}$$

are (bounded) eigenfunctions of the differential operator $Kf = f''$. As for the circle, any nice differential operator commuting with translations and reflection is a polynomial in $K = D^2$.

4. Homomorphisms

The business of homomorphisms of $L^1(R^1)$ is technically more complicated, too. A homomorphism is defined as before, and $e_\gamma \circ f = e_\gamma \hat{f}(\gamma)$, but e_γ is not summable, so the old proof fails. But the *fact* is still true:

$$j_\gamma(f) = \hat{f}(\gamma) = \int f e_\gamma{}^*, \qquad \gamma \in R^1,$$

is a complete list of the homomorphisms of $\mathsf{L}^1(R^1)$.

PROOF. The map $f \to f_y$ interacts with the homomorphism j as follows:

$$j(f_y)\, j(k) = j(f_y \circ k) = j(f \circ k_y) = j(f)\, j(k_y)$$

for summable f and k. Pick k so as to make $j(k) = 1$. Then

$$j(f_y) = e(y)\, j(f)$$

for every summable f, with a universal function

$$e(y) = j(k_y).$$

This function is a character. To begin with,

$$e(x+y)\, j(f) = j(f_{x+y}) = e(x)\, j(f_y) = e(x)\, e(y)\, j(f)$$

---

[1] Adapted from Kac [1965].

so e is multiplicative. It is also bounded:

$$|e(y)| = |j(k_y)| \leqslant \text{constant} \times \|k\|_1.$$

$|e| \equiv 1$ follows readily, and from $e(0) = 1$, you conclude that

$$e(y) = e^{2\pi i \gamma y} = e_\gamma(y)$$

for some real γ. Now $j(f)$ may actually be evaluated as

$$j(f) = j(k)\, j(f) = j(k \circ f)$$

$$= j\left(\int k_{-y} f(y)\, dy\right)$$

$$= \int j(k_{-y}) f(y)\, dy$$

$$= \int f(y)\, e^{-2\pi i \gamma y}\, dy$$

$$= \hat{f}(\gamma).$$

The only point at issue is the passage from line 2 to line 3 which may be justified by picking $k \in C_\downarrow^\infty(R^1)$, still with $j(k) = 1$, and noting that the length

$$\left\| k \circ f - \sum_{|i| \leqslant ln} k_{-i/n} \int_{i/n}^{(i+1)/n} f(y)\, dy \right\|_1$$

tends to 0 as $n \uparrow \infty$ and $l \uparrow \infty$, in that order, for then

$$j(k \circ f) - \sum_{|i| \leqslant ln} j(k_{-i/n}) \int_{i/n}^{(i+1)/n} f(y)\, dy,$$

which is bounded by a constant multiple of this length, also tends to 0, and therefore

$$j(f) = \lim_{l \uparrow \infty} \lim_{n \uparrow \infty} \sum_{|i| \leqslant ln} e_\gamma^*(i/n) \int_{i/n}^{(i+1)/n} f(y)\, dy = \int f e_\gamma^*.$$

The proof is finished.

EXERCISE 3. Give a second proof that $j_\gamma(f) = \hat{f}(\gamma)$ is a complete list of homomorphisms in the style of Exercise 4.2.8.

EXERCISE 4. Redo Subsections 1–4 for R^n ($n \geqslant 2$).

EXERCISE 5. $L^1[0, \infty)$ inherits from $L^1(R^1)$ the product

$$f_1 \circ f_2(x) = \int_0^x f_1(x - y) f_2(y)\, dy.$$

Check that

$$j_\gamma(f) = \hat{f}(\gamma) = \int_0^\infty f(x)\, e^{-\gamma x}\, dx$$

is a homomorphism for any $\gamma \geqslant 0$. What is the most general homomorphism?

EXERCISE 6.   Check that the "Laplace transform" $f \to \hat{f}$, as defined in Exercise 5, is $1:1$ for $f \in L^2[0, \infty)$ ($\gamma = 0$ is now excluded). How do you actually compute f from $\hat{f}$? *Answer:*

$$f(x) = \lim_{a \downarrow 0} \frac{1}{2\pi} \int_{-\infty}^\infty \hat{f}(a + ib)\, e^{ibx}\, db.$$

Compute $[(1 + x^2)^{-1}]^\wedge$, $[\sin x]^\wedge$, $[x^4]^\wedge$. Can you apply your inversion formula to these special cases?

This kind of transform is specially well-suited to problems of electrical circuits as in Subsection 2.7.4, for instance. Heaviside introduced it for that purpose about the turn of the century, though the idea is much older; for additional information and applications, see Doetsch [1958].

EXERCISE 7.[2]   Define a new product for summable functions on the line by the recipe

$$f_1 \mathbin{\square} f_2(x) = f_1(x) \int_{-\infty}^x f_2(y)\, dy + f_2(x) \int_{-\infty}^x f_1(y)\, dy.$$

Check that
(a) $\square$ is commutative and associative;
(b) $\|f_1 \mathbin{\square} f_2\|_1 \leqslant \|f_1\|_1 \|f_2\|_1$;
(c)   every nontrivial homomorphism for the $\square$-product can be expressed as

$$j(f) = \int_{-\infty}^y f$$

for some $-\infty < y \leqslant \infty$, *Hint for* (c): The $\square$-product of the indicator functions 1_I and 1_J of intervals I and J of length $|I|$ and $|J| < \infty$ can be expressed $1_I \mathbin{\square} 1_J = |I|\, 1_J$ if I lies to the left of J, so $j(1_I)\, j(1_J) = |I|\, j(1_J)$, and if $j(1_J) \neq 0$, then $j(f) = \int f$ for every f that lives to the left of J. Try a second proof using Exercise 1.5.6 in the style of Exercise 4.2.8.

By (c), the inversion formula for the $\square$-product is simply the "fundamental theorem of calculus": namely, the transform $\hat{f}(y) = \int_{-\infty}^y f$ is inverted by

---

[2] Adapted from Lardy [1966].

differentiation $f^{\vee} = f'$! This kind of transform has been systematically exploited for solving combinatorial problems by Rota [1964]. A typical example is the "Möbius inversion formula" of number theory. This has to do with functions on the positive integers: the transform is

$$\hat{f}(n) = \sum_{d \text{ dividing } n} f(d),$$

and you invert by use of

$$f(n) = \sum_{d \text{ dividing } n} e(n/d) \hat{f}(d),$$

in which e is the "Möbius function":

$$e(n) = \begin{cases} 1 & \text{if } n = 1, \\ (-1)^m & \text{if } n > 1 \text{ is the product} \\ & \text{of } m \text{ unequal primes}, \\ 0 & \text{if } n > 1 \text{ is not a product} \\ & \text{of unequal primes}. \end{cases}$$

EXERCISE 8. Check the Möbius inversion formula. *Hint:* The binomial identity $\sum_{k=0}^{n} \binom{n}{k}(-1)^k = (1-1)^n = 0$ implies that $\sum_{k \text{ dividing } d} e(k) = 0$ if $d > 1$.

EXERCISE 9. Check that $f \to \hat{f}(n)$ is a homomorphism for the product

$$f_1 \circ f_2(n) = \sum_{(i,j) = n} f_1(i) f_2(j),$$

(i,j) being the least common multiple of i and j.

4.4 FINITE COMMUTATIVE GROUPS

The Fourier idea is seen in its simplest form on a finite commutative group. The purpose of this section is to develop the necessary facts about such groups. The Fourier series themselves occupy Section 4.5; an application to number theory will be found in Section 4.6.

Let G be a finite commutative group, and let H be a subgroup. The coset space G/H is always a group [see Exercise 4.1.5]: the so-called *factor group* $\#(G)$ is the cardinality of G. For $g \in G$, $\#_G(g) = \#(g)$ is the smallest integer $n \geqslant 1$ such that $g^n = 1$. The following exercises have to do with these notions.

EXERCISE 1. Check that $\#(G) = \#(G/H) \#(H)$. *Hint:* Two cosets gH are either identical or disjoint.

EXERCISE 2. Check the following facts: $\#(g)$ divides $\#(G)$ for every $g \in G$. $g^n = 1$ iff n is an integral multiple of $\#(g)$. $\#_{G/H}(gH)$ divides $\#_G(g)$. *Hint:* Use Exercise 1 with $H = (g^k: 0 \leqslant k < \#(g))$ to prove the first assertion.

The "product" $G_1 \times G_2$ of the groups G_1 and G_2 is the set-theoretical product, made into a group by use of componentwise multiplication: If $g = (g_1, g_2)$ and $h = (h_1, h_2)$, then $gh = (g_1 h_1, g_2 h_2)$. A simple [infinite] example is provided by Z^2 [the lattice of integral points in the plane] which is the direct product of two copies of $Z^1 \colon Z^2 = Z^1 \times Z^1$.

BASIS THEOREM. *Any finite commutative group G is isomorphic to a direct product of "counters"*

$$Z_m^+ = \textit{the integers under addition modulo } m,$$
$$\textit{alias the multiplicative group of mth roots}$$
$$\textit{of unity } e^{2\pi i k/m} \colon 0 \leqslant k < m,$$

i.e., G is isomorphic to

$$Z_{m_1}^+ \times \cdots \times Z_{m_n}^+$$

for some $1 \leqslant n < \infty$ and some integral $m_i \colon i \leqslant n$.

The m_i's need not be primes, but you can choose them to be powers of a prime. You may take this on faith if you like and pass directly on to Section 4.5. The proposition goes back to Gauss; the proof presented below is adapted from Speiser [1945, pp. 46–49].

Step 1: Check that G is isomorphic to $Z_p^+ \times \cdots \times Z_p^+$ if $\#(g)$ is a fixed prime p for every $g \in G$, excepting $g = 1$.

PROOF. Pick $g_i \colon i \leqslant n$ from G such that

$$g_1^{e_1} \cdots g_n^{e_n} \neq 1$$

for any $0 \leqslant e_i < p$, excepting $e_i \equiv 0$, and make n as big as possible subject to this condition. These products fill out a subgroup H of G which is isomorphic to $Z_p^+ \times \cdots \times Z_p^+$ (n-fold), and the statement is that $H = G$; if not, you could find $g \notin H$, and you could augment n (against the assumption), as follows from the fact that $\#_{G/H}(g) > 1$, since gH is not the identity in G/H, and therefore $\#_{G/H}(g) = p$, since it divides $\#_G(g) = p$ [see Exercise 2].

Step 2: Check that G is isomorphic to a product of counters if $\#(g)$ is a power of a fixed prime p for every $g \in G$.

PROOF. Put

$$\max_{G} \#(g) = p^m.$$

The proposition is proved by induction on m; Step 1 is the case $m = 1$, and you assume that everything works for any integer less than m for $m \geqslant 2$. Now $G' = (g^p : g \in G)$ is a subgroup of G, and

$$\max_{G'} \#(g') = p^{m-1},$$

so G' is a product of counters. Therefore, you can pick $n \geqslant 1$ and g_i': $i \leqslant n$ out of G' so that every $g' \in G'$ can be expressed in precisely one way as

$$g' = (g_1')^{e_1'} \cdots (g_n')^{e_n'}, \qquad 0 \leqslant e_i' < \#(g_i') = p^{f_i}.$$

Because $g_i' \in G'$, it is a pth power of some $g_i \in G$, and

$$\#(g_i) = p\#(g_i') = p^{f_i+1}.$$

The claim is that

$$g_1^{e_1} \cdots g_n^{e_n} \neq 1$$

for any $0 \leqslant e_i < p^{f_i+1}$, excepting $e_i \equiv 0$. In the opposite case,

$$g_1^{a_1} \cdots g_n^{a_n} = 1$$

for some $0 \leqslant a_i < p^{f_i+1}$, and a_1 would have to be an integral multiple of $\#_{G/K}(g_1 K)$, K being the subgroup

$$(g_2^{b_2} \cdots g_n^{b_n} : 0 \leqslant b_i < p^{f_i+1}).$$

But $\#_{G/K}(g_1 K)$ divides $\#_G(g_1)$ and is therefore a power of p, so that a_1 (and likewise every one of the other exponents a_i) is a power of p, and the offending identity $g_1^{a_1} \cdots g_n^{a_n} = 1$ can be expressed in the forbidden form

$$(g_1')^{e_1'} \cdots (g_n')^{e_n'} = 1, \qquad 0 \leqslant e_i' < p^{f_i}!$$

This contradiction shows that the subgroup

$$H = (g_1^{e_1} \cdots g_n^{e_n} : 0 \leqslant e_i < p^{f_i+1})$$

is isomorphic to

$$Z_{p^{f_1+1}}^+ \times \cdots \times Z_{p^{f_n+1}}^+.$$

Now if $H = G$, you are finished. If not, pick $g \notin H$. Then $g^{-p} \in G'$; as such it is the pth power of some $h \in H$, and $\#(gh) = p$ since $(gh)^p = 1$ but $gh \neq 1$. By the now familiar argument

$$g_1^{e_1} \cdots g_n^{e_n} (gh)^{e_{n+1}} \neq 1$$

for any $0 \leqslant e_i < p^{f_i + 1}$ and $e_{n+1} < p$, excepting $e_i \equiv 0$. Put $g_{n+1} = gh$ and $f_{n+1} = 0$. Then the subgroup

$$H_1 = (g_1^{e_1} \cdots g_{n+1}^{e_{n+1}} : 0 \leqslant e_i < p^{f_i + 1})$$

is isomorphic to $H \times Z_p^+$, it contains g, and if you have not already exhausted (either yourself or) the whole of G, you can go on in this way adjoining factors Z_p^+ until you have done so. The bulk of the proof is finished.

Step 3:   You have only to check that G splits into a direct product of groups of the kind disposed of in Step 2.

PROOF.   Pick a prime p dividing $\#(g)$ for some $g \neq 1$ from G. Both

$$G_1 = (g_1 : \#(g_1) \text{ is a power of } p)$$

and

$$G' = (g' : \#(g_1') \text{ is not divisible by } p)$$

are subgroups of G, as is plain from the fact that $\#(g_1 g_2)$ divides $\#(g_1) \#(g_2)$. Also, G_1 contains a nontrivial power of g so that $\#(G_1) \geqslant 2$, and if you now grant that G is isomorphic to $G_1 \times G'$, the rest will follow by induction: G' splits into $G_2 \times G''$ so as to make $\#(g_2)$ a power of a fixed prime p_2 for every $g_2 \in G_2$ and $\#(G_2) \geqslant 2$, and so on. The point at issue is whether or not every $g \in G$ can be expressed as a product $g = g_1 g'$ in precisely one way. The fact that you can have at most one such splitting is self-evident from $G_1 \cap G' = 1$. To prove the existence of such a splitting, put $\#(g) = p^f q$ with q not divisible by p and pick integers i and j so that $ip^f + jq = 1$. Then

$$g = g^{ip^f} g^{jq}$$

is the desired splitting: Namely, for the second factor,

$$(g^{jq})^{p^f} = (g^{p^f q})^j = 1,$$

so that $\#(g^{jq})$ divides p^f and is therefore a power of p, which is to say that $g^{jq} \in G_1$, while, for the first factor,

$$(g^{ip^f})^q = (g^{p^f q})^i = 1,$$

so that $\#(g^{ip^f})$ divides q and is therefore not divisible by p, which is to say that $g^{ip^f} \in G'$. The proof is finished.

4.5   FOURIER SERIES ON A FINITE COMMUTATIVE GROUP

By Section 4.4, you may as well suppose that the group in question is a product of counters:

$$G = Z_{m_1}^+ \times \cdots \times Z_{m_n}^+.$$

The usefulness of this splitting may now be seen in the following computation of the dual group $\hat{G}$.

Z_m^+ is identified with the integers $0 \leqslant k < m$ under addition modulo m, and $g \in G$ is identified with a point $k = (k_1, \cdots, k_n)$ of the additive group $Z_{m_1}^+ \times \cdots \times Z_{m_n}^+$, that is to say, you identify g with $k_1 g_1 + \cdots + k_n g_n$, in which $g_l = (0, \cdots, 1, \cdots, 0)$ contains a 1 in the lth place and 0's elsewhere. A character $e \in \hat{G}$ is now seen to split as

$$e(g) = e(g_1)^{k_1} \cdots e(g_n)^{k_n},$$

and since

$$e(g_l)^{m_l} = e(g_l^{m_l}) = e(1) = 1,$$

you see that the numbers $e(g_l)$ are m_lth roots of unity:

$$e(g_l) = e^{2\pi i j_l / m_l}$$

for some $0 \leqslant j_l < m_l$. But then the character e is completely specified by

$$j = (j_1, \cdots, j_n) \in Z_{m_1}^+ \times \cdots \times Z_{m_n}^+,$$

and the map $e \to j$ establishes an isomorphism between the dual group $\hat{G}$ and the group $Z_{m_1}^+ \times \cdots \times Z_{m_n}^+$; in particular, G is self-dual!

A function f on G may be expanded into a sum of characters, or "Fourier series." The chief point is that under the inner product

$$(f_1, f_2) = \sum_{g \in G} f_1(g) f_2(g)^*,$$

the characters form a perpendicular family:

$$(e_1, e_2) = \#(G) \quad or \quad 0 \qquad according \ as \ e_1 = e_2 \ or \ not.$$

PROOF.

$$e_1(g_0)(e_1, e_2) = \sum_G e_1(g_0) e_1(g) e_2(g)^*$$

$$= \sum_G e_1(g_0 g) e_2(g)^*$$

$$= \sum_G e_1(g) e_2(g_0^{-1} g)^*$$

$$= \sum_G e_1(g) e_2(g_0^{-1})^* e_2(g)^*$$

$$= e_2(g_0)(e_1, e_2),$$

for any $g_0 \in G$, so either $e_1 \equiv e_2$ or else $(e_1, e_2) = 0$. The evaluation

$$\|e\|^2 = (e, e) = \sum_G |e(g)|^2 = \#(G)$$

is automatic from $|e(g)| = 1$. The proof is finished.

EXERCISE 1.   Check that if H is a subgroup of G, then

$$\sum_H e(h) = \#(H) \quad \text{or} \quad 0 \qquad \text{according as } e \equiv 1 \text{ on } H \text{ or not.}$$

Hint: Think of the sum as the inner product of e with the identity of $\hat{H}$.

Every $g \in G$ can be viewed as a character on $\hat{G}$, i.e., as an element in $G^{\wedge\wedge}$, by defining

$$g(e) \equiv e(g).$$

Clearly g is multiplicative:

$$g(e_1 e_2) \equiv (e_1 e_2)(g) = e_1(g)e_2(g) = g(e_1)g(e_2),$$

and of modulus 1. Moreover in the present circumstances [G isomorphic to $\hat{G}$] distinct elements in G give rise to distinct elements in $G^{\wedge\wedge}$ and therefore you see that

$$\sum_{\hat{G}} e(g_1)e(g_2)^* = \#(\hat{G}) \quad \text{or} \quad 0 \qquad \text{according as } g_1 = g_2 \text{ or not.}$$

This leads at once to the

PLANCHEREL THEOREM.   *Any function f on G can be expanded into a Fourier series*

$$f = \sum_{\hat{G}} \hat{f}(e)\, e$$

with coefficients

$$\hat{f}(e) = \#(G)^{-1}(f, e) = \#(G)^{-1} \sum_G f(g)e(g)^*,$$

and there is a Plancherel identity:

$$\|f\|^2 = \sum_G |f(g)|^2 = \#(G) \sum_{\hat{G}} |\hat{f}(e)|^2 = \#(G)\|\hat{f}\|^2.$$

PROOF.   To see that $f = \sum \hat{f}(e)\, e$, just compute the sum as follows:

$$\sum_{\hat{G}} \hat{f}(e)e(g_0) = \sum_{\hat{G}} e(g_0)\#(G)^{-1} \sum_G f(g)e(g)^*$$

$$= \sum_G f(g)\#(G)^{-1} \sum_{\hat{G}} e(g_0)e(g)^*$$

$$= f(g_0).$$

The proof of the Plancherel identity is just as easy.

The next topic is the Poisson summation formula, but first a brief aside. Given a subgroup H of G, let $(G/H)'$ be the class of characters e of G which are trivial on H: $e(h) \equiv 1$. A character of this kind is a function of cosets gH and so can be thought of as a character of the factor group G/H. This correspondence goes the other way too: any character e' of G/H can be lifted up to G by the rule $e(g) = e'(gH)$ and so can be thought of as belonging to $(G/H)'$. To sum up, $(G/H)'$ and $(G/H)^\wedge$ are isomorphic, and you may take the liberty of confusing the two. The Poisson summation formula may now be stated as

$$\sum_H f(h) = \#(H) \sum_{(G/H)^\wedge} \hat{f}(e).$$

PROOF.   Bring in the function

$$f^0(g) = \sum_H f(gh)$$

and compute

$$\#(G) f^{0\wedge}(e) = \sum_G f^0(g) e(g)^*$$

$$= \sum_G \sum_H f(gh) e(g)^*$$

$$= \sum_H \sum_G f(g) e(gh^{-1})^*$$

$$= \sum_H e(h) \sum_G f(g) e(g)^*$$

$$= \#(H) \#(G) \hat{f}(e)$$

if $e \equiv 1$ on H, and 0 otherwise, in accordance with Exercise 1. But then

$$f^0(g) = \sum_{\hat{G}} f^{0\wedge}(e) e(g)$$

$$= \#(H) \sum_{(G/H)^\wedge} \hat{f}(e) e(g),$$

and Poisson's formula drops out upon putting $g = 1$.

This formula should be compared with the Poisson summation formula for functions on the line:

$$\sum_{Z^1} f(n) = \sum_{Z^1} \hat{f}(n)$$

[see Subsection 2.7.5]. The similarity is plain; in fact, if G is the additive group R^1 and if H is the subgroup Z^1, then G/H is the circle group $S^1 = [0, 1)$, and $(G/H)^\wedge = (S^1)^\wedge$ is a copy of Z^1 [see Exercise 4.1.10]. Therefore, apart from the factor $\#(H)$, the formula has the same group-theoretical flavor in both cases.

EXERCISE 2★.[1]   Find the solution of

$$x_0^2 - x_1 x_2 = y_0, \qquad x_1^2 - x_2 x_0 = y_1, \qquad x_2^2 - x_0 x_1 = y_2,$$

for known y. Think of x and y as functions on the additive group Z_3. The dual group $Z_3^\wedge$ can be identified with the multiplicative group of cube roots of unity:

$$\omega = 1, \quad e^{2\pi i/3}, \quad e^{4\pi i/3},$$

via the formula $e(k) = \omega^k$. Check the identity

$$\hat{x}(e^{2\pi i/3}\omega)\,\hat{x}(e^{4\pi i/3}\omega) = \tfrac{1}{3}\hat{y}(\omega^2)$$

and obtain the *Answer:*

$$\hat{x}(\omega) = \pm\,[\hat{y}(\omega^2)]^{-1}\{\tfrac{1}{27}\hat{y}(1)\,\hat{y}(e^{2\pi i/3})\,\hat{y}(e^{4\pi i/3})\}^{\frac{1}{2}},$$

assuming $\hat{y} \neq 0$.

EXERCISE 3.[2]   Check that for any function f on G, the determinant of the $\#(G) \times \#(G)$ matrix $[f(g_1 g_2^{-1})]$ can be expressed as

$$\det[f(g_1 g_2^{-1})] = \prod_{\hat{G}} \#(G)\,\hat{f}(e)$$

and use this to prove a primitive variant of Exercise 4.2.4. *Hint:*

$$f(g_1 g_2^{-1}) = \sum_{\hat{G}} \hat{f}(e)\,e(g_1)\,e(g_2)^*.$$

EXERCISE 4.[3]   Check the following variant of the Poisson summation formula:

$$\sum_{G/H}\left|\sum_{H} f(gh)\right|^2 = \#(G)\,\#(H)\sum_{(G/H)^\wedge} |\hat{f}(e)|^2$$

with the previous convention about $(G/H)^\wedge$. *Hint:* Apply the Plancherel identity to $f^\circ(g) = \sum_H f(gh)$.

EXERCISE 5.[3]   The group $G = Z_2^+ \times \cdots \times Z_2^+$ (n-fold) is placed in $1:1$ correspondence with the set $Q = 0, 1, \cdots, 2^n - 1$ by mapping

$$g = (k_0, \cdots, k_{n-1}) \to \sum_{i=0}^{n-1} k_i 2^i,$$

---

[1] K. Itô, private communication.
[2] Adapted from W. N. Anderson, Jr., private communication.
[3] After Crimmins *et al.* [1969].

in which $k_i = 0$ or 1 for $0 \leqslant i < n$. This permits you to think of Q as a group isomorphic to G. Prove that $j: G \to Q$ is an isomorphism iff

$$j(g) = \sum_{i=0}^{n=1} [1 - e_i(g)] 2^{i-1},$$

in which e_i: $0 \leqslant i < n$ is a basis of the dual group $\hat{G}$, that is to say, every character e can be expressed in precisely one way as a product $e = e_0^{k_0} \cdots e_{n-1}^{k_{n-1}}$ with $0 \leqslant k_i < 2$.

EXERCISE 6.   The *infinite* product $G = Z_2^+ \times Z_2^+ \times \cdots$ is a commutative group; it may be put into correspondence with the interval $0 \leqslant x < 1$ by means of the map

$$j: g = (k_1, k_2, \cdots) \to x = \sum_{i=1}^{\infty} k_i 2^{-i}.$$

The correspondence is not $1:1$ owing to the fact that rational numbers x have ambiguous binary expansions, but they fill up a set of measure 0, only. Check that the action of G on $[0, 1]$ defined by the recipe $gx = j[gj^{-1}(x)]$ preserves the lengths of intervals and therefore the measure of nice sets, also. The so-called "Rademacher function" $e_n(x) = 1 - 2k_n$ is a character of G, as is the "Walsh function"

$$e^i(x) = e_1^{i_1}(x) e_2^{i_2}(x) \cdots$$

for any "tame" $i = (i_1, i_2, \cdots) \in G$, i.e., any string of 0's and 1's with only 0's from some point on. Draw pictures of the first few Rademacher functions. Prove that the family of Walsh functions is a unit-perpendicular basis of $L^2[0, 1]$.

4.6*   GAUSS' LAW OF QUADRATIC RECIPROCITY

Fourier series on a finite commutative group *look* very simple, but the applications can be both complicated and deep. Applications to number theory can be found in Chandrasekharan [1968], Hardy and Wright [1954], and Rademacher [1956]; to statistical mechanics in Ginibre [1970], and McKean [1964]; to coding in Crimmins *et al.* [1969]; and this is only a tiny sample. The present application is to number theory.

Gauss' law of quadratic reciprocity has to do with a problem of arithmetic: *for which integers $0 < n < p$ is it possible to solve the quadratic congruence $x^2 = n$, modulo p, for a fixed prime $p > 2$,* or, what is the same, which of the integers $0 < n < p$ are "quadratic residues" modulo p? To study this problem, bring in the "Legendre symbol"

$$e(n) = \left(\frac{n}{p}\right) = \begin{cases} +1 & \text{if } n \text{ is a quadratic residue modulo } p, \\ -1 & \text{otherwise}. \end{cases}$$

Gauss' law states that for any odd primes p and q

$$\left(\frac{p}{q}\right)\left(\frac{q}{p}\right) = (-1)^{(1/2)(p-1)(1/2)(q-1)}.$$

The purpose of this section is to prove this fact by Fourier methods via the so-called *Gaussian sum* defined by the recipe

$$G_q(e^{2\pi in/q}) = \sum_{k=0}^{q-1} e^{2\pi ik^2n/q}$$

for $0 \leqslant n < q$ and any odd integer q, prime or not. This is one of the simpler though not the most elementary proofs of the law of quadratic reciprocity. Gauss himself gave eight different proofs; see Nagell [1951, p. 144], and, for additional information on this beautiful circle of ideas, Bachmann [1907] and/or Rademacher [1964]. The first steps of the proof are contained in Exercises 1 and 2. $Z_p^{\times}$ is the group of integers $0 < n < p$ under multiplication modulo p [see Exercise 4.1.1]. Q is the class of quadratic residues $0 < n < p$, and Q' the complementary class of quadratic nonresidues.

EXERCISE 1.   Check that the integers $1^2, 2^2, \cdots, (p-1)^2$, considered modulo p, provide a twofold list of Q; in particular, both Q and Q' contain $\frac{1}{2}(p-1)$ integers, each.

EXERCISE 2.   Check that the Legendre symbol $e(n) = (n/p)$ is a character of Z_p, that is to say, $Q \cdot Q \subset Q$, $Q' \cdot Q \subset Q'$, and $Q' \cdot Q' \subset Q$. *Hint:*   Check $Q \cdot Q \subset Q$ first; the rest follows by counting with the help of Exercise 1.

Here, the proof takes a peculiar twist: You take e, which is a character of $Z_p^{\times}$, extend it to the additive group Z_p^+ by putting $e(0) = 0$, and expand it into a Fourier series on the latter. This is how the Gaussian sums come in. To begin with,

$$e^{\vee}(n) = \sum_{k=0}^{p-1} e(k) e^{2\pi ink/p}$$

$$= \sum_Q e^{2\pi ink/p} - \sum_{Q'} e^{2\pi ink/p}$$

for $0 \leqslant n < p$, in which $Z_p^{+\wedge}$ is identified with the multiplicative group of pth roots of unity. The elementary identity

$$0 = \sum_{k=0}^{p-1} e^{2\pi ink/p} = 1 + \sum_Q e^{2\pi ink/p} + \sum_{Q'} e^{2\pi ink/p}$$

and Exercise 1 are now applied to identify $e^\vee$ as a Gaussian sum:

$$e^\vee(n) = 1 + 2\sum_Q e^{2\pi ink/p}$$

$$= \sum_{k=0}^{p-1} e^{2\pi ink^2/p}$$

$$= G_p(e^{2\pi in/p}).$$

This leads at once to a useful formula: For $0 < n < p$, nk runs through Z_p^{+} once (modulo p) as k runs from 0 to $p-1$, so by Exercise 2,

$$G_p(e^{2\pi in/p}) = e^\vee(n) = \sum_{k=0}^{p-1} e(nk)\, e^{2\pi ink/p}\, e(n)$$

$$= \sum_{k=0}^{p-1} e(k)\, e^{2\pi ik/p}\, e(n)$$

$$= G_p(e^{2\pi i/p})\, e(n),$$

permitting you to express the Legendre symbol for $0 < n < p$ as a ratio of Gaussian sums:

$$e(n) = \frac{G_p(e^{2\pi in/p})}{G_p(e^{2\pi i/p})}.$$

The latter is nothing but an odd way of writing the Z_p^{+} Fourier series for e, modified at $n = 0$ so as to make $e(0) = G_p(e^{2\pi i/p})^{-1}$.

EXERCISE 3.   Deduce that $G_p(e^{2\pi i/p}) = \sqrt{p}$ times a power of i; no computations are needed! *Hint:* $e^{\vee\vee} = G_p(e^{2\pi i/p})^2 e = pe(-\cdot)$, and this implies $G_p^2 = \pm p$.

EXERCISE 4.

$$G_{pq}(e^{2\pi i/pq}) = G_p(e^{2\pi iq/p})\, G_q(e^{2\pi ip/q})$$

for any odd primes p and q. *Hint:* Compute the left-hand side from the definition, using the fact that as $k[j]$ runs once from 0 to $p-1$ $[q-1]$, $kq + jp$ runs once over $0 \leqslant n < pq - 1$, modulo pq.

The law of quadratic reciprocity can now be stated entirely in terms of Gaussian sums:

$$(-1)^{(1/2)(p-1)(1/2)(q-1)} = \left(\frac{p}{q}\right)\left(\frac{q}{p}\right)$$

$$= \frac{G_q(e^{2\pi ip/q})\, G_p(e^{2\pi iq/p})}{G_q(e^{2\pi i/q})\, G_p(e^{2\pi i/p})}$$

$$= \frac{G_{pq}(e^{2\pi i/pq})}{G_p(e^{2\pi i/p})\, G_q(e^{2\pi i/q})}.$$

EXERCISE 5. Check that the law of quadratic reciprocity follows from the evaluation of the Gaussian sum:

$$G_p(e^{2\pi i/p}) = \sqrt{p}\,(i)^{[(p-1)/2]^2}$$

for *any* odd integral p, prime or not. *Hint:* The elementary congruence

$$\left(\frac{pq-1}{2}\right)^2 - \left(\frac{p-1}{2}\right)^2 - \left(\frac{q-1}{2}\right)^2 = \frac{(p-1)(q-1)}{2} \quad \text{modulo } 4$$

is helpful.

The proof is finished by the actual evaluation of the Gaussian sum, based upon the deeper formula of Landsberg and Schaar:

$$\frac{1}{\sqrt{p}} \sum_{n=0}^{p-1} e^{2\pi i n^2 q/p} = \frac{e^{\pi i/4}}{\sqrt{2q}} \sum_{n=0}^{2q-1} \exp\left(\frac{-\pi i n^2 p}{2q}\right)$$

for any integral p and $q \geq 1$.

PROOF.[1] The proof is based upon the Jacobi identity for the theta-function:

$$\sum_{n=-\infty}^{\infty} \exp(-\pi n^2 t) = t^{-\frac{1}{2}} \sum_{n=-\infty}^{\infty} \exp(-\pi n^2/t)$$

[see Subsection 1.7.5], which was proved for $t > 0$, but is actually valid in the open right-hand half-plane, both sides being analytic in that region. Replace $t > 0$ by $t - 2iq/p$ and make $t\downarrow 0$. The left-hand side of Jacobi's identity is

$$\sum e^{-\pi n^2 t} \exp(2\pi i n^2 q/p) = t^{-\frac{1}{2}}\left[(1/p)\sum_{n=0}^{p-1}\exp(2\pi i n^2 q/p) + o(1)\right],$$

since $\exp(2\pi i n^2 q/p)$, as a function of n, is of period p, and

$$\sum \exp(-\pi n^2 t) = t^{-\frac{1}{2}}[1 + o(1)].$$

Besides, the right-hand side is

$$\frac{1}{(t-2iq/p)^{\frac{1}{2}}} \sum \exp\left(-\frac{\pi n^2 t}{t^2 + 4q^2/p^2}\right)\exp\left(-\frac{2\pi i n^2 q/p}{t^2 + 4q^2/p^2}\right)$$

$$= \frac{1}{(-2iq/p)^{\frac{1}{2}}}\left(\frac{4q^2}{tp^2}\right)^{\frac{1}{2}}\left[\frac{1}{2q}\sum_{n=0}^{2q-1}\exp\left(\frac{-\pi i n^2 p}{2q}\right) + o(1)\right]$$

for similar reasons. A comparison of the two expressions produces the Landsberg–Schaar identity.

---

[1] Adapted from Bellman [1961].

The evaluation of the Gaussian sum is now achieved by putting $q = 1$:

$$G_p(e^{2\pi i/p}) = \left(\frac{p}{2}\right)^{1/2} e^{\pi i/4} [1 + e^{-\pi i p/2}]$$

$$= \sqrt{p}(i)^{[(p-1)/2]^2},$$

as you can easily check by looking at the cases $p = 1$ modulo 4 and $p = 3$ modulo 4, separately.

EXERCISE 6.   Use the Landsberg–Schaar identity to prove the so-called "supplementary theorems":

(a) $\left(\dfrac{-1}{p}\right) = (-1)^{[(p-1)/2]^2}$

(b) $\left(\dfrac{2}{p}\right) = (-1)^{(p^2-1)/8}$.

Hint: (b) is proved by putting $q = 2$; as to (a), look at

$$\left(\frac{-1}{p}\right) G_p(e^{2\pi i/p}) = G_p(e^{-2\pi i/p}) = G_p(e^{2\pi i/p})^*.$$

EXERCISE 7.   Check that the Möbius function of Exercise 4.3.8 can be expressed as

$$e(n) = \sum_{\substack{1 \leqslant k < n \\ (k,n)=1}} e^{2\pi i k/n}.$$

$(k, n) = 1$ signifies that k and n have no common primes. Hint: Prove first that the sum $[f(n)]$ is "multiplicative" $[f(ij) = f(i)f(j)$ if $(i,j) = 1]$ with the help of the trick suggested for use in Exercise 4. Then evaluate $f(n)$ for n a prime power.

4.7 NONCOMMUTATIVE GROUPS

The rest of this chapter is devoted to a number of special but important noncommutative groups.

A finite noncommutative group G cannot have enough characters to do Fourier series: If it did, then every function f on G could be expanded as a sum

$$f(g) = \sum_{\hat{G}} \hat{f}(e) e(g),$$

and you would have

$$f(g_1 g_2) = \sum \hat{f}(e) e(g_1) e(g_2) = f(g_2 g_1),$$

for every g_1 and g_2 in G. But that cannot happen unless G is commutative. What to do?

To overcome this obstacle Frobenius introduced the idea of a group "representation," towards the end of the 19th century. This is a map e from G into the group of $n \times n$ nonsingular complex matrices (for some $1 \leqslant n < \infty$) which respects the multiplication of G:

$$e(g_1 g_2) = e(g_1) e(g_2).$$

EXERCISE 1.   A one-dimensional representation of a finite group is a character. *Hint:* $e(g^n) = e(g)^n$ is bounded in modulus, by a finite constant which is independent of n.

An n-dimensional representation e is called "reducible" if, after a common similarity $e \to jej^{-1}$ with nonsingular j, the matrices $e(g)$ split into a $k \times k$ block e' and an $(n-k) \times (n-k)$ block e'' for some fixed $0 < k < n$:

$$jej^{-1} = \begin{pmatrix} e' & 0 \\ 0 & e'' \end{pmatrix}.$$

The Fourier transform of a function f on G is now defined by the recipe

$$\hat{f}(e) = \#(G)^{-1} \sum_G f(g) e(g)^{\#}$$

for *irreducible* representations e, $e^{\#}$ being the conjugate transpose of e. The transform may be inverted by the rule

$$f(g) = \sum_{\substack{\text{irreducible} \\ \text{representations}}} \text{sp}[\hat{f}(e) e(g)] |e|,$$

in which sp stands for the "spur" or "trace" of the matrix inside the brackets and $|e|$ is a positive "weight" ascribed to e. There is even a Plancherel identity:

$$\|f\|^2 = \sum_{\substack{\text{irreducible} \\ \text{representations}}} \text{sp}[\hat{f}(e) \hat{f}(e)^{\#}] |e|.$$

The initial discouragement at having insufficiently many characters is now dispelled: The group did not have "enough" one-dimensional representations, but it *does* have enough several-dimensional ones. In fact, the whole Fourier apparatus can be reconstructed, though it becomes much more cumbersome than before. A nice account of the representations of the group of permutations of n letters can be found in Weyl [1931]. A second important class for which everything is known is the crystallographic groups; see, for example, van der Waerden [1948]. Besides, the whole thing can be carried over to a wide class of continuous noncommutative groups. The best elementary account of the whole field (finite groups included) is Boerner [1963]; for a deeper and quite personal account, see Weyl [1939]. The finite groups are now left to the interested student, and the rest of the chapter is devoted to three noncommutative groups of prime geometrical importance:

SO(3):   the group of proper rotations of three-dimensional space, that is to say, the real 3×3 orthogonal matrices of determinant $+1$.

M(2):   the group of proper rigid motions of the plane, that is to say, the sense-preserving Euclidean congruences.

SL(2, R):   the group of proper rigid motions of the hyperbolic plane, that is to say, 2×2 real matrices of determinant $+1$.

4.8   THE ROTATION GROUP

A real 3×3 matrix g is "orthogonal" if its transpose $g^{\sharp}$ is also its inverse $[g^{\sharp}g = gg^{\sharp} = 1]$.[1] The class of all such matrices is a group, the so-called "orthogonal group" O(3). A member g of O(3) acts upon R^3 according to the rule

$$(gx)_i = \sum_{j=1}^{3} g_{ij} x_j.$$

This action is such as to preserve inner products

$$\sum_{i=1}^{3} x_i y_i = x \cdot y = x \cdot g^{\sharp} g y = gx \cdot gy$$

and, especially, lengths

$$\sum_{i=1}^{3} x_i^2 = |x|^2 = |gx|^2,$$

i.e., g is a "rigid motion" of R^3. The determinant $\det(g)$ of $g \in O(3)$ is ± 1:

$$[\det(g)]^2 = \det(g) \det(g^{\sharp}) = \det(gg^{\sharp}) = \det(1) = 1.$$

This lets you classify $g \in O(3)$ as *proper* $[\det(g) = +1]$ or *improper* $[\det(g) = -1]$. The geometrical significance of this distinction is that a proper rotation preserves the orientation ("handedness") of R^3, while an improper one reverses it. The subgroup of proper rotations is the "special orthogonal group" $G = SO(3)$, which is the subject of the next few sections.

---

[1] The superscript $\sharp$ still means "conjugate transpose" but since g is real the conjugation is of no account.

EXERCISE 1. Explain why SO(3) is a subgroup. Do the improper rotations form a subgroup, too? A typical improper rotation is the reflection

$$k = \begin{pmatrix} 0 & 1 & 0 \\ 1 & 0 & 0 \\ 0 & 0 & 1 \end{pmatrix} : (x_1, x_2, x_3) \rightarrow (x_2, x_1, x_3).$$

The orthogonal group O(3), pictured as a "submanifold" of R^9, comes in two disconnected pieces SO(3) and kSO(3). The adjective "disconnected" means that it is impossible to move from a point of SO(3) to a point of kSO(3) along a continuous path in O(3). Why?

The rest of this section is devoted to elementary facts about $G = $ SO(3) needed later; for additional information, see Gelfand *et al.* [1963].

1.   SO(3) Is Noncommutative

The proof is easy: If a, b, c are the points shown in Fig. 1 and if g_1 is the "east pole" rotation mapping c into a leaving b fixed, while g_2 is the "north pole" rotation mapping a into b leaving c fixed, then $g_1 g_2 c = g_1 c = a$, whereas $g_2 g_1 c = g_2 a = b$. The noncommutativity of SO(3) is in fact so bad that $e \equiv 1$ *is the only character of* SO(3), as will be proved in Subsection 4.

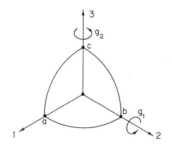

FIGURE 1

2.   Any Rotation Has an Axis

Every $g \in $ SO(3) has an axis $x \in R^3 [gx = x]$. Relative to this axis, g is simply a two-dimensional rotation in the plane perpendicular to x, as pictured in Fig. 2. The point is that if y is perpendicular to x, then so is gy:

$$gy \cdot x = gy \cdot gx = y \cdot x = 0.$$

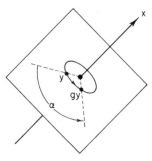

FIGURE 2

PROOF.   The existence of the axis comes down to checking that 1 is an eigenvalue of $g \in SO(3)$, which is the same as to say that $\det(g-1)=0$. But $\det(g)=\det(g^{\sharp})=1$, so

$$\begin{aligned} \det(g-1) &= \det(g^{\sharp})\det(g-1) = \det[g^{\sharp}(g-1)] \\ &= \det(1-g^{\sharp}) = \det[(1-g)^{\sharp}] = \det(1-g) \\ &= (-1)^3\det(g-1) = -\det(g-1), \end{aligned}$$

which does the trick. The fact that the dimension is *odd* is essential; for example, the only rotation of R^2 that has an axis is $g=1$.

EXERCISE 2. $g \in SO(3)$ has two (independent) axes iff it is the identity. *Hint:* $\det(g)$ is the product of the eigenvalues of g.

3.   Projective Space and Exponential Map

The group $SO(3)$ can be pictured as a three-dimensional "surface" in R^9: $g \in SO(3)$ has nine entries and $g^{\sharp}g = 1$ cuts the dimension down by 6 degrees of freedom. But this is not a good way to look at it. A better picture is provided by the closed three-dimensional ball $B: |x| \leqslant \pi$ with antipodal identifications on the surface $|x| = \pi$; this is the so-called "three-dimensional projective space."

EXERCISE 3*.   Try to picture the two-dimensional projective space in R^3, i.e., take a disk $|x| \leqslant 1$, say, and try to paste together antipodal points of the periphery $|x| = 1$. You will quickly get into a mess! For more information see Courant and Robbins [1941]. The pasting *can* be carried out in R^4; for the three-dimensional projective space, you need 5 dimensions to do the pasting.

The identification of $SO(3)$ with B comes about as follows. For $g \neq 1$, let e be a unit vector pointing along the axis of g and pick the direction of e

so that g appears as a counterclockwise rotation in the plane perpendicular to e, through an angle $0 \leqslant \alpha \leqslant \pi$, in accordance with the "right-hand screw rule"; see Fig. 2. The desired map of SO(3) onto B is $g \to x = \alpha e$, with the understanding that $g = 1$ maps into $x = 0$. The antipodal identifications on the surface of B reflect the fact that antipodal points represent the *same* rotation; this makes the map SO(3) $\to B$ be 1:1. The inverse map $B \to$ SO(3) is the so-called "exponential map"; it is not needed later, but the formula is pretty and provides a nice series of exercises if you feel like exploring a little.

EXERCISE 4★. The exponential sum $e^h = \sum_{n=0}^{\infty} h^n/n!$ converges for any real 3×3 matrix h. Check that $\exp(h_1 + h_2) = \exp(h_1) \exp(h_2)$ if h_1 commutes with h_2; in particular, verify that $[\exp(h)]^{-1} = \exp(-h)$.

EXERCISE 5★. Check that $\det[\exp(h)] = \exp[\mathrm{sp}(h)]$, where "sp" stands for "spur" or "trace."

EXERCISE 6★. Check that $g = \exp(h)$ belongs to SO(3) if h is any real 3×3 skewsymmetric matrix [i.e., if $h^{\sharp} = -h$].

EXERCISE 7★ Check that

$$h_1 = \begin{pmatrix} 0 & 0 & 0 \\ 0 & 0 & -1 \\ 0 & 1 & 0 \end{pmatrix}, \qquad h_2 = \begin{pmatrix} 0 & 0 & 1 \\ 0 & 0 & 0 \\ -1 & 0 & 0 \end{pmatrix}, \qquad h_3 = \begin{pmatrix} 0 & -1 & 0 \\ 1 & 0 & 0 \\ 0 & 0 & 0 \end{pmatrix}$$

is a basis (over the reals) for the class of real 3×3 skew-symmetric matrices, and especially, that the map

$$x \to g = \exp(x \cdot h) = \exp(x_1 h_1 + x_2 h_2 + x_3 h_3)$$

is a mapping of $|x| \leqslant \pi$ into SO(3).

EXERCISE 8★. Check that $g = \exp(x \cdot h)$ has x as an axis. *Hint:* Compute the action of $x \cdot h = x_1 h_1 + x_2 h_2 + x_3 h_3$ on x and use the exponential sum.

EXERCISE 9★. For any $k \in$ SO(3) and $g = \exp(x \cdot h)$ as in Exercises 7 and 8, check that

$$kgk^{\sharp} = \exp[k(x \cdot h)k^{\sharp}] = \exp[(kx) \cdot h].$$

Hint: $k(x \cdot h)k^{\sharp}$ is skew-symmetric: as such it can be expressed as $y \cdot h$ for some $y \in R^3$. Use Exercises 2 and 8 to check that $y = \varepsilon(kx) \cdot h$ for some real number ε. Now check

$$\mathrm{sp}[(x \cdot h)^2] = -2|x|^2$$

and

$$-\varepsilon^2 2|x|^2 = \varepsilon^2 \mathrm{sp}[(kx \cdot h)^2]$$
$$= \mathrm{sp}[k(x \cdot h)^2 k^*]$$
$$= -2|x|^2,$$

and conclude that $\varepsilon = \pm 1$. The sign is independent of the choice of k. Why? Think of what happens if you change k slightly. $\varepsilon = +1$ follows.

EXERCISE 10★.   Check that for $x = (0, 0, x_3)$, $g = \exp(x \cdot h) = \exp(x_3 h_3)$ is a counterclockwise rotation about the north pole through the angle x_3. Hint: Compute $(x_3 h_3)^n/n!$ by hand and sum.

EXERCISE 11★.   Check that $g = \exp(x \cdot h)$ is a counterclockwise rotation through the angle $|x|$ about the axis x, and so verify that the mapping $x \to \exp(x \cdot h)$ carries B onto SO(3). This is the inverse of the map $g \to x$ from SO(3) onto B. Why? Hint: By Exercise 9, $\exp(x \cdot h) = k^* \exp[(kx) \cdot h] k$ for any $k \in$ SO(3). Pick k so as to make $kx = (0, 0, |x|)$ and use Exercise 10.

4.   SO(3) Has No Nontrivial Characters

Now you are in a position to prove this fact, first cited in Subsection 1.

PROOF.   A character e of SO(3) depends only upon the angle of rotation and not upon the axis: Indeed, if $g \in$ SO(3) is a rotation of angle $|x|$ about the axis x and if $h \in$ SO(3) swings x into the direction of the north pole, then $k = hgh^{-1}$ is a rotation through the same angle about the north pole, and

$$e(g) = e(h) e(g) e(h)^* = e(hgh^{-1}) = e(k).$$

Now pick a north pole rotation k and an "east pole" rotation j, both of angle α. Then kj^{-1} is a rotation through a certain angle β, and as α runs from 0 to π, so does β. The fact that $e \equiv 1$ is now evident:

$$e(\beta) = e(kj^{-1}) = e(k) e(j^{-1}) = e(\alpha) e(\alpha)^* = 1.$$

5.   G/K Is a Sphere

Let $K \subset G = $ SO(3) be the subgroup of rotations k about the north pole:

$$k = \begin{pmatrix} \cos\theta & -\sin\theta & 0 \\ \sin\theta & \cos\theta & 0 \\ 0 & 0 & 1 \end{pmatrix}$$

with $0 \leqslant \theta < 2\pi$; it is isomorphic to the group SO(2) of (proper) plane rotations (alias the circle group). The class G/K of cosets $gK: g \in G$ is *not* a group: what it looks like is the surface $S^2 = (x \in R^3 : |x| = 1)$ of the three-dimensional ball, and the identification can be made in such a way that the natural action of $h \in G$ on $x = gK \in S^2$ is the *same* as the coset action $h: gK \to (hg)K$.

PROOF. The obvious correspondence to try is the map f that sends $gK \in G/K$ into the point $x = gn \in S^2$, n being the north pole $(0, 0, 1)$. f is a map of cosets $[gn = gkn$ for any $k \in K]$, and the only point at issue is whether f is $1:1$. But $g_1 n = g_2 n$ only if $g_2^{-1} g_1 n = n$, i.e., only if $g_2^{-1} g_1 \in K$, and that is the same as to say that $g_1 K = g_2 K$.

6. Double Cosets

The "double coset space" $K/G/K$ is the class of "orbits" KgK into which $S^2 = G/K$ breaks up under the action of K. The orbit $(kx: k \in K)$ of a point $x \in S^2$ is simply a horizontal circle, or what is the same, a locus of constant latitude, and may be labeled in a natural way by the central angle φ (co-latitude), as in Fig. 3. This permits you to identify $K/G/K$ with the closed interval $0 \leqslant \varphi \leqslant \pi$.

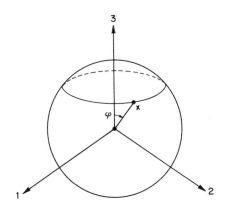

FIGURE 3

EXERCISE 12. Check that g and g^{-1} have the same double coset. *Hint:* A picture will convince you that gn and $g^{-1}n$ lie at the same latitude; even simpler is to check that $(g^{\#}n, n) = (gn, n)$.

7. Integration on *G*

An important ingredient of Fourier series and integrals is the use of a special kind of integral. The group SO(3) is no exception to this rule. The identification of $G = SO(3)$ with the projective space B makes it natural to identify a function $f(g)$ on G with a function $f^\bullet(x)$ on B and to try to express the integral as

$$\int_G f(g)\, dg = \int_B f^\bullet(x)\, \Delta(x)\, d^3 x$$

with a suitable nonnegative "weight" Δ. The problem is to pick Δ so as to make the integral "left-invariant" in the sense that

$$\int_G f(hg)\, dg = \int_G f(g)\, dg$$

for every $h \in G$ and every "nice" f; in such a case the volume element $dg = \Delta(x)\, d^3 x$ is also called left invariant. As it happens, this is not the simplest format for proving the existence of such an integral, and the actual value of Δ:

$$\Delta(x) = \frac{1 - \cos|x|}{4\pi^2\, |x|^2}$$

is not computed until Exercise 4.14.4. Before embarking upon the proof that such an integral actually exists, it may be helpful to look at some examples for simpler groups.

EXAMPLE 1.   The integral $\int f(x)\, dx$ is left- (and right-) invariant for the circle or the line: $\int f(x+y)\, dx = \int f(x)\, dx$.

EXAMPLE 2.   The integral $\int f(k)\, dk$ is left- (and right-) invariant for $K = SO(2)$ if you think of $f(k)$ as a function of the (counterclockwise) rotation $0 \leqslant \theta < 2\pi$ and put $dk = (2\pi)^{-1}\, d\theta$.

EXAMPLE 3.   The sum $\sum_G f(g)$ is left- (and right-) invariant for a finite commutative (or noncommutative) group G:

$$\sum_G f(hg) = \sum_G f(g) = \sum_G f(gh).$$

EXERCISE 13.   Check that the similarities of the line:

$$g: x \to ax + b, \qquad a > 0, \qquad -\infty < b < \infty$$

form a noncommutative group G under composition. Find a left-invariant volume element. How many are there? How about a right-invariant volume element? *Answer:* $dg = a^{-2} da\, db$ is left-invariant, $dg = a^{-1} da\, db$ is right-invariant, and up to positive constant multiples, these are the only possibilities. *Hint for the left integral:* Identify the function $f(g)$ on G with a new function $f^{\bullet}(a, b)$ on $(0, \infty) \times R^1$; the objective is to find a weight Δ so as to make

$$\int_G f(g)\, dg = \int_0^{\infty} da \int_{-\infty}^{\infty} f^{\bullet}(a, b)\, \Delta(a, b)\, db$$

$$= \int_G f(hg)\, dg = \int_0^{\infty} da \int_{-\infty}^{\infty} f^{\bullet}(a'a, a'b + b')\, \Delta(a, b)\, db$$

for every $h: x \to a'x + b'$. Derive an identity for Δ and solve it.

Back to $G = SO(3)$.

PROOF OF EXISTENCE.   Pick any nice function f on $G = SO(3)$ and let

$$f^0(g) = \int_K f(gk)\, dk,$$

in which $K = SO(2)$ is the subgroup of rotations about the north pole and $dk = (2\pi)^{-1} d\theta$ is its (invariant) volume element; see Subsection 5 and Example 2. Now for any $k' \in K$,

$$f^0(gk') = \int_K f(gk'k)\, dk = \int_K f(gk)\, dk = f^0(g),$$

so f^0 is really a function of the coset $x = gK \in S^2$ only, by Subsection 5. This function may now be integrated over S^2 with respect to the element of surface area $do = (4\pi)^{-1} d\theta \sin \varphi \, d\varphi$, defining an integral on the full group $G = SO(3)$ by the recipe

$$\int_G f(g)\, dg = \int_{S^2} f^0(x)\, do.$$

To see that the volume element $dg = do\, dk$, as defined above, is left-invariant notice first that $h \in SO(3)$ is a volume-preserving map on R^3 since the Jacobian, $\det(h)$, is 1. Because h is also length-preserving, it also preserves the element of surface area do, and so

$$\int_G f(hg)\, dg = \int_{S^2} f^0(hx)\, do = \int_{S^2} f^0(x)\, do = \int_G f(g)\, dg.$$

The proof (of existence) is finished.

PROOF OF UNIQUENESS. $dg = do \, dk$ is actually the *only* left-invariant volume element on G with total volume $\int_G dg = 1$. To help check this, bring in the notation $f^{-1}(g) = f(g^{-1})$. Then

$$\int_G f(g)\, dg = \int_G dh \int_G f(h^{-1}g)\, dg$$

$$= \int_G dg \int_G f^{-1}(g^{-1}h)\, dh$$

$$= \int_G dg \int_G f^{-1}(h)\, dh$$

$$= \int_G f^{-1}(h)\, dh$$

$$= \int_G f(g^{-1})\, dg\,,$$

and this implies that $\int_G f(g)\, dg$ is also right-invariant:

$$\int_G f(gh)\, dg = \int_G f^{-1}(h^{-1}g^{-1})\, dg$$

$$= \int_G f^{-1}(h^{-1}g)\, dg$$

$$= \int_G f^{-1}(g)\, dg$$

$$= \int_G f(g)\, dg\,.$$

Now let $\Delta(g)\, dg$ be a second left-invariant volume element on G with total volume $\int \Delta(g)\, dg = 1$. Then

$$\int_G f(g)\, \Delta(g)\, dg = \int_G dh \int_G f(hg)\, \Delta(g)\, dg$$

$$= \int_G \Delta(g)\, dg \int_G f(hg)\, dh$$

$$= \int_G \Delta(g)\, dg \int_G f(h)\, dh$$

$$= \int_G f(h)\, dh\,,$$

as advertised.

4.9   THREE CONVOLUTION ALGEBRAS

The study of functions on the group $G = SO(3)$ now begins with a look at the class $L^1(G)$ of (summable) functions f on G with

$$\|f\|_1 = \int_G |f(g)|\, dg < \infty.$$

The product

$$f_1 \circ f_2(g) = \int_G f_1(gh^{-1}) f_2(h)\, dh$$

makes $L^1(G)$ into an algebra, as follows from the next exercise.

EXERCISE 1.   Check that $\|f_1 \circ f_2\|_1 \leqslant \|f_1\|_1 \|f_2\|_1$.

EXERCISE 2.   Show that the product in $L^1(G)$ is associative but not commutative.

$L^1(G/K)$ is the class of functions $f \in L^1(G)$ which are constant on cosets gK [$K = SO(2)$ is the subgroup of north-pole rotations]. A function of this kind may be thought of as a function $f^0(x)$ of the point $x = gn$ of S^2 identified with gK, and since

$$\|f\|_1 = \int_G |f(g)|\, dg = \int_{S^2} |f^0(x)|\, do = \|f^0\|_1,$$

you may identify $L^1(G/K)$ with $L^1(S^2, do)$. $L^1(G/K)$ "inherits" a product from $L^1(G)$ and is an algebra in its own right: Namely, if f_1 and f_2 belong to $L^1(G/K)$, then the product $f_1 \circ f_2$ is also a function of cosets, as you can see from

$$f_1 \circ f_2(gk) = \int_G f_1(gkh^{-1}) f_2(h)\, dh$$

$$= \int_G f_1(gh^{-1}) f_2(hk)\, dh$$

$$= \int_G f_1(gh^{-1}) f_2(h)\, dh$$

$$= f_1 \circ f_2(g).$$

$L^1(G/K)$ is still *noncommutative*.

PROOF.   To see this, think of the x_1 coordinate of gn as a function of $g \in G$. Because $gn = gkn$, this is really a function on the sphere, and for any $f \in L^1(G/K)$, you have

$$f \circ x_1(g) = \int_G f(gh^{-1}) x_1(h) \, dh$$

$$= \int_G f(gh^{-1}k) x_1(k^{-1}h) \, dh$$

$$= \int_G f(gh^{-1}) x_1(k^{-1}h) \, dh,$$

independently of $k \in K$. As k runs over K, $k^{-1}hn$ sweeps out a horizontal circle, and

$$\int x_1(k^{-1}h) \, dk = 0.$$

Therefore,

$$f \circ x_1(g) = \int_K dk \int_G f(gh^{-1}) x_1(k^{-1}h) \, dh = 0.$$

Now suppose that $L^1(G/K)$ is commutative. Then you would also have

$$0 = x_1 \circ f(g) = \int_G x_1(gh^{-1}) f(h) \, dh.$$

Pick $f \geqslant 0$ with $\int f = 1$ and $f = 0$ outside a small spherical cap D about the north pole. As a function on G, $f = 0$ except near K, and as you shrink D, $x_1 \circ f$ approximates

$$\int_K x_1(gk^{-1}) \, dk = x_1(g) \int_K dk = x_1(g).$$

But then $x_1(g)$ would be 0, and that is not the case.

$L^1(K/G/K)$ is the class of functions $f \in L^1(G)$ which are constant on double cosets KgK; such a function can be viewed as a function $f^\circ(\cos \varphi)$ of the colatitude φ of the horizontal circle $Kgn \subset S^2$, and since

$$\|f\|_1 = \int_G |f(g)| \, dg = \int_0^\pi f^\circ(\cos \varphi) \tfrac{1}{2} \sin \varphi \, d\varphi,$$

you may identify $L^1(K/G/K)$ with $L^1([0, \pi], \tfrac{1}{2} \sin \varphi \, d\varphi)$. The product $\circ$ is "inherited" by $L^1(K/G/K)$ much as before, only now you get a *commutative* algebra!

PROOF. By Exercise 4.8.12, g and g^{-1} have the same double coset, so that $f(g) = f(g^{-1})$ for any function of double cosets, and for the product of two such functions, you find

$$f_1 \circ f_2(g) = \int_G f_1(gh^{-1}) f_2(h) \, dh$$

$$= \int_G f_1(hg^{-1}) f_2(h) \, dh$$

$$= \int_G f_1(h) f_2(hg) \, dh$$

$$= \int_G f_2(g^{-1}h^{-1}) f_1(h) \, dh$$

$$= f_2 \circ f_1(g^{-1})$$

$$= f_2 \circ f_1(g),$$

as advertised. The fact that $f_1 \circ f_2$ is a function of double cosets is used in the last line.

4.10   HOMOMORPHISMS OF $L^1(K/G/K)$

Because $L^1(K/G/K)$ is commutative, it can be expected to have a fair number of homomorphisms. The next step toward a Fourier series for SO(3) is to find out what they are. A partial answer is that *a map j of* $L^1(K/G/K)$ *into the complex numbers is a homomorphism iff*

$$j(f) = \int_G f(g) \, p(g) \, dg = \int_0^\pi f(\cos \varphi) \, p(\cos \varphi) \, \tfrac{1}{2} \sin \varphi \, d\varphi,$$

in which p is a "spherical" function, that is,

(a) $p \in C^\infty(K/G/K)$,

(b) $|p(g)| \leqslant p(1) = 1$,

(c) $p(g_1) \, p(g_2) = \int_K p(g_1 k g_2) \, dk$.

$C^\infty(K/G/K)$ is the class of infinitely differentiable functions on the sphere which depend upon colatitude only. The actual computation of the spherical functions will be carried out in Section 4.11.

PROOF.   Define a map of $L^1(K/G/K)$ into itself by the rule

$$f \to f^h(g) = \int_K f(gkh^{-1}) \, dk.$$

Then

$$\int_G f_1^h(g) f_2(h) \, dh = \int_G f_2(h) \, dh \int_K f_1(gkh^{-1}) \, dk$$

$$= \int_K dk \int_G f_1(gkh^{-1}) f_2(h) \, dh$$

$$= \int_K f_1 \circ f_2(gk) \, dk$$

$$= f_1 \circ f_2(g),$$

so

$$j(f_1) j(f_2) = j(f_1 \circ f_2) = j\left(\int_G f_1^h f_2(h) \, dh \right)$$

$$= \int_G j(f_1^h) f_2(h) \, dh,$$

modulo a few technical flourishes, and if you now pick f_1 so as to make $j(f_1) = 1$ and write f in place of f_2 you will have

$$j(f) = \int_G f(g) \, p(g) \, dg$$

with

$$p(g) = j(f_1^g).$$

Because f_1^g depends upon g only via its double coset, p is a function of colatitude alone, and (a) follows by taking $f_1 \in C^\infty(K/G/K)$, as you may. To prove (c), notice first that

$$\int_{G \times G} f_1(g_1) f_2(g_2) \, p(g_1) \, p(g_2) \, dg_1 \, dg_2$$

$$= \int_G f_1 \, p \, dg \times \int_G f_2 \, p \, dg$$

$$= j(f_1) j(f_2)$$

$$= j(f_1 \circ f_2)$$

$$= \int f_1 \circ f_2 \, p \, dg$$

$$= \int_{G \times G} f_1(g_1 g_2^{-1}) f_2(g_2) \, p(g_1) \, dg_1 \, dg_2$$

$$= \int_{G \times G} f_1(g_1) f_2(g_2) \, p(g_1 g_2) \, dg_1 \, dg_2.$$

This holds for all f_1 and f_2 from $L^1(K/G/K)$, and it is tempting to conclude that $p(g_1)p(g_2) = p(g_1 g_2)$, but this is not correct because $p(g_1 g_2)$ is *not* a function of the double cosets of g_1 and g_2, separately! The difficulty is easily overcome: The final integral is unchanged if you replace $p(g_1 g_2)$ by $p(g_1 k g_2)$, and if you now average over K, replacing $p(g_1 g_2)$ by

$$\int_K p(g_1 k g_2)\, dk$$

which *is* a function of the separate double cosets, you will find that (c) must hold. As to (b), $\|p\|_\infty < \infty$ since $p \in C^\infty(K/G/K)$ and $K/G/K = [0, \pi]$ is compact. Therefore,

$$|p(g)|^2 \leqslant \int_K |p(gkg)|\, dk \leqslant \|p\|_\infty,$$

so that $\|p\|_\infty \leqslant 1$, and the fact that $p(1) = 1$ follows from $p \neq 0$ and

$$p(g)\,p(1) = \int_K p(gk)\, dk = p(g).$$

This finishes the proof that every homomorphism comes from a spherical function. The proof of the converse is simpler and is left to you as an exercise.

EXERCISE 1.   Give a second proof of (c) in the style of Exercise 4.2.8.

The identity (c) says that the spherical function p is trying its best to act like a character. It cannot actually do so unless it is $\equiv 1$ as $G = SO(3)$ has no other characters [Subsection 4.8.4], but the hope is that if you help things along by averaging over K, you will find a considerable number of non-constant (spherical) functions which factor in this "character-like" way. To make (c) more concrete, you can write it out, thinking of p as a function of colatitude $0 \leqslant \varphi \leqslant \pi$, or better still, as a function of $\cos \varphi$:

$$p(\cos \alpha)\, p(\cos \beta) = \frac{1}{2\pi} \int_0^{2\pi} p(\sin \alpha \sin \beta \cos \theta + \cos \alpha \cos \beta)\, d\theta.$$

PROOF.   Bring in $\cos \alpha = \cos(\text{colatitude } g_1 n) = (g_1 n) \cdot n$, and similarly $\cos \beta = (g_2 n) \cdot n$. Then (c) says that

$$p(\cos \alpha)\, p(\cos \beta) = p(g_1)\, p(g_2) = p(g_1^{-1})\, p(g_2)$$

$$= \int_K p(g_1^{-1} k g_2)\, dk$$

$$= \int_K p[\cos(\text{colatitude of } g_1^{-1} k g_2)]\, dk.$$

The argument of the spherical function in line 3 is the inner product

$$kg_2 n \cdot g_1 n,$$

between $g_1^{-1} kg_2 n$ and n,
and if you adjust g_1 and g_2 so as to make

$$g_1 n = (\sin \alpha, 0, \cos \alpha) \quad \text{and} \quad g_2 n = (\sin \beta, 0, \cos \beta),$$

as you may without prejudice to the integral, you will find

$$kg_2 n \cdot g_1 n = \begin{pmatrix} \cos\theta & -\sin\theta & 0 \\ \sin\theta & \cos\theta & 0 \\ 0 & 0 & 1 \end{pmatrix} \begin{pmatrix} \sin\beta \\ 0 \\ \cos\beta \end{pmatrix} \cdot \begin{pmatrix} \sin\alpha \\ 0 \\ \cos\alpha \end{pmatrix}$$

$$= \begin{pmatrix} \sin\beta\cos\theta \\ \sin\beta\sin\theta \\ \cos\beta \end{pmatrix} \cdot \begin{pmatrix} \sin\alpha \\ 0 \\ \cos\alpha \end{pmatrix}$$

$$= \sin\alpha \sin\beta \cos\theta + \cos\alpha \cos\beta.$$

The proof is finished.

EXERCISE 2. Find polynomial solutions $p = p_n(x)$ of (c) of degree n in $x = \cos \varphi$ for $n = 0, 1, 2, 3$. *Answer:* $p_0 = 1$, $p_1 = x$, $p_2 = \frac{1}{2}(3x^2 - 1)$, $p_3 = \frac{1}{2}(5x^3 - 3x)$. Apart from multiplicative constants, these are the first four Legendre polynomials of Exercise 1.3.7.

EXERCISE 3. Check that for functions of double cosets the product $f_1 \circ f_2$ can be expressed as

$$f_1 \circ f_2 (\cos \alpha)$$

$$= \frac{1}{4\pi} \int_0^\pi \sin\beta \, d\beta \int_0^{2\pi} f_1(\sin\alpha \sin\beta \cos\theta + \cos\alpha \cos\beta) f_2(\cos\beta) \, d\theta.$$

Use this formula to give a direct proof that the product is commutative. *Hint for the formula:*

$$f_1 \circ f_2(g) = \int_G f_1^h(g) f_2(h) \, dh.$$

4.11   SPHERICAL FUNCTIONS ARE EIGENFUNCTIONS OF THE LAPLACIAN

The next step is to compute the spherical functions p of Section 4.10 and to verify that they span $L^2(K/G/K) = L^2([0, \pi], \frac{1}{2}\sin\varphi \, d\varphi)$. The chief thing you need to know is that p, as a function of $x = gn$, is an eigenfunction of the spherical Laplace operator

$$\Delta = \frac{1}{\sin\varphi} \frac{\partial}{\partial\varphi} \sin\varphi \frac{\partial}{\partial\varphi} + \frac{1}{\sin^2\varphi} \frac{\partial^2}{\partial\theta^2}.$$

This is proved in the present section; the actual identification of the spherical functions occupies Section 4.12.

Step 1:   Check that Δ commutes with the action of $G = SO(3)$ on the spherical surface S^2; this means that if $f^g(x) = f(gx)$, then $\Delta(f^g) = (\Delta f)^g$.

PROOF.   The three-dimensional Laplace operator

$$\Delta^3 = \frac{\partial^2}{\partial x_1{}^2} + \frac{\partial^2}{\partial x_2{}^2} + \frac{\partial^2}{\partial x_3{}^2}$$

commutes with the action of G on R^3. To see this, pick $f \in C_\downarrow^\infty(R^3)$ and look at

$$(\Delta^3 f)^\wedge = \int (\Delta^3 f)(x) e^{-2\pi i \gamma \cdot x} d^3 x = -4\pi^2 |\gamma|^2 \hat{f}.$$

Because $\det(g) = 1$, you have

$$(f^g)^\wedge = \int f(gx) e^{-2\pi i \gamma \cdot x} d^3 x$$

$$= \int f(x) e^{-2\pi i \gamma \cdot g^\# x} d^3 x$$

$$= \int f(x) e^{-2\pi i g\gamma \cdot x} d^3 x$$

$$= \hat{f}^g,$$

and so

$$\Delta^3 f^g = [(\Delta^3 f^g)^\wedge]^\vee$$

$$= [-4\pi^2 |\gamma|^2 (f^g)^\wedge]^\vee$$

$$= [-4\pi^2 |\gamma|^2 \hat{f}^g]^\vee$$

$$= [(-4\pi^2 |\gamma|^2 \hat{f})^g]^\vee$$

$$= [(\Delta^3 f)^\wedge]^{g\vee}$$

$$= [(\Delta^3 f)^\wedge]^{\vee g}$$

$$= (\Delta^3 f)^g,$$

as advertised.

EXERCISE 1.   Check that $\Delta^3(f^g) = (\Delta^3 f)^g$ by direct computation.

EXERCISE 2.   Compute Δ^3 in spherical polar coordinates

$$x_1 = r \sin \varphi \cos \theta, \qquad x_2 = r \sin \varphi \sin \theta, \qquad x_3 = r \cos \varphi.$$

Answer:

$$\Delta^3 = \frac{\partial^2}{\partial r^2} + \frac{2}{r}\frac{\partial}{\partial r} + \frac{1}{r^2}\Delta.$$

EXERCISE 3.   Finish the proof of Step 1.

EXERCISE 4.   Check that $(\Delta f, f) = \int \Delta f f^* \, do \leq 0$ for $f \in C^\infty(S^2)$.

EXERCISE 5.   Check that $(\Delta f_1, f_2) = (f_1, \Delta f_2)$ for functions of class $C^\infty(S^2)$.

Step 2:   Check that any spherical function p is an eigenfunction of Δ [$\Delta p = \gamma p$] with eigenvalue $\gamma \leq 0$.

PROOF.   A spherical function p is of class $C^\infty(S^2)$ and solves

$$p(g_1)\,p(g_2) = \int_K p(g_1 k g_2) \, dk.$$

This identity may be written in the form

$$p(g)\,p(x) = \int_K p(gkx) \, dk$$

with $x = g_2 n$, and you can apply Δ to both sides to obtain

$$p(g)\,\Delta p(x) = \int_K \Delta p^{gk}(x) \, dk$$

$$= \int_K (\Delta p)^{gk}(x) \, dk$$

$$= \int_K \Delta p(gkx) \, dk,$$

or, what is the same,

$$p(g_1)\,\Delta p(g_2) = \int_K \Delta p(g_1 k g_2) \, dk.$$

But Δp is a function of colatitude only since p is such, so

$$p(g_1)\Delta p(g_2) = p(g_1^{-1})\,\Delta p(g_2^{-1})$$

$$= \int_K \Delta p(g_1^{-1}kg_2^{-1})\,dk$$

$$= \int_K \Delta p(g_2 k^{-1}g_1)\,dk$$

$$= \int_K \Delta p(g_2 kg_1)\,dk$$

$$= p(g_2)\,\Delta p(g_1),$$

and putting $g_1 = 1$ and $g_2 = g$, you see that p is an eigenfunction of Δ:

$$\Delta p(g) = p(1)\Delta p(g) = \Delta p(1)\,p(g).$$

The eigenvalue $\gamma = \Delta p(1)$ is less than or equal to 0 by Exercise 4:

$$0 \geqslant (\Delta p, p) = \gamma \|p\|_2^2,$$

and the proof is finished.

4.12   SPHERICAL FUNCTIONS ARE LEGENDRE POLYNOMIALS

For fixed colatitude $0 \leqslant \varphi \leqslant \pi$, the function $(1 - 2\gamma \cos\varphi + \gamma^2)^{-\frac{1}{2}}$ is analytic in the open unit disk $|\gamma| < 1$ and so can be expanded into a power series

$$(1 - 2\gamma \cos\varphi + \gamma^2)^{-\frac{1}{2}} = \sum_{n=0}^{\infty} p_n(\cos\varphi)\,\gamma^n$$

for $|\gamma| < 1$. The coefficient $p_n(\cos\varphi)$ is a (real) polynomial in $\cos\varphi$ of exact degree n. The goal of this section is to verify that the spherical functions of Section 4.10 are precisely these so-called "Legendre polynomials" $p_n: n \geqslant 0$; see also Exercise 4.10.2.

Step 1: $p_n(1) = 1$.

PROOF. $\sum_{n=0}^{\infty} p_n(1)\gamma^n = (1 - 2\gamma + \gamma^2)^{-\frac{1}{2}} = (1 - \gamma)^{-1} = \sum_{n=0}^{\infty}\gamma^n.$

Step 2: $\Delta p_n = (\sin\varphi)^{-1}(\sin\varphi p_n')' = -n(n+1)\,p_n$, the prime signifying differentiation by the colatitude φ.

PROOF. The polynomial p_n is viewed as a function of $x \in S^2$, which is constant on circles of fixed latitude, and $(1 - 2r \cos \varphi + r^2)^{-\frac{1}{2}}$ is interpreted as the reciprocal of the distance $R = |n - x|$ between the north pole $n = (0, 0, 1)$ and the point

$$x = (x_1, x_2, x_3) = r(\sin \varphi \cos \theta, \sin \varphi \sin \theta, \cos \varphi).$$

Because Δ^3 commutes with translations, you have

$$0 = \left[\frac{\partial^2}{\partial R^2} + \frac{2}{R} \frac{\partial}{\partial R} \right] R^{-1} = \Delta^3 R^{-1}$$

$$= \sum_{n=0}^{\infty} \Delta^3 p_n(\cos \varphi) r^n$$

$$= \sum_{n=0}^{\infty} \left[\frac{\partial^2}{\partial r^2} + \frac{2}{r} \frac{\partial}{\partial r} + \frac{1}{r^2} \Delta \right] p_n(\cos \varphi) r^n$$

$$= \sum_{n=0}^{\infty} [n(n+1) + \Delta] p_n(\cos \varphi) r^{n-2}.$$

The proof is finished.

Step 3: The polynomials p_n are perpendicular in $L^2([0, \pi], \frac{1}{2} \sin \varphi \, d\varphi)$, and

$$\|p_n\|_2^2 = \int_0^\pi p_n^2(\cos \varphi) \tfrac{1}{2} \sin \varphi \, d\varphi = (2n+1)^{-1}.$$

PROOF. The perpendicularity is automatic from Exercise 4.11.5 and the fact that $n(n+1) \neq m(m+1)$ for $n \neq m$. By this result,

$$\sum_{n=0}^{\infty} \|p_n\|_2^2 \gamma^{2n} = \frac{1}{2} \int_0^\pi \frac{\sin \varphi}{1 - 2\gamma \cos \varphi + \gamma^2} \, d\varphi$$

for real γ between -1 and $+1$, and this is an elementary integral:

$$\frac{1}{2} \int_0^\pi \frac{\sin \varphi}{1 - 2\gamma \cos \varphi + \gamma^2} \, d\varphi = \frac{1}{2} \int_{-1}^{+1} \frac{dx}{1 - 2\gamma x + \gamma^2}$$

$$= \left. \frac{-1}{4\gamma} \log(1 - 2\gamma x + \gamma^2) \right|_{-1}^{+1}$$

$$= \frac{1}{2\gamma} \log \frac{1 + \gamma}{1 - \gamma}$$

$$= \sum_{n=0}^{\infty} \frac{\gamma^{2n}}{2n+1}.$$

The proof is finished by matching coefficients of like powers.

Step 4: The polynomials p_n: $n \geqslant 0$ span $L^2([0, \pi], \frac{1}{2} \sin \varphi \, d\varphi)$.

PROOF. This is the same as to say that the polynomials $p_n(x)$ $[x = \cos \varphi]$ span $L^2[-1, +1]$ since $dx = -\sin \varphi \, d\varphi$. But that is self-evident: $p_n(x)$ is of exact degree n, so the Weierstrass approximation theorem of Subsection 1.7.3 does the trick.

EXERCISE 1. Check that the polynomials $p_n(x)$: $n \geqslant 0$, $-1 \leqslant x \leqslant 1$ are the same, up to multiplicative constants, as the polynomials of Exercise 1.3.7, constructed by applying the Gram–Schmidt recipe to the powers x^n: $n \geqslant 0$ in $L^2[-1, +1]$.

EXERCISE 2. Check that the polynomials p_n are the only eigenfunctions of Δ of class $L^2([0, \pi], \frac{1}{2} \sin \varphi \, d\varphi)$. *Hint:* Expand the eigenfunction f as a sum $\sum_{n=0}^{\infty} \hat{f}(n) \, p_n$.

Step 5: The Legendre polynomials are spherical functions.

PROOF. The polynomial $p_n = p_n(\cos \varphi)$ is of class $C^{\infty}(K/G/K)$ and $p_n(1) = 1$, so it is enough to check the identity

$$p_n(g_1) \, p_n(g_2) = \int_K p_n(g_1 \, k g_2) \, dk,$$

thinking of p_n as a function on $G = SO(3)$ which is constant on double cosets. Fix $g_1 = g$ and think of the integral on the right as a function $f(x)$ of $x = g_2 n$. Then

$$\Delta f = \int_K \Delta(p_n^{gk}) \, dk$$

$$= \int_K (\Delta p_n)^{gk} \, dk$$

$$= -n(n+1) \int_K p_n^{gk} \, dk$$

$$= -n(n+1) f,$$

and since f is actually a function of colatitude only, it has to be a constant multiple of p_n, by Exercise 2:

$$f(x) = \int_K p_n(g_1 \, k g_2) = c(g_1) \, p_n(g_2).$$

Because f depends upon g_1 only via its double coset, the same is true of c, and

$$c(g_1)\, p_n(g_2) = \int_K p_n(g_1^{-1} k g_2^{-1})\, dk$$

$$= \int_K p_n(g_2 k^{-1} g_1)\, dk = c(g_2)\, p_n(g_1);$$

especially, for $g_2 = 1$, you find

$$c(g) = c(g)\, p_n(1) = c(1)\, p_n(g),$$

and for $g_1 = g_2 = 1$, you see that

$$c(1) = c(1)\, p_n(1) = \int_K p_n(k)\, dk = p_n(1) = 1.$$

Therefore,

$$\int_K p_n(g_1 k g_2) = c(g_1)\, p_n(g_2) = p_n(g_1)\, p_n(g_2),$$

as advertised.

EXERCISE 3.   Prove the converse of Step 5: Every spherical function is a Legendre polynomial. *Hint:*   The bulk of the work is done in Exercise 2.

EXERCISE 4.   Check the formula:

$$p_n(\cos \varphi) = \frac{1}{2\pi} \int_0^{2\pi} (i \sin \varphi \, \cos \theta + \cos \varphi)^n \, d\theta.$$

Hint:

$$\sum_{n=0}^{\infty} \frac{1}{2\pi} \int_0^{2\pi} (i \sin \varphi \, \cos \theta + \cos \varphi)^n \, d\theta \, \gamma^n = \frac{1}{2\pi} \int_0^{2\pi} \frac{d\theta}{A - iB \cos \theta}$$

with $A = 1 - \gamma \cos \varphi$ and $B = \gamma \sin \varphi$. The integral may be evaluated by use of Cauchy's formula after expressing it as an integral around the unit circle:

$$\frac{1}{2\pi i} \int \frac{1}{A - i(B/2)(\gamma + \gamma^{-1})} \frac{d\gamma}{\gamma}.$$

Before going on to bigger and better Fourier series on $G/K = S^2$ and $G = SO(3)$, it will be best to stop for a moment to sum up the present state of things for $K/G/K = [0, \pi]$ and to relate it to earlier ideas.

For (very) general commutative groups G, there is a 1:1 correspondence between characters $e \in \hat{G}$ and homomorphisms $f \to \hat{f}(e)$ of $L^1(G)$ expressed as an integral [or a sum]

$$\hat{f}(e) = \int_G f(g) e^*(g) \, dg \qquad \left[\hat{f}(e) = \#(G)^{-1} \sum_G f(g) e^*(g) \right],$$

and there are sufficiently many characters to recover the function f by an inverse integral [or sum]:

$$f(g) = \hat{f}^{\vee} = \int_{\hat{G}} \hat{f}(e) e(g) \, de \qquad \left[f(g) = \sum_{\hat{G}} \hat{f}(e) e(g) \right].$$

The space at hand, $K/G/K = [0, \pi]$, is not a group, so it does not have any characters, but $L^1(K/G/K)$ is commutative, and it has enough homomorphisms [spherical functions or Legendre polynomials] to enable you to expand any function of class $L^2(K/G/K)$ into a nice Fourier-type series:

$$f = \sum_{n=0}^{\infty} \hat{f}(n) \, p_n$$

with coefficients

$$(2n+1) \hat{f}(n) = \int_G f p_n \, dg = \int_{S^2} f p_n \, do = \int_0^{\pi} f p_n \tfrac{1}{2} \sin \varphi \, d\varphi,$$

thinking of f as a function on $G = SO(3)$, $G/K = S^2$, and $K/G/K = [0, \pi]$, successively.

4.13 SPHERICAL HARMONICS

The next step is to "lift" the Fourier–Legendre series of Section 4.12 from $K/G/K = [0, \pi]$ up to $G/K = S^2$, so as to obtain some kind of Fourier series for functions of class $L^2(G/K)$. This is done by means of "spherical harmonics." A spherical harmonic "of weight n" is a member of the span $M_n \subset L^2(G/K)$ of the left translates

$$p_n^g(x) = p_n(gx)$$

of the spherical function p_n. The most important properties of these functions will now be developed in a series of short articles.

1. Addition Formula

Pick a unit-perpendicular basis $e_n^l \colon l = 0, \pm 1, \ldots$ of M_n, and for fixed $g \in G$, expand $p_n^g(x)$ into a Fourier series

$$p_n^g(x) = \sum_l (p_n^g)^{\wedge}(l) \, e_n^l(x)$$

with coefficients

$$(p_n^g)^\wedge(l) = \int_G p_n^g(h)\, e_n^l(h)^*\, dh.$$

Now for any g and g' from $G = SO(3)$ and any north pole rotation k,

$$\int_G p_n^g p_n^{g'} = \int_G p_n(gh)\, p_n(g'h)\, dh$$

$$= \int_G p_n(k^{-1}gh)\, p_n(g'h)\, dh$$

$$= \int_G p_n(h)\, p_n(g'g^{-1}kh)\, dh$$

$$= \int_K dk \int_G p_n(h)\, p_n(g'g^{-1}kh)\, dh$$

$$= \int_G p_n(h)\, dh \int_K p_n(g'g^{-1}kh)\, dk$$

$$= \int_G p_n(h)\, dh\, p_n(g'g^{-1})\, p_n(h)$$

$$= \|p_n\|^2 p_n(g'g^{-1})$$

$$= (2n+1)^{-1} p_n^{g'}(g^{-1})$$

by Step 3 of Section 4.12. Therefore,

$$\int_G p_n^g f = (2n+1)^{-1} f(g^{-1})$$

for any $f \in M_n$. To spell this out a little more, the formula is true for any translate f of p_n, and since these span M_n while the left-hand side is continuous in g for *any* $f \in L^2(G)$, you see that each $f \in M_n$ can be modified so as to be continuous on G, and that with this modification, the formula holds *identically*. The case $f = e_n^{l*}$ is of special interest as it permits you to evaluate the coefficients $(p_n^g)^\wedge$:

$$(p_n^g)^\wedge(l) = \int_G p_n^g e_n^{l*} = (2n+1)^{-1} e_n^l(g^{-1})^*,$$

and if you put this back into the Fourier series for p_n^g, you will obtain *the addition formula for spherical harmonics:*

$$p_n(g^{-1}h) = (2n+1)^{-1} \sum_l e_n^l(g)^* e_n^l(h).$$

EXERCISE 1. The sum converges in $L^2(G)$ for fixed $g \in G$. Use the formula $\int p_n^h f = (2n+1)^{-1} f(h^{-1})$ to check that convergence also takes place pointwise.

EXERCISE 2. Use the formula for $(p_n^g)^\wedge$ to check that

$$\int_K e_n^l(g_1 kg_2)\, dk = e_n^l(g_1)\, p_n(g_2).$$

2. Dimension of M_n

An immediate consequence of the addition formula, as amplified by Exercise 1, is that $\dim M_n = 2n+1$, as you can see by putting $g = h$ and integrating both sides over G:

$$1 = p_n(1) = \int_G p_n(g^{-1}g)\, dg = (2n+1)^{-1} \sum_l \int_G |e_n^l(g)|^2\, dg = (2n+1)^{-1} \sum_l 1.$$

This means that there are only $2n+1$ functions e_n^l. They are now indexed in the natural way with l running from $-n$ to $+n$, inclusive.

3*. The Zeeman Effect

The Zeeman effect of atomic physics must be mentioned here, however briefly. The wave function ("state") of the hydrogen atom is an eigenfunction ψ of class $L^2(R^3)$ of the differential operator $\Delta^3 + |x|^{-1}$, the negatives of the eigenvalues being the possible energy levels. For fixed energy, ψ splits into the product of a radial function and a spherical harmonic e. The "quantum number" n which is the "weight" of e is related to the total angular momentum; and for fixed n, there are $2n+1$ (experimentally indistinguishable) states, one to each choice of $|l| \leqslant n$. This is the mathematical explanation of the Zeeman effect in which each spectral line of hydrogen "splits" into an odd number of neighboring lines when you switch on a magnetic field: each of the $2n+1$ formerly indistinguishable states now carries a different energy and can therefore be "seen." The details of this fascinating story can be found in a very simple form in Heitler [1945].

4. Eigenfunctions of Δ

By the addition formula,

$$e_n^l(x) = e_n^l(g) = (2n+1) \int_G p_n(g^{-1}h)\, e_n^l(h)\, dh$$

$$= (2n+1) \int_G p_n(h^{-1}x)\, e_n^l(h)\, dh,$$

for $x = g$(north pole),[1] and since Δ commutes with the action of h^{-1} on $G/K = S^2$, you see that every spherical harmonic of weight n is an eigenfunction of Δ with the *same* eigenvalue $-n(n+1)$ as p_n.

EXERCISE 3*. Show that up to a multiplicative constant Δ is the only partial differential operator of degree less than or equal to 2 which commutes with the action of $G = SO(3)$ on the sphere. *Hint:* Check (a) that such an operator, call it L, maps functions of colatitude into themselves and is completely determined by this action, (b) that p_n is an eigenfunction of L for every $n \geqslant 0$, and (c) that $\int f_1 \, Lf_2 = \int f_2 \, Lf_1$ for functions of class $C^\infty(S^2)$. Use (a) and (c) to express Lf as $(\sin \varphi)^{-1}(A \sin \varphi f')'$ for functions $f = f(\cos \varphi)$ with an unknown coefficient A. Then infer from $Lp_1 = \gamma_1 p_1$ that $A = -\frac{1}{2}\gamma_1 + B \sin^{-2}\varphi$ with a constant B, and from $Lp_2 = \gamma_2 p_2$ that $B = 0$.

5. Spherical Harmonics Span $\mathbf{L}^2(G/K)$

$L^2(G/K)$ *is the perpendicular sum* $\bigoplus_{n=0}^{\infty} M_n$ *of the subspaces of spherical harmonics of the several weights* $n \geqslant 0$; *in particular, the family* $e_n^l \colon |l| \leqslant n$, $n \geqslant 0$ *is a unit-perpendicular basis of* $L^2(G/K)$, *so that every function* f *of this class may be expanded into a Fourier series*

$$f = \sum_{n=0}^{\infty} \sum_{|l| \leqslant n} \hat{f}\binom{l}{n} e_n$$

with coefficients

$$\hat{f}\binom{l}{n} = \int_G f e_n^{l*} \, dg = \int_{S^2} f e_n^{l*} \, do \, .$$

PROOF. To begin with, spherical harmonics of unequal weights are perpendicular since they are eigenfunctions of Δ belonging to different eigenvalues, so all you have to do is to check that $f \in L^2(G/K)$ is perpendicular to every spherical harmonic only if $f = 0$. To do this, fix $g \in G$ and let

$$f^0(x) = f^0(h) = \int_K f(gkh) \, dk \quad \text{with} \quad x = h(\text{north pole}) \, .$$

Then $f^0 \in L^2(K/G/K)$ and is prependicular to the spherical functions:

$$\begin{aligned}
(f^0, p_n) &= \int_G \left[\int_K f(gkh) \, dk \right] p_n(h) \, dh \\
&= \int_K dk \int_G f(h) p_n(k^{-1}g^{-1}h) \, dh \\
&= \int_G f(h) p_n(g^{-1}h) \, dh = 0 \, .
\end{aligned}$$

---

[1] g (north pole) is written occasionally in place of gn to clarify the notation.

Therefore, $f^0 = 0$, and since g^{-1} preserves surface area on the sphere,

$$f^0(g^{-1}x) = \int_K f(gkg^{-1}x)\,dk$$

vanishes, too. Now as k runs over $K = \mathrm{SO}(2)$, $y = gkg^{-1}x$ traces out a circle perpendicular to the axis g(north pole), as in Fig. 1, and by varying g and x, it is plain from the picture that the integral of f over every spherical cap must vanish. But this is not possible unless $f = 0$. The proof is finished.

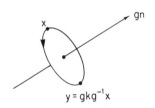

FIGURE 1

EXERCISE 4. Give a full technical proof that the integral of f over every spherical cap is 0 only if $f = 0$. *Hint:* The integral of f over a cap, when suitably normalized, tends to f in $\mathsf{L}^2(S^1)$ as the cap shrinks to a point.

EXERCISE 5. Any closed subspace $\mathsf{M} \subset \mathsf{L}^2(G/K)$ which is invariant under the action of $G = \mathrm{SO}(3)$ is the perpendicular sum of a certain number of the subspaces M_n; especially, the latter are the minimal (closed) invariant subspaces of $\mathsf{L}^2(G/K)$. *Hint:* Check

$$\int_G f^h(g)e_n^{\,l}(h)^*\,dh = \int_G f(h)e_n^{\,l}(h)^*\,dh \times p_n(g)$$

with the help of Exercise 2; Exercises 4.2.2 and 4.2.3 will serve as a model for the rest.

EXERCISE 6. Check that if f is a homogeneous polynomial of degree n, then $\Delta^3 f = 0$ iff the restriction of f to $|x| = 1$ is a spherical harmonic of weight n; for additional information about this approach to spherical harmonics, see Exercise 12 and the introductory articles of Müller [1966] and/or Seeley [1966b]. *Hint:* Express Δ^3 in spherical polar coordinates.

EXERCISE 7. Bring in the "associated Legendre polynomials"

$$p_n^{\,l}(\cos \varphi) = \frac{1}{2\pi} \int_0^{2\pi} (i \sin \varphi \cos \theta + \cos \varphi)^n e^{-il\theta}\,d\theta,$$

for $|l| \leqslant n$ $[p_n^{\,0} = p_n$, by Exercise 4.12.4]. Check that for $x = (\sin \varphi \cos \theta,$

$\sin \varphi \sin \theta, \cos \varphi)$

$$e_n^{\ l}(x) = p_n^{\ l}(\cos \varphi) e^{il\theta}$$

is a spherical harmonic of weight n. This recipe produces $2n+1$ perpendicular spherical harmonics, in agreement with the dimension count of Subsection 2. Why don't you get more from $|l| > n$? *Hint:* Apply Exercise 6 to the function

$$f(x) = \sum_{|l| \leqslant n} p_n^{\ l}(\cos \varphi) e^{il\theta} = (i \sin \varphi \cos \theta + \cos \varphi)^n = (ix_1 + x_3)^n.$$

Now let α be the angle of rotation of k and deduce that

$$e_n^{\ l}(x) = p_n^{\ l}(\cos \varphi) e^{il\theta} = \int_K f^k(x) e^{-il\alpha} dk$$

is a spherical harmonic of the same weight n.

6. Maxwell's Poles

Maxwell [1892] found a very beautiful way of describing spherical harmonics as electrostatic potentials. The following notation will be useful:

$$x = (x_1, x_2, x_3) \in R^3,$$

$$r = |x| = (x_1^2 + x_2^2 + x_3^2)^{1/2},$$

$\mathbf{n}$ is a triple (n_1, n_2, n_3) of nonnegative integers,

$$n = |\mathbf{n}| = n_1 + n_2 + n_3 = \text{the "weight" of } \mathbf{n},$$

$$\partial^{\mathbf{n}} = \left(\frac{\partial}{\partial x_1}\right)^{n_1} \left(\frac{\partial}{\partial x_2}\right)^{n_2} \left(\frac{\partial}{\partial x_3}\right)^{n_3}.$$

The electrostatic potential of a unit charge placed at the origin of R^3 is

$$f^{(0,0,0)}(x) = r^{-1}$$

(after fixing up the units), this being the amount of work needed to bring a like unit charge from ∞ to the place x against the field $\mathfrak{f}(x) = xr^{-3}$. Now the potential of a (vertical) dipole is

$$f^{(0,0,1)}(x) = \partial r^{-1}/\partial x_3 = -x_3 r^{-3},$$

and, in general, the potential of a "multipole" is

$$f^{\mathbf{n}}(x) = \partial^{\mathbf{n}} r^{-1}.$$

This is a homogeneous function of degree $-n-1$, i.e.,

$$f^{\mathbf{n}}(x) = r^{-n-1} f^{\mathbf{n}}(x/r);$$

especially, f^n is the product of r^{-n-1} and a function $e^n = f^n(x/r)$ of co-latitude and longitude only. Maxwell noticed that *e^n is a spherical harmonic of weight n and that, for fixed n, the family e^n: $|\mathbf{n}| = n$ spans out the spherical harmonics of that weight.* The proof is sketched in Exercises 8 through 10.

EXERCISE 8.   Check that $\Delta e^n = -n(n+1)e^n$. *Hint:* $\Delta^3 f^n = 0$; now write out Δ^3 in spherical polars; compare Exercise 6.

EXERCISE 9.   Verify that $e^{(0,0,n)} = n!(-1)^n p_n(\cos\varphi)$. *Hint:*

$$[x_1^2 + x_2^2 + (x_3-\gamma)^2]^{-\frac{1}{2}} = \sum_{n=0}^{\infty} e^{(0,0,n)}(-\gamma)^n/n!$$

on $|x| = 1$.

EXERCISE 10.   Finish the proof of Maxwell's statement. *Hint:*   The span of e^n for fixed $|\mathbf{n}| = n$ is closed under the action of $G = SO(3)$ on $|x| = 1$.

EXERCISE 11.   Check by a picture that the dipoles

$$e^{(1,0,0)} = \partial r^{-1}/\partial x_1, \qquad e^{(0,1,0)} = \partial r^{-1}/\partial x_2, \qquad e^{(0,0,1)} = \partial r^{-1}/\partial x_3$$

are perpendicular on $|x| = 1$.

EXERCISE 12.   Prove the converse of Exercise 6: Every spherical harmonic of weight n is the restriction to $|x| = 1$ of a homogeneous polynomial f of degree n with $\Delta^3 f = 0$. *Hint:* $f = r^{2n+1}f^n = r^n e^n$.

EXERCISE 13.   Discuss the analog of Maxwell's poles for the plane, replacing $1/r$ by $\log(1/r)$.

4.14*   REPRESENTATIONS OF SO(3)

Now comes the final step: to "lift" up from $L^2(G/K)$ to the full space $L^2(G)$!

Bring in the basis e_n^l: $|l| \leqslant n$ of the spherical harmonics of weight n and for $g \in SO(3)$, define a (complex) $(2n+1) \times (2n+1)$ matrix $o_n(g)$ by the rule

$$e_n^i(gx) = \sum_{|j| \leqslant n} o_n^{ij}(g)e_n^j(x).$$

Then o_n is "multiplicative":

$$o_n(g)o_n(h) = o_n(gh),$$

as you see from

$$\sum_{|k| \leqslant n} o_n^{ik}(g) \sum_{|j| \leqslant n} o_n^{kj}(h) e_n^j(x) = \sum_{|k| \leqslant n} o_n^{ik}(g) e_n^k(hx)$$

$$= e_n^i(ghx)$$

$$= \sum_{|j| \leqslant n} o_n^{ij}(gh) e_n^j(x).$$

EXERCISE 1.   Check that o_n is nonsingular; in fact, it is a "unitary" matrix, meaning that $o_n^{-1} = o_n^{\#}$ (the conjugate transpose of o_n). *Hint:*   Compute

$$\int_{S^2} e_n^i(x) e_n^j(x)^* \, do = \int_{S^2} e_n^i(gx) e_n^j(gx)^* \, do,$$

using the definition of o_n.

The moral is that the map o_n from SO(3) to $(2n+1) \times (2n+1)$ unitary matrices is a "representation" of SO(3), as introduced in Section 4.7. The goal of all the work done for $G = $ SO(3) is now before you: $G = $ SO(3) *has sufficiently many (irreducible) representations to do Fourier series in the manner suggested in Section 4.7; specifically, any function $f \in L^2(G)$ can be expanded into a "Fourier series"*

$$f(g) = \sum_{n=0}^{\infty} (2n+1) \operatorname{sp}[\hat{f}(n) o_n(g)]$$

with $(2n+1) \times (2n+1)$ coefficients

$$\hat{f}(n) = \int_G f(g) o_n(g)^{\#} \, dg \qquad \left[\text{i.e.}, \hat{f}(n)^{ij} = \int_G f(g) o_n^{ji}(g)^* \, dg\right],$$

and there is a Plancherel identity:

$$\|f\|_2^2 = \int_G |f(g)|^2 \, dg = \sum_{n=0}^{\infty} (2n+1) \operatorname{sp}[\hat{f}(n) \hat{f}(n)^{\#}],$$

in which sp *means "spur" or "trace," as usual.* The proof is broken up into a series of easy steps. Additional information about the representations of $G = $ SO(3) is given by Boerner [1963]; see also Gelfand *et al.* [1963], Hammermesh [1962], and Müller [1969] for applications to the Zeeman effect, crystals, Maxwell's equations, and the like. An excellent account of the representations of the "classical groups," including SO(3), and of the special functions connected with them will be found in Vilenkin [1968]. Now to the proof.

Step 1: The functions o_n^{ij}: $|i| \leqslant n$, $|j| \leqslant n$, $n \geqslant 0$ form a perpendicular family in $L^2(G)$, and

$$\int_G |o_n^{ij}(g)|^2 \, dg = (2n+1)^{-1}.$$

PROOF.

$$o_n^{ij}(g) = \int_{S^2} e_n^{i}(gx) e_n^{j}(x)^* \, do = \int_G e_n^{i}(gh) e_n^{j}(h)^* \, dh,$$

so

$$\int_G o_n^{ik}(g) o_m^{jl}(g)^* \, dg$$

$$= \int_G dg \int_G e_n^{i}(gh_1) e_n^{k}(h_1)^* \, dh_1 \int_G e_m^{j}(gh_2)^* e_m^{l}(h_2) \, dh_2$$

$$= \int_G e_n^{k}(h_1)^* \, dh_1 \int_G e_m^{l}(h_2) \, dh_2 \int_G e_n^{i}(gh_1) e_m^{j}(gh_2)^* \, dg.$$

The inner integral is computed with the help of Exercise 4.13.2:

$$\int_G e_n^{i}(gh_1) e_m^{j}(gh_2)^* \, dg = \int_G e_n^{i}(g) e_m^{j}(gh_1^{-1}h_2)^* \, dg$$

$$= \int_G e_n^{i}(g) \, dg \int_K e_m^{j}(gkh_1^{-1}h_2)^* \, dk$$

$$= p_m(h_1^{-1}h_2) \int_G e_n^{i}(g) e_m^{j}(g)^* \, dg.$$

Now you see that the inner product is 0 unless $i = j$ and $n = m$, in which case the addition formula of Subsection 4.13.1 comes to the rescue:

$$\int_G o_n^{ik}(g) o_n^{il}(g)^* \, dg$$

$$= \int_G e_n^{k}(h_1)^* \, dh_1 \int_G e_n^{l}(h_2) \, p_n(h_2^{-1}h_1) \, dh_2$$

$$= (2n+1)^{-1} \sum_{|j| \leqslant n} \int_G e_n^{k}(h_1)^* e_n^{j}(h_1) \, dh_1 \int_G e_n^{l}(h_2) e_n^{j}(h_2)^* \, dh_2$$

$$= (2n+1)^{-1} \int_G e_n^{k}(h_1)^* e_n^{l}(h_1) \, dh_1$$

$$= (2n+1)^{-1} \quad \text{if} \quad k = l, \quad \text{and} \quad 0 \quad \text{otherwise}.$$

Step 2: This step involves a number of new definitions. $f \in L^2(G)$ is a "class function" if $f(g)$ depends upon the "conjugacy class" $(hgh^{-1}: h \in G)$ only. This is the same as to say that f is an even function of the angle $0 \leqslant \alpha \leqslant \pi$ through which g rotates the plane perpendicular to its axis. An example of a class function is the "character" of the representation o_n:

$$\mathrm{ch}_n(g) = \mathrm{sp}[o_n(g)];$$

in fact,

$$\mathrm{sp}[o_n(hgh^{-1})] = \mathrm{sp}[o_n(hg)o_n(h^{-1})] = \mathrm{sp}[o_n(h^{-1})o_n(hg)] = \mathrm{sp}[o_n(g)].$$

Warning: Do not be confused by the nomenclature: ch_n is *not* a character of G as defined in Subsection 4.1.3 unless the representation is one-dimensional $(n = 0)$.

The purpose of Step 2 is to compute ch_n, *with a view to showing that the characters span the class functions.* Note that the characters form a unit perpendicular family in $L^2(G)$ by Step 1.

To begin with,

$$\mathrm{ch}_n(g) = \sum_{|l| \leqslant n} o_n^{ll}(g)$$

$$= \sum_{|l| \leqslant n} \int_{S^2} e_n^l(gx) e_n^l(x)^* \, do$$

$$= (2n+1) \int_{S^2} p_n(x \cdot gx) \, do$$

by the addition formula of Subsection 4.13.1, $x \cdot gx$ being the cosine of the angle between x and gx. Because this is a class function, it depends only upon the angle α of g, so you can replace g by the north-pole rotation

$$k = \begin{pmatrix} \cos\alpha & -\sin\alpha & 0 \\ \sin\alpha & \cos\alpha & 0 \\ 0 & 0 & 1 \end{pmatrix}$$

of angle α and express the character as

$$\mathrm{ch}_n(\cos\alpha) = (2n+1) \int_{S^2} p_n(\cos\alpha(x_1^2 + x_2^2) + x_3^2) \, do$$

$$= (2n+1)/2 \int_0^\pi p_n(\cos\alpha \sin^2\varphi + \cos^2\varphi) \sin\varphi \, d\varphi$$

$$= (2n+1) \int_0^1 p_n(\cos\alpha(1 - x^2) + x^2) \, dx.$$

Now for any $0 \leqslant \gamma < 1$,

$$\sum_{n=0}^{\infty} \frac{\mathrm{ch}_n(\cos\alpha)}{2n+1} \gamma^n = \int_0^1 \sum_{n=0}^{\infty} P_n(\cos\alpha(1-x^2)+x^2)\gamma^n\,dx$$

$$= \int_0^1 \frac{dx}{\{1-2\gamma[\cos\alpha(1-x^2)+x^2]+\gamma^2\}^{\frac{1}{2}}}$$

$$= \int_0^1 \frac{dx}{(a^2-b^2x^2)^{\frac{1}{2}}}$$

with

$$a^2 = 1 - 2\gamma\cos\alpha + \gamma^2 > b^2 = 2\gamma(1-\cos\alpha),$$

and this is an elementary integral:

$$\int_0^1 \frac{dx}{(a^2-b^2x^2)^{\frac{1}{2}}} = \frac{1}{b}\int_0^{b/a} \frac{dy}{(1-y^2)^{\frac{1}{2}}} = \frac{1}{b}\sin^{-1}\frac{b}{a}.$$

Think of both sides as functions f of γ and compute $2\gamma f' + f$ to get rid of the factors $(2n+1)^{-1}$ in the left-hand sum. After a long but elementary computation [Do it!], you find

$$\sum_{n=0}^{\infty} \mathrm{ch}_n(\cos\alpha)\gamma^n = \frac{1+\gamma}{1-2\gamma\cos\alpha+\gamma^2}$$

$$= \frac{1}{2i\sin(\alpha/2)}\left[\frac{e^{i\alpha/2}}{1-\gamma e^{i\alpha}} - \frac{e^{-i\alpha/2}}{1-\gamma e^{-i\alpha}}\right]$$

$$= \text{the imaginary part of } \frac{1}{\sin(\alpha/2)}\frac{e^{i\alpha/2}}{1-\gamma e^{i\alpha}}$$

$$= \text{the imaginary part of } \frac{1}{\sin(\alpha/2)}\sum_{n=0}^{\infty} e^{i(2n+1)\alpha/2}\gamma^n$$

$$= \sum_{n=0}^{\infty} \frac{\sin[(2n+1)\alpha/2]}{\sin(\alpha/2)}\gamma^n.$$

Therefore, the character of o_n is

$$\mathrm{ch}_n(g) = \mathrm{ch}_n(\cos\alpha) = \frac{\sin[(2n+1)\alpha/2]}{\sin(\alpha/2)},$$

and as this is a polynomial in $\cos\alpha$ of exact degree n [Why?], it follows from the Weierstrass approximation theorem [Subsection 1.7.3] that the characters span the class functions.

EXERCISE 2.   Give a second proof that the characters span the class functions based on the fact that ch_n is the Dirichlet kernel D_n of Section 1.4, adapted to $[-\pi, \pi)$.

EXERCISE 3.   Show that the integral of a class function $f(g) = f^\bullet(\cos \alpha)$ can be expressed as

$$\int_G f(g)\, dg = \int_0^\pi f^\bullet(\cos \alpha)\,\frac{1 - \cos \alpha}{\pi}\, d\alpha.$$

Hint: The characters span, so it is enough to check the formula for $f = \mathrm{ch}_n$. You can do this by showing that $\Delta(\alpha) = (1 - \cos \alpha)/\pi$ is the only nonnegative summable weight relative to which the characters

$$\frac{\sin\left[(2n+1)\alpha/2\right]}{\sin(\alpha/2)} : \quad n \geqslant 0$$

form a unit-perpendicular family on the interval $0 \leqslant \alpha \leqslant \pi$.

EXERCISE 4.   Show that the integral for $G = \mathrm{SO}(3)$ can be expressed in the projective space picture of Subsection 4.8.3 as

$$\int_G f(g)\, dg = \frac{1}{4\pi^2} \int_B f^\bullet(x)\,\frac{1 - \cos |x|}{|x|^2}\, d^3 x,$$

in which $f^\bullet$ is f viewed as a function of $x \in B$. *Hint:* $\int_G f(hgh^{-1})\, dh$ is a class function with the same integral as f; now use Exercise 3.

Step 3:   The functions o_n^{ij} span $L^2(G)$.

PROOF.   The span of the functions o_n^{ij} contains the translates (from both sides) of any class function f. To see this, expand f into a sum of characters:

$$f(g) = \sum_{n=0}^\infty \hat{f}(n)\,\mathrm{ch}_n(g)$$

and observe that

$$f(gh) = \sum_{n=0}^\infty \hat{f}(n)\,\mathrm{sp}\left[o_n(g)o_n(h)\right] = f(hg)$$

belongs to the span of the o_n^{ij} as a function of either argument. Now bring in the special class functions

$$f_n(g) = e^{-n(1-\cos \alpha)} \times \left[\int_G e^{-n(1-\cos \alpha)}\, dg\right]^{-1}$$

for $n \geqslant 1$. f_n is positive, $\int f_n = 1$, and

$$\lim_{n \uparrow \infty} \int_G f_n(gh^{-1}) f(h) \, dh = f(g)$$

for $f \in C(G)$ and fixed $g \in G$ since the angle of gh^{-1} vanishes only if $gh^{-1} = 1$. But the left-hand side is bounded by $\|f\|_\infty$, independently of $n \geqslant 1$, so the convergence also takes place in $L^2(G)$. The proof is finished by observing that $\int f_n(gh^{-1}) f(h) \, dh$ belongs to the span of translates of class functions.

Step 4:   Check the formula

$$f(g) = \sum_{n=0}^{\infty} (2n+1) \operatorname{sp} [\hat{f}(n) o_n(g)]$$

for the Fourier series of $f \in L^2(G)$ and the Plancherel identity

$$\|f\|_2^2 = \sum_{n=0}^{\infty} (2n+1) \operatorname{sp} [\hat{f}(n) \hat{f}(n)^*]$$

using the fact that the unit-perpendicular family $(2n+1)^{1/2} o_n^{ij}$ spans $L^2(G)$, as established in Step 3. This is just a matter of unwinding the sums: for example,

$$\sum_{n=0}^{\infty} (2n+1) \operatorname{sp} [\hat{f}(n) \hat{f}(n)^*]$$

$$= \sum_{n=0}^{\infty} (2n+1) \sum_{|i| \leqslant n} \sum_{|j| \leqslant n} \int_G f(g) o_n^{ji}(g)^* \, dg \int_G f(g)^* o_n^{ji}(g) \, dg$$

$$= \sum_{n=0}^{\infty} \sum_{|i| \leqslant n} \sum_{|j| \leqslant n} \left| \int_G f(g) (2n+1)^{1/2} o_n^{ji}(g)^* \, dg \right|^2$$

$$= \|f\|_2^2 .$$

4.15*   THE EUCLIDEAN MOTION GROUP

The purpose of this section is to parallel the discussion of $SO(3)$ [Sections 4.8–4.13] for the group $G = M(2)$ of (proper) rigid Euclidean motions of the plane, up to the Fourier integral for functions on G/K; the latter turns out to be an old friend. Unlike $SO(3)$, G is noncompact; that is why integrals come out instead of series. To obtain a Fourier integral on the whole group, you must deal with ∞-dimensional representations, and that gets pretty complicated, as you can imagine; see Gelfand *et al.* [1969] for information on this subject in the context of $G = SL(2, R)$. A brief discussion of the group $M(1)$ of rigid Euclidean motions of the line will set the stage.

1. The Group $M(1)$

The one-dimensional group $G = M(1)$ comprises translations $[x \to x+y]$, reflection $[x \to -x]$, and products of these, that is to say, G is the class of maps

$$g: x \in R^1 \to kx + y \quad \text{with} \quad k = \pm 1.$$

EXERCISE 1. Check that G is a (noncommutative) group.

The subgroup K, comprising the reflection and the identity, plays the role of $SO(2)$. G/K can be identified as the line and $K/G/K$ as the half-line $R^+ = [0, \infty)$. G has a nice integral:

$$\int_G f(g)\, dg = \int_{G/K} \frac{1}{2}\left[\sum_K f^{\bullet}(k,y)\right] dy = \int_{-\infty}^{\infty} \tfrac{1}{2}[f^{\bullet}(+1,y) + f^{\bullet}(-1,y)]\, dy$$

which is invariant under the group action, and you may introduce a product for summable functions on G, or G/K, or $K/G/K$, as before. $L^1(G/K)$ may be identified with $L^1(R^1, dx)$. The product of two functions of that class is

$$f_1 \circ f_2(x) = \tfrac{1}{2}\int f_1(x-y)f_2(y)\, dy + \tfrac{1}{2}\int f_1(x+y)f_2(y)\, dy$$

$$= \tfrac{1}{2}\int f_1(x-y)\,[f_2(y) + f_2(-y)]\, dy,$$

and this is noncommutative. $L^1(K/G/K)$ can be identified with $L^1(R^+, 2\,dx)$, alias even functions of class $L^1(R^1)$, and the product *is* commutative there! The homomorphisms of $L^1(K/G/K)$ are

$$f \to \hat{f}(\gamma) = 2\int_0^{\infty} f(x) \cos 2\pi\gamma x\, dx$$

for $\gamma \geq 0$; in particular, you recover the "cosine transform" of Exercise 2.3.5. The identity

$$p(g_1)\,p(g_2) = \int_K p(g_1 k g_2)\, dk$$

for the "spherical function" $p = \cos 2\pi\gamma x$ is just a learned way of stating the elementary trigonometrical identity

$$\cos 2\pi\gamma x \cos 2\pi\gamma y = \tfrac{1}{2}\cos 2\pi\gamma(x-y) + \tfrac{1}{2}\cos 2\pi\gamma(x+y),$$

and you even have an "addition formula"

$$\cos 2\pi\gamma(x-y) = \cos 2\pi\gamma x \cos 2\pi\gamma y + \sin 2\pi\gamma x \sin 2\pi\gamma y,$$

which shows that the associated "spherical harmonics" are simply $\cos 2\pi\gamma x$ and $\sin 2\pi\gamma x$, $\gamma = 0$ excepted.

EXERCISE 2.   What is the counterpart for SO(3) of the formulas

$$\cos x = \tfrac{1}{2}(e^{ix}+e^{-ix}), \qquad \sin x = (2i)^{-1}(e^{ix}-e^{-ix}) \, ?$$

Hint: Express these identities as integrals over K.

The expansion of $f \in L^2(G/K)$ into "spherical harmonics" is now seen to be the standard one-dimensional Fourier transform pair, expressed by means of the sine and cosine transforms of Exercise 2.3.5, namely,

$$f(x) = f_{\text{even}}(x) + f_{\text{odd}}(x) = 2\int_0^\infty \hat{f}_{\text{even}}(\gamma)\cos 2\pi\gamma x \, d\gamma + 2\int_0^\infty \hat{f}_{\text{odd}}(\gamma)\sin 2\pi\gamma x \, d\gamma \, .$$

The goal of the discussion of M(2) is to make a similar group-theoretical explanation of the two-dimensional Fourier transform.

2.   The Group $M(2)$

An "affine" transformation of the plane is a map

$$g \colon x \in R^2 \to kx + y,$$

in which k is a real 2×2 matrix and $y \in R^2$: It is a "rigid motion" if it preserves distances [$k \in O(2)$]; it is "proper" if it preserves the orientation ("handedness") of R^2, which is to say $\det(k) = +1$ [$k \in SO(2)$]. The Euclidean motion group $G = M(2)$ is the (noncommutative) group of proper rigid motions of the plane. $K = SO(2)$ is a subgroup. The coset space G/K is identified with the plane via the map

$$g \to g(0) = y \in R^2 \, .$$

Under this map, the double coset KgK appears as a circle [$ky \colon k \in SO(2)$] of radius $|y|$ about the origin, so you may identify $K/G/K$ with the half-line R^+ via the map

$$g \to |g(0)| = |y| \, .$$

This identification shows that g and g^{-1} have the same double coset: Namely, $g^{-1}x = k^{-1}(x-y)$, so

$$|g^{-1}(0)| = |-k^{-1}(y)| = |y| = |g(0)| \, .$$

3. Integration on $M(2)$

The (left) invariant integral for G is easily computed. As for SO(3),

$$f^0(g) = \int_K f(gk) \, dk$$

is a function of the coset gK, only. As such it can be viewed as a function of $g(0) = y \in R^2$, and as you might expect

$$\int_G f(g) \, dg = \int_{R^2} f^0(y) \, d^2 y = \int_{G/K} d^2 y \int_K f(gk) \, dk$$

is a (left-) invariant integral since the action of G on R^2 is area-preserving; for practical computation, it is preferable to think of $f(g)$ as a function of k and y and to express the volume element directly as $dg = d^2 y \, dk$.

EXERCISE 3.   Check that $\int_G f(g) \, dg$ is also invariant under the right action of G. *Warning:* $\int_G 1 \, dg = \infty$, so you cannot copy the SO(3) proof, nevertheless the computation is easy.

EXERCISE 4.   Check that $\int_G f(g) \, dg$, defined as above, is the only left-invariant integral on G up to multiplicative constants.

4. Spherical Functions

As for SO(3), each of the spaces

$$L^1(G) \supset L^1(G/K) = L^1(R^2, d^2 x)$$

$$\supset L^1(K/G/K) = L^1(R^+, 2\pi r \, dr)$$

is associative under the product

$$f_1 \circ f_2(g) = \int_G f_1(gh^{-1}) f_2(h) \, dh,$$

though only the smallest one $L^1(K/G/K)$ is commutative; as before, the proof of commutativity hinges upon the fact that g and g^{-1} have the same double coset. The homomorphisms of $L^1(K/G/K)$ may now be computed. They come from "spherical functions" p, subject to

(a) $p \in C^\infty(K/G/K)$

(b) $|p| \leqslant p(0) = 1$[1]

(c) $p(g_1) \, p(g_2) = \int_K p(g_1 k g_2) \, dk,$

---

[1] 0 is the double coset of the identity! This is something of a notational misfortune, but will not lead to confusion if you keep it in mind.

just as for SO(3). The identity (c) can be expressed more concretely if you identify the double coset of $g_1 \, kg_2$ with

$$|g_1 \, kg_2(0)| = |k_1 \, ky_2 + y_1|.$$

As k runs over SO(2), this runs through the values

$$(a^2 + b^2 - 2ab \cos \theta)^{1/2} \quad \text{with} \quad a = |y_1|, \quad b = |y_2|, \quad \text{and} \quad 0 \leqslant \theta < 2\pi,$$

as you can easily see from the law of cosines and a picture, and if you now think of the spherical function $p(g)$ as a function of the radius $|g(0)|$, you can express the identity (c) as

$$p(a) \, p(b) = \frac{1}{2\pi} \int_0^{2\pi} p([a^2 + b^2 - 2ab \cos \theta]^{1/2}) \, d\theta.$$

EXERCISE 5.   Find a similar formula for the product of functions of double cosets. *Answer:* $\int_0^{2\pi} \int_0^\infty f_1((a^2 + b^2 - 2ab \cos \theta)^{1/2}) f_2(b) \, b \, db \, d\theta$.

EXERCISE 6.   Check that up to a multiplicative constant the Laplace operator

$$\Delta = \frac{\partial^2}{\partial x_1{}^2} + \frac{\partial^2}{\partial x_2{}^2} = \frac{\partial^2}{\partial r^2} + \frac{1}{r} \frac{\partial}{\partial r} + \frac{1}{r^2} \frac{\partial^2}{\partial \theta^2}$$

is the only partial differential operator of degree less than or equal to 2 which commutes with the action of $G = M(2)$.

EXERCISE 7.   Check that any spherical function p is a (radial) eigen-function of Δ:

$$\Delta p = p'' + r^{-1} p' = \gamma p.$$

in which $\gamma = \Delta p(0)$, and the prime stands for differentiation by $r = (x_1{}^2 + x_2{}^2)^{1/2}$.

EXERCISE 8.   Show that the eigenvalues γ of Exercise 7 are always less than or equal to zero. *Hint:* As it happens, $\int |p|^2 \, dg = \infty$ [See Exercise 9], so the SO(3) proof is no good. But

$$0 = \int_{R^2} f(\Delta - \gamma) \, p = \int_{R^2} p(\Delta - \gamma) f$$

for any $f \in C_{\uparrow}^\infty(R^2)$, and if γ is not less than or equal to 0, then $\Delta - \gamma$ maps $C_{\uparrow}^\infty(R^2)$ *onto* itself, as you can see by looking at $[(\Delta - \gamma)f]^\wedge$. This contradicts $p \neq 0$.

5. Spherical Functions Are Bessel Functions

The content of this article is that if J_0 is the Bessel function

$$J_0(r) = \frac{1}{2\pi} \int_0^{2\pi} e^{ir\cos\theta}\, d\theta$$

then

$$p(r) = J_0(2\pi \varkappa r), \qquad \varkappa \geq 0,$$

is a complete list of the spherical functions of $G = M(2)$. The proof is divided into simple steps.

Step 1:   Check that $J_0(2\pi\varkappa r)$ is a (radial) eigenfunction of Δ with eigenvalue $-4\pi^2\varkappa^2 \leq 0$.

PROOF.   The first point is to notice that

$$J_0(2\pi\varkappa r) = \int_K e^{2\pi i y\,\cdot\, kx}\, dk = \int_K e^{2\pi i k^\# y\,\cdot\, x}\, dk,$$

in which $|x| = r$ and $|y| = \varkappa$. Now apply the Laplace operator Δ to both sides:

$$\Delta J_0(2\pi\varkappa r) = \int_K (2\pi i)^2 |k^\# y|^2\, e^{2\pi i k^\# y\,\cdot\, x}\, dk$$

$$= -4\pi^2\varkappa^2 J_0(2\pi\varkappa r).$$

Step 2:   Up to a multiplicative constant, $J_0(2\pi\varkappa r)$ is the only bounded radial eigenfunction of Δ with eigenvalue $-4\pi^2\varkappa^2$; in particular, by Exercises 7 and 8, *every spherical function is of the form* $J_0(2\pi\varkappa r)$ *for some* $\varkappa \geq 0$.

PROOF.   Put $J_0(2\pi\varkappa r) = f_1$, and let f_2 be a second solution of $f'' + r^{-1}f' = -4\pi^2\varkappa^2 f$. Then

$$[r(f_1'f_2 - f_1 f_2')]' = (rf_1'' + f_1')f_2 - f_1(rf_2'' + f_2') = 0,$$

so that the Wronskian

$$f_1'f_2 - f_1 f_2' = \text{constant} \times r^{-1},$$

and consequently *either* f_2 is proportional to f_1, *or else* the constant is different from 0. But in the latter case, f_2 cannot stay bounded as $r \downarrow 0$ because

$$f_1(0) = 1,$$

$$f_1'(0+) = \frac{1}{2\pi} \int_0^{2\pi} i2\pi\varkappa \cos\theta\, d\theta = 0,$$

and if f_2 were bounded you would have

$$f_2' = [\text{constant} + o(1)] \times r^{-1},$$

which is contradictory.

Step 3:   Check that every $J_0(2\pi\varkappa r)$ is a spherical function.

PROOF.   By Subsection 4, it suffices to check the identity

$$J_0(a) J_0(b) = \frac{1}{2\pi} \int_0^{2\pi} J_0([a^2 + b^2 - 2ab \cos\theta]^{\frac{1}{2}}) \, d\theta,$$

and this is easy to do if you express it in a little more group-theoretical way. Take x and y from R^2 with $|x| = a$ and $|y| = b$ and let $n = (1,0)$. Then

$$\frac{1}{2\pi} \int_0^{2\pi} J_0([a^2 + b^2 - 2ab \cos\theta]^{\frac{1}{2}}) \, d\theta$$

$$= \int_K J_0(|x + ky|) \, dk$$

$$= \int_K dk \frac{1}{2\pi} \int_0^{2\pi} e^{i|x + ky||n| \cos\theta} \, d\theta$$

$$= \int_K dk \int_K e^{i(x + ky) \cdot k'n} \, dk'$$

$$= \int_K e^{ix \cdot k'n} \, dk' \int_K e^{iy \cdot k^{-1}k'n} \, dk$$

$$= \frac{1}{2\pi} \int_0^{2\pi} e^{i|x| \cos\theta} \, d\theta \frac{1}{2\pi} \int_0^{2\pi} e^{i|y| \cos\theta} \, d\theta$$

$$= J_0(a) J_0(b),$$

as advertised. The proof is finished.

6.   The Bessel Transform for $\mathbf{L^2}(K/G/K)$

The spherical functions $p(r) = J_0(2\pi\varkappa r)$ may now be used to make a Fourier-type integral for functions of class $L^2(K/G/K) = L^2(R^+, 2\pi r \, dr)$, in a very perfect analogy with the Legendre series of Section 4.12 for functions of colatitude on S^2. To do this, bring in the transform

$$\hat{f}(\varkappa) = \int_0^\infty f(r) J_0(2\pi\varkappa r) 2\pi r \, dr.$$

This is just the Bessel transform of Section 2.10, alias the conventional two-dimensional Fourier transform specialized to radial functions. The function f is recovered by the recipe

$$f(r) = \int_0^\infty \hat{f}(\varkappa) J_0(2\pi\varkappa r) 2\pi\varkappa \, d\varkappa,$$

and there is a Plancherel identity

$$\|f\|_2^2 = \int_0^\infty |f(r)|^2 2\pi r \, dr = \int_0^\infty |\hat{f}(\varkappa)|^2 2\pi\varkappa \, d\varkappa = \|\hat{f}\|_2^2,$$

as you already know from Section 2.10.

7. Spherical Harmonics: Formal Discussion

If you now try to "lift" this Fourier integral up to $L^2(G/K) = L^2(R^2)$ via "spherical harmonics," by looking at the span of the left translates p^g of the spherical function $p(r) = J_0(2\pi\varkappa r)$ under the action of $G = M(2)$, you run into trouble at once:

$$\|p\|_2^2 = \int_0^\infty |J_0(2\pi\varkappa r)|^2 2\pi r \, dr = \infty,$$

EXERCISE 9.   Confirm that $\int_0^\infty |J_0(r)|^2 r \, dr = \infty$. *Hint:*   It is enough to check $\int_0^\infty |J_0|^2 \, dr = \infty$; now apply the standard one-dimensional Plancherel identity to the integral formula for J_0.

The interpretation of the "span" looks like a delicate technical point; not to prejudice it, let us try another tack.

By analogy with $G = SO(3)$, the class of "spherical harmonics of weight $\varkappa$" associated with the spherical function $p(r) = J_0(2\pi\varkappa r)$ is (or ought to be) the "nice" eigenfunctions of Δ with eigenvalue $-4\pi^2\varkappa^2$. As in Exercise 4.13.7, it is easy to see that if J_l is the "associated" Bessel function

$$J_l(r) = \frac{1}{2\pi} \int_0^{2\pi} e^{ir\cos\theta} e^{-il\theta} \, d\theta,$$

and if $x = r(\cos\theta, \sin\theta)$, then

$$e_\varkappa^l(x) = J_l(2\pi\varkappa r) e^{il\theta}$$

is such an eigenfunction for any $|l| < \infty$. The proof will be self-evident from the fact that

$$\sum_{|l| < \infty} e_\varkappa^l(x) = e^{i2\pi\varkappa r\cos\theta} = e^{i2\pi\varkappa x \cdot (1,0)}$$

is also an eigenfunction of Δ with the desired eigenvalue $-4\pi^2\varkappa^2$ if you will recollect Exercise 4.13.7. The adjective "associated" is not customarily used in this context, but it should be inasmuch as J_l plays the same role for $G = M(2)$ that the associated Legendre polynomials do for $SO(3)$. The hope that these eigenfunctions are bona fide "spherical harmonics" is reinforced by *Graf's addition formula*:

$$J_0([a^2+b^2-2ab\cos\theta]^{1/2}) = \sum_{l=-\infty}^{\infty} J_l(a)J_l(b)e^{il\theta},$$

which is a perfect analogue of the addition formula of Subsection 4.13.1. To see this analogy, let $g_1 K = x = a(\cos\alpha, \sin\alpha)$ and $g_2 K = y = b(\cos\beta, \sin\beta)$. Then by Graf's formula

$$
\begin{aligned}
p(g_1^{-1}g_2) &= J_0(2\pi\varkappa|x-y|) \\
&= J_0\big(2\pi\varkappa[a^2+b^2-2ab\cos(\alpha-\beta)]^{1/2}\big) \\
&= \sum_{|l|<\infty} [J_l(2\pi\varkappa a)e^{il\alpha}]^* J_l(2\pi\varkappa b)e^{il\beta} \\
&= \sum_{|l|<\infty} e_\varkappa^{\,l}(x)^* e_\varkappa^{\,l}(y).
\end{aligned}
$$

The fact that J_l is real is used in line 3.

PROOF OF GRAF'S ADDITION FORMULA. The proper tool is the Parseval identity for the circle, which justifies the summation

$$
\begin{aligned}
\sum_{|l|<\infty} & J_l(a)^* J_l(b)\,e^{il\theta} \\
&= \sum_{|l|<\infty}\left[\frac{1}{2\pi}\int_0^{2\pi} e^{ia\cos\alpha}\,e^{-il\alpha}\,d\alpha\right]^* \frac{1}{2\pi}\int_0^{2\pi} e^{ib\cos\beta}\,e^{-il\beta}\,d\beta\,e^{il\theta} \\
&= \sum_{|l|<\infty}\left[\frac{1}{2\pi}\int_0^{2\pi} e^{ia\cos\alpha}\,e^{-il\alpha}\,d\alpha\right]^* \frac{1}{2\pi}\int_0^{2\pi} e^{ib\cos(\beta+\theta)}\,e^{-il\beta}\,d\beta \\
&= \frac{1}{2\pi}\int_0^{2\pi} e^{-i[a\cos\psi - b\cos(\psi+\theta)]}\,d\psi\,.
\end{aligned}
$$

To reduce this to the required form

$$\frac{1}{2\pi}\int_0^{2\pi} \exp[i(a^2+b^2-2ab\cos\theta)^{1/2}\cos\psi]\,d\psi = J_0([a^2+b^2-2ab\cos\theta]^{1/2}),$$

you have only to notice that

$$
\begin{aligned}
a\cos\psi - b\cos(\psi+\theta) &= (a-b\cos\theta)\cos\psi + b\sin\theta\sin\psi \\
&= (a^2+b^2-2ab\cos\theta)^{1/2}\cos(\psi + \text{a suitable phase angle}).
\end{aligned}
$$

The proof is finished.

8.   Spherical Harmonics as They Should Be

Both on grounds of Graf's addition formula and the fact that

$$e_x{}^l = J_l(2\pi \varkappa r)\, e^{il\theta}$$

is an eigenfunction of Δ with the proper eigenvalue $-4\pi^2 \varkappa^2$, the class $M_\varkappa$ of formal sums

$$f = \sum_{|l| < \infty} \hat{f}(l)\, e_x{}^l$$

with

$$\|f\|_\varkappa^2 = \sum_{|l| < \infty} |\hat{f}(l)|^2 < \infty$$

is now *declared* to be the "spherical harmonics of weight $\varkappa$." The inner product for $M_\varkappa$ is

$$(f_1, f_2)_\varkappa = \sum_{|l| < \infty} \hat{f}_1(l)\, \hat{f}_2(l)^*,$$

from which it is clear that the family $e_x{}^l : |l| < \infty$ is a unit-perpendicular basis. Graf's addition formula makes sense in this context:

$$\sum |e_x{}^l(x)|^2 = \sum e_x{}^l(x)^* e_x{}^l(x) = p(1) = 1 < \infty,$$

and for $x \in R^2$ and $gK = y \in R^2$ the formula

$$p(g^{-1}x) = \sum e_x{}^l(y)^* e_x{}^l(x) = \sum e_x{}^l(y)\, e_x{}^l(x)^*$$

is just the Fourier series of the translated spherical function $p^{g^{-1}}$; in particular, every translate of p is a spherical harmonic of weight $\varkappa$.

EXERCISE 10.   Check that $M_\varkappa$ is the span of the translates of p and infer that it is invariant under the action of $G = M(2)$. *Hint:*

$$\int_K p(g^{-1}kx)\, e^{-il\alpha}\, dk = e_x{}^l(x)\, J_l(|y|),$$

in which α is the angle of rotation of $k \in K = SO(2)$; compare Exercise 4.13.7.

Besides this, it is very satisfactory to find that the formal sums $f = \sum \hat{f}(l)\, e_x{}^l$ inhabiting $M_\varkappa$ actually converge to bona fide eigenfunctions of Δ with the proper eigenvalue $-4\pi^2 \varkappa^2$. The easiest way to see this is to infer from $\sum |\hat{f}(l)|^2 < \infty$ that $\hat{f}(l)$ can be expressed as the Fourier coefficient of a bona fide function f^* of class $L^2(K)$:

$$\hat{f}(l) = \frac{1}{2\pi} \int_0^{2\pi} f^*(\theta)\, e^{-il\theta}\, d\theta$$

and to use this to sum the formal series for the "spherical harmonic" f:

$$
\begin{aligned}
f(x) &= \sum_{|l|<\infty} \hat{f}(l)\, J_l(2\pi\varkappa r)\, e^{il\theta} \\
&= \sum_{|l|<\infty} \frac{1}{2\pi} \int_0^{2\pi} f^{\#}(\alpha)\, e^{-il\alpha}\, d\alpha\, \frac{1}{2\pi} \int_0^{2\pi} e^{i2\pi\varkappa r\cos\beta}\, e^{-il\beta}\, d\beta\, e^{il\theta} \\
&= \frac{1}{2\pi} \int_0^{2\pi} e^{i2\pi\varkappa r\cos(\theta-\psi)} f^{\#}(\psi)\, d\psi \\
&= \int_K e^{i2\pi\varkappa x\cdot kn} f^{\#}(k)\, dk, \qquad x = r(\cos\theta, \sin\theta), \quad n = (1,0).
\end{aligned}
$$

Δ can now be applied under the integral to check that $\Delta f = -4\pi^2\varkappa^2 f$.

9. Fourier Transforms for $\mathbf{L}^2(G/K)$

Now you can "project" a function $f \in L^2(G/K) = L^2(R^2)$ onto the space of spherical harmonics of weight $\varkappa$ by the formal recipe

$$
\begin{aligned}
f \to f_\varkappa(x) &= \sum_{|l|<\infty} e_\varkappa^l(x) \int_{R^2} f(y)\, e_\varkappa^l(y)^*\, d^2 y \\
&= \int_{R^2} J_0(2\pi\varkappa\,|x-y|)\, f(y)\, d^2 y,
\end{aligned}
$$

and you may hope to add these "slices" back together to reconstitute the original function. A formal computation substantiates this:

$$
\begin{aligned}
\int_0^\infty f_\varkappa(x)\, 2\pi\varkappa\, d\varkappa &= \int_0^\infty 2\pi\varkappa\, d\varkappa \int_{R^2} J_0(2\pi\varkappa\,|x-y|)\, f(y)\, d^2 y \\
&= \int_{R^2} f(y)\, d^2 y \int_0^\infty J_0(2\pi\varkappa\,|x-y|)\, 2\pi\varkappa\, d\varkappa \\
&= \int_{R^2} f(y)\, d^2 y \int_0^\infty \int_0^{2\pi} e^{2\pi i\varkappa|x-y|\cos\theta}\, d\theta\, \varkappa\, d\varkappa \\
&= \int_{R^2} f(y)\, d^2 y \int_{R^2} e^{2\pi i\gamma\cdot(x-y)}\, d^2\gamma \\
&= \int_{R^2} \hat{f}(\gamma)\, e^{2\pi i\gamma\cdot x}\, d^2\gamma \\
&= f(x),
\end{aligned}
$$

in which $|\gamma|$ is identified with $\varkappa$ in line 4, and $\hat{f}(\gamma) = \int f(y)\, e^{-2\pi i\gamma\cdot y}\, d^2 y$ is the customary two-dimensional Fourier transform. The circle is completed by translating the Plancherel identity into the new language. To do this, compute $f_\varkappa$ in a new way:

$$f_\varkappa(x) = \int_{R^2} J_0(2\pi\varkappa|x-y|)f(y)\,d^2y$$

$$= \int_{R^2} f(y)\,d^2y \int_K e^{i2\pi\varkappa(x-y)\cdot kn}\,dk$$

$$= \int_K e^{i2\pi\varkappa x\cdot kn}\hat{f}(\varkappa kn)\,dk\,.$$

By comparison with the last formula of Subsection 8, you see that $f_\varkappa^{\#}(k) = \hat{f}(\varkappa kn)$, so

$$\|f_\varkappa\|_\varkappa^2 = \sum_{|l| < \infty} |f_\varkappa^\wedge(l)|^2 = \int_K |f_\varkappa^{\#}(k)|^2\,dk = \int_K |\hat{f}(\varkappa kn)|^2\,dk\,,$$

and

$$\int_0^\infty \|f_\varkappa\|_\varkappa^2\, 2\pi\varkappa\,d\varkappa = \int_0^\infty 2\pi\varkappa\,d\varkappa \int_K |\hat{f}(\varkappa kn)|^2\,dk$$

$$= \int_{R^2} |\hat{f}(\gamma)|^2\,d^2\gamma = \|\hat{f}\|_2^2 = \|f\|_2^2 = \int_{R^2} |f(x)|^2\,d^2x\,,$$

by the conventional Plancherel theorem for $L^2(R^2)$.

To put this formal development on a solid mathematical footing, you may phrase everything in terms of a perpendicular "sum" (or better, an "integral") of the spaces of spherical harmonics. This refers to the class M of measurable functions $f = f_\varkappa(x)$ of the weight $\varkappa \geq 0$ and $x \in R^2$ such that

(a) $f_\varkappa$ is a spherical harmonic (of the indicated weight) for almost every $\varkappa \geq 0$, and

(b) $\|f\|^2 = \int_0^\infty \|f_\varkappa\|_\varkappa^2\, 2\pi\varkappa\,d\varkappa < \infty.$

EXERCISE 11.   Show that M is a Hilbert space.

EXERCISE 12.   Check that the map

$$f \to f_\varkappa(x) = \int_{R^2} J_0(2\pi\varkappa|x-y|)f(y)\,d^2y$$

is an isomorphism between $L^2(R^2)$ and M. *Hint:*   Do everything for $C_\downarrow^\infty(R^2)$ first, and then carry it over to $L^2(R^2)$, using Section 2.3 as a model.

The isomorphism can be expressed in an informal but suggestive way:

$$\mathsf{L}^2(G/K) = \mathsf{L}^2(R^2) = \int_0^\infty M_\varkappa\, 2\pi\varkappa\, d\varkappa,$$

in perfect analogy with the perpendicular splitting

$$\mathsf{L}^2(G/K) = \mathsf{L}^2(S^2) = \oplus M_n$$

for $G = SO(3)$.

EXERCISE 13.   Repeat the discussion of Subsections 2–5 for dimension $n = 3$. Check that

$$f_1 \circ f_2(a) = \int_0^\infty \left(\tfrac{1}{2}\int_0^\pi f_1([a^2+b^2-2ab\cos\varphi]^{1/2})\sin\varphi\, d\varphi\right)f_2(b)\, 4\pi b^2\, db$$

for functions of double cosets, and verify that this is the same as

$$af_1 \circ f_2(a) = 2\pi \int\int_{\substack{|x-y|\,\leqslant\, a\,\leqslant\, x+y \\ x\,\geqslant\, 0,\, y\,\geqslant\, 0}} f_1(x)f_2(y)\, xy\, dx\, dy.$$

Check that

$$\hat{f}(\varkappa) = \int_0^\infty f(r)\frac{\sin 2\pi\varkappa r}{2\pi\varkappa r}\, 4\pi r^2\, dr, \qquad \varkappa \geqslant 0$$

is a complete list of homomorphisms of $\mathsf{L}^1(K/G/K)$; see Section 2.10 for the function $r^{-1}\sin r = j_3(r)$. Similar convolutions are discussed, in a different context, by Leblanc [1968].

4.16★ SL(2, R) AND THE HYPERBOLIC PLANE

As a final example of the group-theoretical approach to Fourier integrals, let us sketch what happens for the group $SL(2, R)$ of real 2×2 matrices of determinant $+1$.

$SL(2, R)$ is the group of proper [sense-preserving] rigid [distance-preserving] motions of the hyperbolic plane H. The latter may be pictured as the open upper half plane $x+iy\colon y>0$ equipped with the length element $y^{-1}(dx^2+dy^2)^{1/2}$ instead of the usual $(dx^2+dy^2)^{1/2}$. This makes H into a surface of constant negative curvature -1, so called, as opposed to the sphere [curvature $+1$] or the plane [curvature 0]: at each place H looks like a mountain pass with two ridges rising up and two valleys falling away. The following articles will explain (mostly without proof) as much of the geometry of H as you need; for proofs and additional information, see Auslander [1967] or Carathéodory [1954].

1.   SL(2, R) as Rigid Motions of H

$G = SL(2, R)$ acts on H according to the rule

$$g = \begin{pmatrix} a & b \\ c & d \end{pmatrix}: \quad z = x + iy \to \frac{az+b}{cz+d}.$$

This is a 1:1 map of H onto itself which is "proper" [sense-preserving] and "rigid" [distance-preserving].

EXERCISE 1.   Check that distances really are preserved. *Hint:*
$$y^{-1}(dx^2 + dy^2)^{\frac{1}{2}} = y^{-1}|dz|.$$
Compute $(y')^{-1}|dz'|$ for $z' = (cz+d)^{-1}(az+b)$.

Every proper rigid motion of H arises in this way. The correspondence is 1:2 because

$$g = \begin{pmatrix} a & b \\ c & d \end{pmatrix} \quad \text{and} \quad -g = \begin{pmatrix} -a & -b \\ -c & -d \end{pmatrix}$$

produce the same motion of H. The multiplication of $G = SL(2, R)$ corresponds to composition of motions, and the actual motion group is isomorphic to G/Z, in which Z is the "central subgroup" of G:

$$Z = (\pm 1) = (g': \quad g'g = gg' \text{ for every } g \in G).$$

$G = SL(2, R)$ is therefore a "two-fold covering" of the motion group.

EXERCISE 2.   Check the isomorphism, assuming only that the map from G to motions is onto.

2.   Hyperbolic Distance

The curve of shortest (hyperbolic) distance between two points of H is the arc of a circle which meets the line $y = 0$ perpendicularly; under "circle" is understood also vertical line. A rigid motion maps such circular arcs into one another.

EXERCISE 3.   What is the hyperbolic distance l between i and $e^x i$? *Answer:* $l = \alpha$. *Hint:* l is the infemum of $\int y^{-1}(dx^2 + dy^2)^{\frac{1}{2}}$ over all curves joining i and $e^x i$.

EXERCISE 4.   Check that the hyperbolic distance between two points on the semicircle $(x^2 + y^2)^{\frac{1}{2}} = R$, $y > 0$ obeys the rule

$$\cosh(\text{hyperbolic distance}) = 1 + \frac{1}{2} \frac{(\text{Euclidean distance})^2}{\text{product of heights}}.$$

3.   Coset Spaces

The subgroup $K = SO(2) \subset G$ of matrices

$$k = k(\theta) = \begin{pmatrix} \cos\theta & -\sin\theta \\ \sin\theta & \cos\theta \end{pmatrix}$$

is the "stability group" $(g \in G: gi = i)$ of the imaginary unit i. The coset space G/K can now be identified with H via the map $g \to gi$. The orbit $(kz: k \in SO(2))$ of a point $z \in H$ is a circle with center at $i\cosh\alpha$ and radius $\sinh\alpha$, α being the hyperbolic distance from i to z. Each point in H lies on precisely one such circle. Check this! As k runs over $K = SO(2)$, the circle is covered *twice* because, as a map of H, k is of period π in its dependence on $0 \le \theta < 2\pi$: if you increase θ by π, k changes sign. This reflects the previously cited fact that $G = SL(2, R)$ is a double covering of the motion The above description of the action of $K = SO(2)$ on H permits you to identify $K/G/K$ with the half-line $R^+ = [0, \infty)$ via the map

$$g \to \text{the hyperbolic distance between } gi \text{ and } i,$$

and it is easy to see that g and g^{-1} have the same double coset.

4.   Polar Coordinates

The point

$$\frac{i\cosh(\alpha/2) + \sinh(\alpha/2)}{i\sinh(\alpha/2) + \cosh(\alpha/2)}$$

lies at hyperbolic distance α from i, and as α runs from 0 to ∞, it sweeps out the positive quadrant of the unit circle $|z| = 1$. This permits you to introduce "polar coordinates" $0 \le \theta < 2\pi$ and $0 \le \alpha < \infty$ in H by the recipe

$$z = \begin{pmatrix} \cos\theta/2 & -\sin\theta/2 \\ \sin\theta/2 & \cos\theta/2 \end{pmatrix} \begin{pmatrix} \cosh\alpha/2 & \sinh\alpha/2 \\ \sinh\alpha/2 & \cosh\alpha/2 \end{pmatrix} i.$$

The distance α is identified with the double coset $(kz: k \in SO(2))$ in accordance with the map of Subsection 3.

EXERCISE 5.   Check that for fixed $0 \le \theta \le 2\pi$,

$$\begin{pmatrix} \cos\theta/2 & -\sin\theta/2 \\ \sin\theta/2 & \cos\theta/2 \end{pmatrix} \begin{pmatrix} \cosh\alpha/2 & \sinh\alpha/2 \\ \sinh\alpha/2 & \cosh\alpha/2 \end{pmatrix} i: 0 \le \alpha \le \infty$$

is a circular arc starting at i and cutting the line $y = 0$ perpendicularly in the point $x = (1 + \sin\theta)^{-1}\cos\theta$. Use this to draw a picture of polar coordinates in the quadrant $x \ge 0$, $y \ge 0$.

EXERCISE 6.   Use polar coordinates to prove that every $g \in G =$ SL(2, R) can be expressed in precisely one way as a product $k'tk''$ with k' and k'' from K and t of the form

$$t = \left(\begin{array}{cc} \cosh \alpha & \sinh \alpha \\ \sinh \alpha & \cosh \alpha \end{array} \right)$$

with $\alpha \geqslant 0$. *Hint:*   Find k and t so that $t^{-1}k^{-1}gi = i$.

5.   Integration on $G = $ SL(2, R)

The element $y^{-2} \, dx \, dy$ of (hyperbolic) area is invariant under the action of $G = $ SL(2, R) on H.

EXERCISE 7.   Prove it.

EXERCISE 8.   Think of $\int_K f(gk) \, dk$ as a function of the coset gK, identified with the point gi of H. Check that the integral

$$\int_G f(g) \, dg = \int_H y^{-2} \, dx \, dy \int_K f(gk) \, dk$$

is insensitive to the (left) action of $G = $ SL(2, R). Up to a multiplicative constant, it is the only such. The proof is a little harder than for $G = $ M(2).

EXERCISE 9.   Show that in the polar coordinates of Subsection 4 the area element $y^{-2} \, dx \, dy$ may be expressed as $\sinh \alpha \, d\alpha \, d\theta$.

6.   Laplace's Operator

As for SO(3) and M(2), there is a "Laplace operator"

$$\Delta = y^2 \left(\frac{\partial^2}{\partial x^2} + \frac{\partial^2}{\partial y^2} \right) = \frac{1}{\sinh \alpha} \frac{\partial}{\partial \alpha} \sinh \alpha \frac{\partial}{\partial \alpha} + \frac{1}{\sinh^2 \alpha} \frac{\partial^2}{\partial \theta^2}$$

which commutes with the action of $G = $ SL(2, R) on H; up to a multiplicative constant it is the only partial differential operator of degree less than or equal to 2 that does so.

7.   Spherical Functions

The stage is set. The rest is in a very perfect analogy with SO(3) and M(2).

$L^1(K/G/K) = L^1(R^+, 2\pi \sinh \alpha \, d\alpha)$ is commutative. The spherical functions $p = p(\cosh \alpha)$ are the solutions of

$$p(\operatorname{ch}\alpha)\, p(\operatorname{ch}\beta) = \frac{1}{2\pi} \int_0^{2\pi} p(\operatorname{sh}\alpha \operatorname{sh}\beta \cos\theta + \operatorname{ch}\alpha \operatorname{ch}\beta)\, d\theta$$

with $|p| \leqslant p(1)$,[1] and much as in Section 4.15, they turn out to be the nice "radial" eigenfunctions of Δ with nonpositive eigenvalues. Now the radial eigenfunctions of Δ are the so-called "conical functions"

$$p_{-\frac{1}{2}+ix}(\operatorname{ch}\alpha) = \frac{1}{2\pi} \int_0^{2\pi} (\operatorname{sh}\alpha \cos\theta + \operatorname{ch}\alpha)^{-\frac{1}{2}+ix}\, d\theta.$$

The corresponding eigenvalue is $-(\frac{1}{4}+\varkappa^2)$, and for this to be less than or equal to 0, you have to restrict $\varkappa$ *either* to the "principal series" R^1 *or* to the "supplementary series" $i[-\frac{1}{2}, +\frac{1}{2}]$; actually, $p_{-\frac{1}{2}+ix} = p_{-\frac{1}{2}-ix}$, so this is a double listing of spherical functions ($p_{-\frac{1}{2}}$ excepted), and you may as well take $[0, \infty)$ as the principal series and $i(0, \frac{1}{2}]$ as the supplementary series.

8.   An Anomaly

As a matter of fact, the supplementary series does *not* figure in the Fourier integral for $L^2(K/G/K)$ or $L^2(G/K)$! This anomaly may be explained by the fact that the supplementary series gives rise to eigenvalues between $-\frac{1}{4}$ and 0, while the spectrum of Δ lies to the left of $-\frac{1}{4}$ in the sense that

$$(\Delta f, f) \leqslant -\frac{1}{4}\|f\|_2^2$$

for every smooth $f \in L^2(G/K)$. [You may take it as an article of faith that what is good for Δ is good for $L^2(K/G)$!] To prove the stated upper bound to the spectrum of Δ, let $f \in C^\infty(H)$ be compact and think of it as a function of $0 < y < \infty$ for fixed x. Then

$$\frac{1}{4}\left(\int_0^\infty |f|^2 y^{-2}\, dy\right)^2 = \frac{1}{4}\left|\int_0^\infty y^{-1}\left(f\frac{\partial f^*}{\partial y} + f^*\frac{\partial f}{\partial y}\right) dy\right|^2$$

$$\leqslant \int_0^\infty |f|^2 y^{-2}\, dy \int_0^\infty \left|\frac{\partial f}{\partial y}\right|^2 dy,$$

---

[1] sh[ch] is short for sinh[cosh].

and if you divide both sides by $\int_0^\infty |f|^2 y^{-2} \, dy$ and integrate over x, you will get

$$\frac{1}{4}\|f\|_2^2 = \frac{1}{4}\int_{-\infty}^\infty dx \int_0^\infty |f|^2 y^{-2} \, dy \leqslant \int_{-\infty}^\infty dx \int_0^\infty \left|\frac{\partial f}{\partial y}\right|^2 dy$$

$$\leqslant \int_{-\infty}^\infty dx \int_0^\infty \left(\left|\frac{\partial f}{\partial x}\right|^2 + \left|\frac{\partial f}{\partial y}\right|^2\right) dy$$

$$= -\int_{-\infty}^\infty dx \int_0^\infty f^* \left(\frac{\partial^2 f}{\partial x^2} + \frac{\partial^2 f}{\partial y^2}\right) dy$$

$$= -\int_{-\infty}^\infty dx \int_0^\infty f^* \Delta f y^{-2} \, dy,$$

$$= -(\Delta f, f),$$

as advertised.

9.   The Mehler Transform for $L^2(K/G/K)$

The spherical functions of the principal series lead to a nice Fourier transform for $L^2(K/G/K)$. This is the so-called "Mehler transform"

$$\hat{f}(\varkappa) = 2\pi \int_0^\infty f(\operatorname{ch}\alpha)\, p_{-\frac{1}{2}+i\varkappa}(\operatorname{ch}\alpha)\operatorname{sh}\alpha\, d\alpha$$

with the inverse transform[2]

$$f(\operatorname{ch}\alpha) = \frac{1}{2\pi}\int_0^\infty \hat{f}(\varkappa)\, p_{-\frac{1}{2}+i\varkappa}(\operatorname{ch}\alpha)\,\varkappa\operatorname{th}\pi\varkappa\, d\varkappa$$

and the Plancherel identity

$$\int_H |f|^2 y^{-2}\, dx\, dy = 2\pi\int_0^\infty |f(\operatorname{ch}\alpha)|^2 \operatorname{sh}\alpha\, d\alpha = \frac{1}{2\pi}\int_0^\infty |\hat{f}(\varkappa)|^2 \varkappa\operatorname{th}\pi\varkappa\, d\varkappa.$$

10.   Spherical Harmonics

As in Section 4.15, this "radial" transform can be "lifted" up to a Fourier transform on $G/K = H$ by means of "spherical harmonics." To do this, you bring in

$$e_\varkappa^l(x) = p_{-\frac{1}{2}+i\varkappa}(\operatorname{ch}\alpha)\, e^{il\theta},$$

_____

[2] th is short for tanh.

in which

$$p^l_{-\frac{1}{2}+i\varkappa}(\operatorname{ch}\alpha) = \frac{1}{2\pi}\int_0^{2\pi}(\operatorname{sh}\alpha\cos\theta+\operatorname{ch}\alpha)^{-\frac{1}{2}+i\varkappa}e^{-il\theta}\,d\theta,$$

is the "associated" conical function and define "the spherical harmonics of weight $\varkappa$" to be the class of $M_\varkappa$ of formal sums

$$f = \sum_{|l|<\infty}\hat{f}(l)e_\varkappa^l$$

with

$$\|f\|_\varkappa^2 = \sum_{|l|<\infty}|\hat{f}(l)|^2 < \infty.$$

These formal sums are bona fide eigenfunctions of Δ with the same eigenvalue $-(\frac{1}{4}+\varkappa^2)$ as $p_{-\frac{1}{2}+i\varkappa}$. $M_\varkappa$ is the span of $p_{-\frac{1}{2}+i\varkappa}$ under the action of G, and you have an addition formula

$$p_{-\frac{1}{2}+i\varkappa}(\operatorname{sh}\alpha\operatorname{sh}\beta\cos\theta+\operatorname{ch}\alpha\operatorname{ch}\beta)$$
$$= \sum_{|l|<\infty}p_{-\frac{1}{2}+i\varkappa}(\operatorname{ch}\alpha)\,p_{-\frac{1}{2}+i\varkappa}(\operatorname{ch}\beta)\,e^{il\theta},$$

or, what is the same,

$$p_{-\frac{1}{2}+i\varkappa}(g_1^{-1}g_2) = \sum_{|l|<\infty}e_\varkappa^l(g_1)^*e_\varkappa^l(g_2).$$

The optimist will now believe that

$$L^2(G/K) = L^2(H, y^{-2}\,dx\,dy) = \frac{1}{2\pi}\int_0^\infty M_\varkappa\,\varkappa\operatorname{th}\pi\varkappa\,d\varkappa,$$

much as in Section 4.15; the proof is similar. The "projection" of $f\in L^2(G/K)$ onto spherical harmonics of weight $\varkappa$ is

$$f\to f_\varkappa(x+iy) = \int_H p_{-\frac{1}{2}+i\varkappa}(\operatorname{ch}\alpha)\,f(x'+iy')\,(y')^{-2}\,dx'\,dy',$$

in which α is the hyperbolic distance between $x+iy$ and $x'+iy'$. The original function f may be recovered by the recipe

$$f(x) = \frac{1}{2\pi}\int_0^\infty f_\varkappa(x)\,\varkappa\operatorname{th}\pi\varkappa\,d\varkappa,$$

and you have the Plancherel formula

$$\|f\|_2^2 = \int_H |f|^2 y^{-2}\,dx\,dy = \frac{1}{2\pi}\int_0^\infty \|f_\varkappa\|_\varkappa^2\,\varkappa\operatorname{th}\pi\varkappa\,d\varkappa.$$

11. Comparison With the Flat Plane

To complete the parallel with Section 4.15, the hyperbolic analogue of the customary "rectilinear" Fourier transform

$$\hat{f}(\gamma) = \int_{R^2} f(x)\, e^{-2\pi i \gamma \cdot x}\, d^2 x,$$

for the flat plane will now be explained.

As k runs over $K = SO(2)$, the point $k0 = -\tan\theta$ runs (twice) over the line $y = 0$, so you can regard K as a double covering of the line and parametrize accordingly. Given $z = x + iy$ from H, let $z \cdot k$ be the hyperbolic distance from i to the circle passing through z which is tangent to $y = 0$ at k, affixing to this distance a plus or a minus sign according as i falls outside or inside the circle; see Fig. 1. The "rectilinear" transform $\hat{f}$ is

$$\hat{f}(\varkappa, k) = \int_H f(z)\, e^{(-\frac{1}{2} - i\varkappa) z \cdot k}\, y^{-2}\, dx\, dy$$

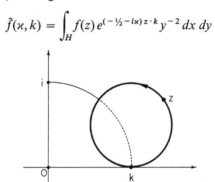

FIGURE 1

which is inverted by

$$f(z) = \frac{1}{2\pi} \int_0^\infty \int_K \hat{f}(\varkappa, k)\, e^{(-\frac{1}{2} + i\varkappa) z \cdot k}\, \varkappa\, \text{th}\, \pi\varkappa\, d\varkappa\, dk,$$

and there is a Plancherel identity

$$\|f\|_2^2 = \int_H |f(z)|^2 y^{-2}\, dx\, dy = \|\hat{f}\|_2^2 = \frac{1}{2\pi} \int_0^\infty \int_K |\hat{f}(\varkappa, k)|^2 \varkappa\, \text{th}\, \pi\varkappa\, d\varkappa\, dk.$$

The rectilinear transform is related to the projections $f \to f_\varkappa$ of Subsection 10 via the formulas

$$p_{-\frac{1}{2} + i\varkappa}(\text{ch}\,\alpha) = \int_K e^{(-\frac{1}{2} + i\varkappa) z \cdot k}\, dk,$$

$$f_\varkappa(z) = \int_K e^{-(\frac{1}{2} + i\varkappa) z \cdot k} \hat{f}(\varkappa, k)\, dk,$$

much as in Subsection 4.15.9. The complicated exponent $(-\frac{1}{2}+i\varkappa)z\cdot k$ looks much more natural if you compare Fig. 1 with what happens in the plane R^2, as indicated in Fig. 2. The heavy circle of Fig. 1 becomes a straight line through the point $x \in R^2$, perpendicular to the direction $k \in SO(2)$, which is pictured as a unit arrow issuing from $x = 0$. The distance between this line and the origin, with sign affixed as for Fig. 1, is simply the inner product $x\cdot k$, so [give or take a $-\frac{1}{2}$] the analog of $(-\frac{1}{2}+i\varkappa)z\cdot k$ is $i\varkappa x\cdot k = ix\cdot\gamma$ with $\gamma = \varkappa k \in R^2$.

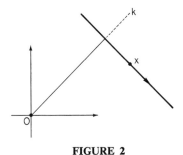

FIGURE 2

A nice introduction to the foregoing [without proofs] is given by Helgason [1967]. Gelfand *et al.* [1969] describe the representations of $G = SL(2, R)$, but it is very complicated; see also Rühl [1969, 1970]. An excellent intermediate presentation is that of Vilenkin [1968]. Robin [1959, Vol. 3, pp. 137–182] is best for information about the conical functions.

Additional Reading

The reader may wish to supplement the present text with additional material; the following brief list will provide some guidance. The simplest (and also the best) introductions are Hardy and Rogozinski [1944] and Seeley [1966a]. The first is purely mathematical and deals with series only; the second deals with integrals, too, and important applications. Bochner and Chandrasekharan [1949], Goldberg [1961], Carslaw [1930], and Papoulis [1962] provide intermediate treatments of series and/or integrals; the second pair has lots of applications, the first has none. Carleman [1944], Paley and Wiener [1934], Titchmarsh [1937], Wiener [1933], and Zygmund [1959] are advanced and purely mathematical, in the style of "hard" analysis; for the "soft" side, Loomis [1953] is recommended; Rudin [1963] and Hewitt and Ross [1963, 1970] should also be consulted. Edwards [1967] and especially Katznelson [1968] provide an excellent introduction to questions of current mathematical interest; see also Coppel [1969] for a quick survey of history and highlights. A nice introduction to applications in analytic number theory is given by Chandrasekharan [1968]. The historically minded will enjoy Hobson [1926], from which the "Historical Introduction" is adapted, and, above all, the great work itself: Fourier [1878] (the original French edition dates back to 1822).

Bibliography

AKHIEZER, N. I. "The Theory of Approximation." Ungar, New York, 1956.
AKHIEZER, N. I. "The Calculus of Variations." Ginn (Blaisdell), Boston, Massachusetts, 1962.
AKHIEZER, N. I., and I. M. GLAZMAN. "Theory of Linear Operators in Hilbert Space," 2 Vols. Ungar, New York, 1961–1963.
AUSLANDER, L. "Differential Geometry." Harper, New York, 1967.
BACHMANN, P. "Grundlehren der Neueren Zahlentheorie." Goschen, Leipzig, 1907.
BATEMAN, H. *In* "Tables of Integral Transforms," (A. Erdélyi, ed.). 2 Vols. McGraw-Hill, New York, 1954.
BELLMAN, R. "A Brief Introduction to Theta Functions." Holt, New York, 1961.
BERBERIAN, S. K. "Introduction to Hilbert Space." Oxford Univ. Press, London and New York, 1961.
BEURLING, A. On two problems concerning linear transformations in Hilbert space. *Acta Math.* **81**, 239–255 (1949).
BIRKHOFF, G., and S. MACLANE. "A Survey of Modern Algebra," 3rd ed. Macmillan, New York, 1965.
BIRKHOFF, G., and G.-C. ROTA. "Ordinary Differential Equations." Ginn, Boston, Massachusetts, 1962.
BOCHNER, S., and K. CHANDRASEKHARAN. "Fourier Transforms." Princeton Univ. Press, Princeton New Jersey, 1949.

BOERNER, H. "Representations of Groups with Special Consideration for the Needs of Modern Physics." North-Holland Publ., Amsterdam, 1963.

CARATHÉODORY, C. "Theory of Functions." Chelsea, Bronx, New York, 1954.

CARLEMAN, T. "L'intégrale de Fourier et Questions qui s'y Rattachent." Almqvist & Wiksell, Stockholm, 1944.

CARLESON, L. On the convergence and growth of partial sums of Fourier series. *Acta Math.* **116**, 135–157 (1966).

CARSLAW, H. S. A historical note on Gibbs' phenomenon in Fourier's series and integrals. *Bull. Amer. Math. Soc.* **31**, 420–424 (1925).

CARSLAW, H. S. "Introduction to the Theory of Fourier's Series and Integrals." Macmillan, New York, 1930. Reprinted by Dover, New York, 1952.

CASE, K. M., F. DE HOFFMAN, and G. PLACZEK. "Introduction to the Theory of Neutron Diffusion." U.S. Government Printing Office, Washington, D.C., 1953.

CHANDRASEKHAR, S. Stochastic problems in physics and astronomy. *Rev. Modern Phys.* **15**, 1–89 (1943). Reprinted in "Selected Papers on Noise and Stochastic Processes" (N Wax, ed.), Dover, New York, 1954.

CHANDRASEKHARAN, K. "Introduction to Analytic Number Theory." Springer-Verlag, Berlin and New York, 1968.

COPPEL, W. A. J. B. Fourier—On the occasion of his two hundredth birthday. *Amer. Math. Monthly* **76**, 468–483 (1969).

COPSON, E. T. "An Introduction to the Theory of Functions of a Complex Variable." Oxford Univ. Press (Clarendon), London and New York, 1935.

COURANT, R., and D. HILBERT. "Methods of Mathematical Physics," Vol. 1. Wiley (Interscience), New York, 1953.

COURANT, R., and F. JOHN. "Introduction to Calculus and Analysis." Wiley (Interscience), New York, 1965.

COURANT, R., and H. ROBBINS. "What is Mathematics?" Oxford Univ. Press, London and New York, 1941.

CRIMMINS, T., H. HORWITZ, C. PALERMO, and R. PALERMO. Minimization of mean-square error for data transmitted via group codes. *IEEE Trans.* **IT-15**, 72–73 (1969).

DE BRANGES, L., and J. ROVNYAK. "Square Summable Power Series." Holt, New York, 1966.

DE LA VALLÉE-POUSSIN, C. "Intégrales de Lebesgue," 2nd ed. Gauthier-Villars, Paris, 1950.

DOETSCH, G. "Einfuhrung in die Theorie und Anwendung der Laplace Transformation." Birkhaeuser, Basel, 1958.

DUREN, P. L. "Theory of H^p Spaces." Academic Press, New York, 1970.

EDWARDS, R. E. "Fourier Series, a Modern Introduction," 2 Vols. Holt, New York, 1967.

FEJÉR, L. Untersuchungen über Fouriersche reihen. *Math. Ann.* **58**, 51–69 (1904).

FELLER, W. "An Introduction to Probability Theory and its Applications," Vol. 1, 3rd ed. and Vol. 2. Wiley, New York, 1968a and 1966.

FELLER, W. On Müntz's theorem and completely monotone functions. *Amer. Math. Monthly* **75**, 342–350 (1968b).

FEYNMAN, R. "Lectures in Physics," 3 Vols. Addison-Wesley, Reading, Massachusetts, 1963–1965.

FLANDERS, M., and D. SWAN. At the drop of another hat. Angel Records, #36388, side 1, band 6.

FORD, G. W., and G. E. UHLENBECK. "Lectures in Statistical Mechanics." Amer. Math. Soc., Providence, Rhode Island, 1963.

FOURIER, J. "The Analytical Theory of Heat," translated by A. Freeman. Cambridge Univ. Press, London and New York, 1878. Reprinted by Dover, New York, 1955.

GELFAND, I. M., R. MINLOS, and Z. YA. ŠAPIRO. "Representations of the Rotation and Lorentz Groups and Their Applications." Pergamon, Oxford, 1963.

GELFAND, I. M., M. I. GRAEV, and I. I. PYATETSKII-ŠAPIRO. "Representation Theory and Automorphic Functions." Saunders, Philadelphia, Pennsylvania, 1969.

GIBBS, J. W. *Nature (London)* **59**, 606 (1899).

GINIBRE, J. General formulation of Griffiths' inequalities. *Comm. Math. Phys.* **16**, 310–328 (1970).

GOLDBERG, R. R. "Fourier Transforms." Cambridge Univ. Press, London and New York, 1961.

HAMMERMESH, M. "Group Theory." Addison-Wesley, Reading, Massachusetts, 1962.

HARDY, G. H., and W. ROGOZINSKI. "Fourier Series." Cambridge Univ. Press, London and New York, 1944.

HARDY, G. H., and E. M. WRIGHT. "An Introduction to the Theory of Numbers." Oxford Univ. Press (Clarendon), London and New York, 1954.

HEITLER, W. "Elementary Wave Mechanics." Oxford Univ. Press (Clarendon), London and New York, 1945.

HELGASON, S. Lie groups and symmetric spaces. In "Battelle Rencontres" C. M. DeWitt and J. A. Wheeler, eds., pp. 1–71, 1967. Benjamin, New York, 1968.

HELSON, H. "Lectures on Invariant Subspaces." Academic Press, New York, 1964.

HEWITT, E., and K. A. ROSS. "Abstract Harmonic Analysis," 2 Vols. Springer-Verlag, Berlin and New York, 1963–1970.

HEWITT, E., and K. STROMBERG. "Real and Abstract Analysis." Springer-Verlag, Berlin and New York, 1965.

HOBSON, E. W. "The Theory of Functions of a Real Variable," Vol. 2. Cambridge Univ. Press, London and New York, 1926. Reprinted by Dover, New York, 1957.

HOFFMAN, K. "Banach Spaces of Analytic Functions." Prentice-Hall, Englewood Cliffs, New Jersey, 1962.

HOPF, E. "Mathematical Problems of Radiative Equilibrium." Cambridge Univ. Press, London and New York, 1934.

HOPF, E., and N. WIENER. Über eine klasse singularer integralgleichungen. *Sitzungsber. Deut. Akad. Wiss. Berlin* 696–706 (1931). Reprinted in "Selected Papers of Norbert Wiener." M.I.T. Press, Cambridge, Massachusetts, 1964.

IKEHARA, S. An extension of Landau's theorem in the analytic theory of numbers. *J. Math. and Phys.* **10**, 1–21 (1931).

INGHAM, A. E. "The Distribution of Prime Numbers." Cambridge Univ. Press, London and New York, 1932.

JOHN, F. "Plane Waves and Spherical Means, Applied to Partial Differential Equations." Wiley (Interscience), New York, 1955.

KAC, M. Quelques remarques sur les zéros des intégrales de Fourier. *J. London Math. Soc.* **13**, 128–130 (1938).

KAC, M. A remark on Wiener's Tauberian theorem. *Proc. Amer. Math. Soc.* **16**, 1155–1157 (1965).

KATZNELSON, Y. "An Introduction to Harmonic Analysis." Wiley, New York, 1968.

KINGMAN, J. F. C. Spitzer's identity and its use in probability theory. *J. London Math. Soc.* **37**, 309–316 (1962).

KOLMOGOROV, A. N. Une série de Fourier-Lebesgue divergente partout. *C.R. Acad. Sci. Paris Sér. A-B* **183**, 1327–1328 (1926).

LANDAU, H. J., and H. O. POLLACK. Prolate spheroidal wave functions, Fourier analysis and uncertainty (2). *Bell System Tech. J.* **40**, 65–84 (1961).

LANDAU, H. J., and H. O. POLLACK. Prolate spheroidal wave functions, Fourier analysis and uncertainty (3): The dimension of the space of essentially time- and band-limited signals. *Bell System Tech. J.* **41**, 1295–1336 (1962).

LARDY, L. J. Order convolution on $L^1[a, b]$. *Notices Amer. Math. Soc.* **13**, 374 (1966).

LEBLANC, N. Classification des algèbres de Banach associées aux opérateurs différentials de Sturm-Liouville. *J. Functional Analysis* **2**, 52–72 (1968).

LEE, Y. W. "Statistical Theory of Communication." Wiley, New York, 1960.

LEVINSON, N. A motivated account of an elementary proof of the prime number theorem. *Amer. Math. Monthly* **76**, 225–245 (1969).

LEVINSON, N., and R. REDHEFFER. "Complex Variables." Holden-Day, San Francisco, California, 1970.

LOOMIS, L. "An Introduction to Abstract Harmonic Analysis." Van Nostrand-Reinhold, Princeton, New Jersey, 1953.

LUDWIG, D. The Radon transform on euclidean space. *Comm. Pure Appl. Math.* **19**, 49–81 (1966).

MCKEAN, H. P. Kramers-Wannier duality for the 2 dimensional Ising model as an instance of the Poisson summation formula. *J. Math. Phys.* **5**, 775–776 (1964).

MAXWELL, J. C. "A Treatise on Electricity and Magnetism." Oxford Univ. Press (Clarendon), London and New York, 1892.

MICHAELSON, A. *Nature (London)* **58**, 544 (1898).

MÜLLER, C. "Spherical Harmonics" (Lecture Notes in Math.) No. 17. Springer-Verlag, Berlin and New York, 1966.

MÜLLER, C. "Foundations of the Mathematical Theory of Electromagnetic Waves." Springer-Verlag, Berlin and New York, 1969.

MUNROE, M. E. "Introduction to Measure and Integration." Addison-Wesley, Reading, Massachusetts, 1953.

NAGELL, T. "Introduction to Number Theory." Wiley, New York, 1951.

NEUGEBAUER, O. E. "The Exact Sciences in Antiquity." Princeton Univ. Press, Princeton, New Jersey, 1952.

PALEY, R. E. A. C., and N. WIENER. "Fourier Transforms in the Complex Domain." Amer. Math. Soc., Providence, Rhode Island, 1934.

PAPOULIS, A. "The Fourier Integral and Its Applications." McGraw-Hill, New York, 1962.

POLLACK, H. O., and D. SLEPIAN. Prolate spheroidal wave functions, Fourier analysis and uncertainty (1). *Bell System Tech. J.* **40**, 43–64 (1961).

PÓLYA, G. Über eine aufgabe der wahrscheinlichkeitsrechnung betreffend die irrfahrt im strassennetz. *Math. Ann.* **84**, 149–160 (1921).

RADEMACHER, H. Fourier analysis in number theory (lecture notes). Cornell Univ., Ithaca, New York, 1956.

RADEMACHER, H. "Lectures on Elementary Number Theory." Ginn (Blaisdell), Boston, Massachusetts, 1964.

RIESZ, F., and B. SZ.-NAGY. "Functional Analysis." Ungar, New York, 1955.

ROBIN, L. "Functions Sphériques de Legendre et Functions Sphéroidales," Vol. 3. Gauthier-Villars, Paris, 1959.

ROBINSON, E. A. "Random Wavelets and Cybernetic Systems." Griffin, London, 1962.

ROTA, G.-C. On the foundations of combinatorial theory, 1. Theory of Möbius functions. *Z. Wahrscheinlichkeitstheorie und Verw. Gebiete* **2**, 340–368 (1964).

ROTA, G.-C. Baxter algebras and combinatorial identities, 1 and 2. *Bull. Amer. Math. Soc.* **75**, 325–334 (1969).

ROYDEN, H. L. "Real Analysis." Macmillan, New York, 1963.

RUDIN, W. "Fourier Analysis on Groups." Wiley (Interscience), New York, 1963.

RUDIN, W. "Real and Complex Analysis." McGraw-Hill, New York, 1966.

RÜHL, W. An elementary proof of the Plancherel theorem for the classical groups. *Comm. Math. Phys.* **11**, 297–302 (1969).

RÜHL, W. "The Lorentz Group and Harmonic Analysis." Benjamin, New York, 1970.

SCHWARZ, L. "Étude des sommes d'exponentielles." Herrmann, Paris, 1959.

SEELEY, R. "An Introduction to Fourier Series and Integrals." Benjamin, New York, 1966a.

SEELEY, R. Spherical harmonics. *Amer. Math. Monthly* **73**, 115–121 (1966b).

SHANNON, C. E. Communication in the presence of noise. *Proc. IRE* **37**, 10–21 (1949).

SOMMERFELD, A. "Partial Differential Equations in Physics. "Academic Press, New York, 1949.

SPEISER, A. "Die Theorie der Gruppen von Endlicher Ordnung." Springer-Verlag, Berlin and New York, 1927. Reprinted by Dover, New York, 1945.

SPITZER, F. A combinatorial lemma and its applications to probability theory. *Trans. Amer. Math. Soc.* **82**, 323–339 (1956).

SPITZER, F. "Principles of Random Walk." Van Nostrand-Reinhold, Princeton, New Jersey, 1964.

SZEGÖ, G. Beiträge zur Theorie der Toeplitzchen Formen (Erste Mitteilung), *Math. Z.* **6**, (1920), 167—202.

TITCHMARSH, E. "Introduction to the Theory of Fourier Integrals." Oxford Univ. Press (Clarendon), London and New York, 1937.

VAN DER WAERDEN, B. L. "Gruppen von Linearen Transformationen." Springer-Verlag, Berlin and New York, 1935. Reprinted by Chelsea, Bronx, New York, 1948.

VILENKIN, N. YA. "Special Functions and the Theory of Group Representations." Amer. Math. Soc., Providence, Rhode Island, 1968.

WENDELL, J. G. Spitzer's formula: A short proof. *Proc. Amer. Math. Soc.* **9**, 905–908 (1958).

WEYL, H. Über die gleichvertielung von zahlen mod. eins. *Math. Ann.* **77**, 111–147 (1916).

WEYL, H. "The Theory of Groups and Quantum Mechanics." Dutton, New York, 1931. Reprinted by Dover, New York, 1950.

WEYL, H. "The Classical Groups." Princeton Univ. Press, Princeton, New Jersey, 1939.

WEYL, H. "Symmetry." Princeton Univ. Press, Princeton, New Jersey, 1952.

WIENER, N. "The Fourier Integral and Certain of its Applications." Cambridge Univ. Press, London and New York, 1933. Reprinted by Dover, New York, 1959.

WIGNER, E. "Symmetry and Reflections." Univ. of Indiana Press, Bloomington, Indiana, 1967.

WILBRAHAM, H. On a certain periodic function. *Cambridge and Dublin Math. J.* **3**, 198 (1848).

YOSIDA, K. "Lectures on Differential and Integral Equations." Wiley (Interscience), New York, 1960.

ZYGMUND, A. "Trigonometric Series," 2nd ed., 2 Vols. Cambridge Univ. Press, London and New York, 1959.

Index